Enrico Baccarini
and Kavya Vaddadi

Reverse Engineering Vedic Vimanas
New Light on Ancient Indian Heritage

Technical Part
By
Kavya Vaddadi

About Kavya:

Kavya Vaddadi is born in Hyderabad and lives in UP, India. Works as CFD Expert in ScheDio CAD Solutions (OPC) Pvt. Ltd. In the past, had two years experience as Design Engineer and CFD analyst in Vaddadi Engineering Design and analysis Services (VEDAS) Company, Research Service Provider in Vedic Scientific Research Foundation (VSRF), Founder of Scientific Works on Advanced Space Technology Investigators for Knowledge (SWASTIK) Research Team. In 2015, Kavya's Project Ideology on Mini Rukma vimana Unmanned Air vehicle has been selected to National Aerolympics Project, Aeronautical Society of India and received Best team Award. Her guidance and support to the students has been mentioned in Telugu Newspaper. In 2016, Kavya Vaddadi name is recorded as "Marvellous Design Engineer of Vimana Project" in "Marvellous Records Book of India". Major works on Vimanas include Retranslations of Vimanashasthra chapters, interpretation of Sanskrit shlokas into technical meanings, 2D, 3D modelling, Assemblies of Vedic Vimanas, Computational Fluid Dynamics fluid flow analysis of Vimanas and Heat Transfer Thermal analysis on Vimana materials. Technical researcher on Vimanashasthra focusing the topics such as Manufacturing, Tactical Air Defense Aerial Wars, UFO's, Aliens, Aerodynamics, structures, Propulsion, space mechanics, Antigravity technology and Vimana prototypes.

Contents

Technical Part 3
Contents 4
Author's Preface 13
Acknowledgment 16
I. Introduction: A Journey from Vedic India's 'Vimana' To Modern India's Hyperplane 18
   Definition of Vimana 19
   Spaceplanes: A Synergy of Rocket and Aircraft Technologies 20
   Different Representations of Vedic "Vimanas" 20
   Vimanas for Space Transportation 21
   Varieties of Vimanas 23
   1. Mantrika 23
      Sanskrit Pilot instructions Shloka: 23
      Re-Translation 23
      Interpretation: 24
      Examples of modern Technology: 24
   2. Tantrika 24
      Sanskrit Pilot instructions Shloka: 24
      Re-Translation 25
      Interpretation: 25
      Examples of modern Technology: 25
   3. Kritaka: 25
      Sanskrit Pilot instructions Shloka: 25
      Re-Translation 25
      Interpretation: 26
      Examples of modern Technology: 26
   Spaceplanes of the Modern World 28
   Design Requirements for SSTO Spaceplanes 28
   Promising Contemporary Spaceplane Design Concepts 29
   The UK "Skylon" (late 1980's) 29
   X-43 High Speed Airbreathing Engine (Scramjet) Test Vehicle 31
   The Indian "Hyperplane" or "Avatar" Spaceplane (late 1980's) 31
   General Comment 32
   Scientific Overview 33

II. Aerial Wars, UFO's and Aliens 37
TACTICAL AIR DEFENSE 37
1. Destruction: 37
- Sanskrit Pilot instructions Shloka: 37
- Re-Translation 38
- Devices used: 38
- Interpretation: 38
- Examples of modern Technology: 39

2. Methods bringing terrific shape: 39
- Sanskrit Pilot instructions Shloka: 39
- Re-Translation 40
- Devices used: 40
- Interpretation: 40
- Examples in modern Technology: 40

3. Transforming shape: 41
- Sanskrit Pilot instructions Shloka: 41
- Re-Translation 41
- Interpretation: 41

4. Beautiful shape: 42
- Sanskrit Pilot instructions Shloka: 42
- Re-Translation 42
- Devices used: 42
- Interpretation: 42

5. The secret to sun-like Glow: 42
- Sanskrit Pilot instructions Shloka: 42
- Re-Translation 43
- Devices used: 43
- Interpretation: 43

6. Facing opposite party Vimana 44
- Sanskrit Pilot instructions Shloka: 44
- Re-Translation 44
- Devices used: 44
- Interpretation: 44

7. Stunning, making immobile 45
- Sanskrit Pilot instructions Shloka: 45
- Re-Translation 45
- Devices used: 46
- Interpretation: 46
- Similar Examples of modern Technology: 46

8. Invisibility 48

  Sanskrit Pilot instructions Shloka: 48
  Re-Translation 48
  Interpretation: 49
 9. Presence *49*
  Sanskrit Pilot instructions Shloka: 49
  Re-Translation 49
  Interpretation: 49
 10. Destruction *49*
  Sanskrit Pilot instructions Shloka: 49
  Re-Translation 50
  Interpretation: 50
 11. Radiance *50*
  Sanskrit Pilot instructions Shloka: 50
  Re-Translation 50
  Interpretation: 51
 12. Ultra-sound art of confusing *51*
  Sanskrit Pilot instructions Shloka: 51
  Re-Translation 51
  Interpretation: 51
 13. Snake-like movement *51*
  Sanskrit Pilot instructions Shloka: 51
  Re-Translation 52
  Interpretation: 52
 14. Wavering/unsteady *52*
  Sanskrit Pilot instructions Shloka: 52
  Re-Translation 52
  Interpretation: 52
 15. Facing all sides *53*
  Sanskrit Pilot instructions Shloka: 53
  Re-Translation 53
  Interpretation: 53
 16. Foreign/alien vimana sound observing *53*
  Sanskrit Pilot instructions Shloka: 53
  Re-Translation 53
  Interpretation: 53
  Examples of modern Technology: 53
 17. Visual view attraction *54*
  Sanskrit Pilot instructions Shloka: 54
  Re-Translation 54
  Interpretation: 54

- 18. Action observing 55
  - Sanskrit Pilot instructions Shloka: 55
  - Re-Translation 55
  - Interpretation: 55
- 19. Sky like appearance 55
  - Sanskrit Pilot instructions Shloka: 55
  - Re-Translation 55
  - Interpretation: 56
- 20. Cloud like appearance 56
  - Sanskrit Pilot instructions Shloka: 56
  - Re-Translation 56
  - Interpretation: 56
- 21. Absence 57
  - Sanskrit Pilot instructions Shloka: 57
  - Re-Translation 57
  - Devices used: 57
  - Interpretation: 57
- 22. Darkness trap 57
  - Sanskrit Pilot instructions Shloka: 57
  - Re-Translation 58
  - Devices used: 58
  - Interpretation: 58
- 23. Solar flare Escaping Secrets 58
  - Sanskrit Pilot instructions Shloka: 58
  - Re-Translation 58
  - Devices used: 59
  - Interpretation: 59

III. SPACE AND FLIGHT MECHANICS 60
- *Landing Gears* 60
- *Re-Translations from Vimana shasthra* 60
- 1. Defense/offense 60
  - Sanskrit Pilot instructions Shloka: 60
  - Re-Translation 60
  - Interpretation: 61
- 2. Intermediate Space 61
  - Sanskrit Pilot instructions Shloka: 61
  - Re-Translation 61
  - Interpretation: 61
- 3. Visibility 61
  - Sanskrit Pilot instructions Shloka: 61

    Re-Translation 62
    Devices used: 62
    Interpretation: 62
    Examples of modern Technology: 62
  4. Contraction 63
    Sanskrit Pilot instructions Shloka: 63
    Re-Translation 63
    Devices used: 63
    Interpretation: 64
  5. Expansion 64
    Sanskrit Pilot instructions Shloka: 64
    Re-Translation 64
    Devices used: 64
    Interpretation: 64
  6. Direction display 64
    Sanskrit Pilot instructions Shloka: 64
    Re-Translation 65
    Devices used: 65
    Interpretation: 65
  Vimana Flight and space mechanics: 65
    Rukma Vimana 65

IV. Structures and Flight: 3D modeling Vedic Vimanas 67
  RUKMA VIMANA :- 68
    Structural analysis Rukma Vimana:- 75
  SHAKUNA VIMANA 78
    Structural Analysis Shakuna Vimana:- 85
  SUNDARA VIMANA 88
    Structural Analysis Sundara Vimana:- 91
  TRIPURA VIMANA 94
    Structural analysis of Tripura Vimana:- 97

V. AERODYNAMICS 100
  Aerial Routes 100
  RUKMA VIMANA 101
  SUNDARA VIMANA 103
  SHAKUNA VIMANA 104
  TRIPURA VIMANA 106

VI. Ancient Aerospace Materials 109
  Manufacturing Materials of Vimana 109
  Propulsion materials 115

Nanomaterials - Nanotechnology in Ancient India...........................*121*
Comment..*122*

VII. PROPULSION...123
The Power..*123*
   Interpretation:..*123*
Electric Generator...*123*
   Interpretation:..*124*
The Electric Motor..*125*
   Interpretation:..*125*
Ganapa-yantra..*125*
   Interpretation:..*126*
Electric Dynamo..*126*
   Interpretation:..*126*
Rukma Vimana Propulsion...*130*

Advanced Antigravity Technology..132

VIII. Vimana Prototypes..142
MRV UAV..*142*
   RESULTS AND DISCUSSION *146*
TRIPURA VIMANA PROTOTYPE *147*

IX. Nikola Tesla's Vimana and Yantras 149
Death Ray..*153*
Tesla's Oscillator *153*
Free Electricity System...*154*
The Flying Saucer..*155*
Improved Airships *156*

X. Talpade's Vimana 156
MARUTSAKHA VIMANA..*157*

XI. Hitler's De glocke..159

XII. Marutsakha Vimana, 3D Printing and Wind Tunnel Testing............163

As shown in Ancient Aliens, History Channel............................163
Wind Tunnel Testing in California University Irvine.................*166*

XIII. 3D and 4D Printing of Ancient Indian Vimanas.................169
1. Introduction:...*169*
   1.2: Additive Manufacturing:..*171*
2. Steps involved in 3D Printing of Vimanas:............................*172*
3. Tradational Machining of CNC Vs Additive Manufacturing Process...........*176*

- 3.1 Material Choice ... 177
- 3.2 Speed ... 177
- 3.2 Complexity ... 178
- 3.3 Accuracy ... 178
- 3.4 Geometry ... 179
- 3.5 Additive Manufacturing Materials ... 180
- 3.6 Additive Manufacturing Applications ... 183
- 3.7 Additive Manufacturing Capabilities ... 184
- 3.8 Additive Manufacturing Limitations ... 185
- *4 AM Data Formats & Software's* ... *185*
- 4.1 AM Data Formats ... 185
- 4.2 AM Software's ... 187
- *5. Exporting an STL file from Creo* ... *188*
- 5.1 Deviation control ... 188
- 5.2 Angle control ... 189
- 5.3 Steps to convert to STL file. ... 189
- *6. AM Build Setup Preparation* ... *189*
- 6.1 Cura ... 190
- 6.2 Loading of file in Cura ... 190
- *7. AM Build Processes* ... *198*
- 7.1 Material extrusion ... 199
- 7.2 Material Jetting ... 201
- 7.3 Binder Jetting ... 204
- 7.5 Vat Photopolymerization ... 208
- 7.6 Powder Bed Fusion ... 210
- 7.7 Directed Energy Deposition ... 212
- *8. Future scope in the advancement of manufacturing processes of Ancient vimanas:* ... *214*

RESULTS AND DISCUSSIONS ... 216
- *Mercury Vortex Propulsion* ... *216*

Appendix – 1 ... 223
- *New Sanskrit Translations in Vimana shasthra:* ... *223*
  - Aerial Routes ... 223
  - Aeroplane parts ... 224
  - Manufacturing ... 224
- *Flight Mechanics, Tactical Air Defense and Propulsion* ... *225*

References ... 227
- *Main sources:* ... *227*
- *3D and 4D Printing chapter:* ... *228*

Antigravity chapter: *228*

HISTORICAL PART **231**

Historical Part **232**
Introduction *232*

Chapter I –Vimanas, God's flying cars **234**
Ratha and Vimana, the old flying cars *236*
Other references to flying carts *243*
A Vimana for every era *247*
The Mantra Vimana *250*
The Tantra Vimana *251*
The Krutaka Vimana *251*
Vimana, the confirmation inside the edicts of Ashoka *253*
God Salva's Vimana and his terrible arms *254*
"Chapter 76" *257*

Chapter II –Lost Technologies **331**
The electricity into Vedic's literature *331*
The Saubha, the town on the space *335*
The Saubhikas *337*
The Sanskrit and the artificial intelligence *340*
The gunpowder *341*
Amsu Bodhini *346*
Cosmology of the old India *352*
Atomic philosophies *357*
The Veda's speed of light *360*
Parallel universes in Hinduism *362*
Stealth's materials on the old Vedic texts *366*
The Chumbakamani *368*
The Panchloh, a corrosion-resistant material *370*
Test Tube Babies *372*
Human and Animal cloning *373*

Chapter IV – Astra God's weapons **378**
Astra, God's weapons *378*
Astra's nature *383*
War machines *385*
The Tejas astras, and the energy weapons *399*
The Visvakarma, the Lanka and the Dwarka *405*
Gods's wars *406*
The Rig Veda *413*

Sanskrit literature 413
The Vedas and the other inhabited worlds 417
The Kumaras and the Narada Muni, the travelers of the Worlds 424
The Samarangana Sutradhara 427
Konark and the temples in imitation of the Vimana 432
The Ramayana and the Vimana 434

Author's Preface
By Kavya Vaddadi, kavya.vaddadi@gmail.com

This book is written in an attempt to decode and bring back the ancient scientific knowledge on Aerospace Design and Engineering and to educate its importance to young minds. Merging ancient texts investigation with modern technology, the unique works of re-translations, interpretations with unique 3D modeling and analysis are presented in this book. My observation on ancient technology is that it is advanced and more than 100% efficient, free energy using, and which never harm or disturb nature in any manner.

Most of the sources for my works mentioned in this book are from the writings of Maharshi Bharadwaja, Shri. Subbaraya Shasthry, drawings of Shri. Ellappa and translation of Shri. Josyer. As the present generations have started approaching advanced aerospace engineering technology development, the writings of our India's ancient Vimana shasthra are becoming more understandable. Thanks a lot for the technical works of Talpade, Tesla, Boeing, NASA, and ISRO for which I am able to understand and interpret Vimana shasthra.

I am happy that I am one among the researchers who did attempt to understand Vimana shasthra in a scientific manner for the first time in history. The Sanskrit shlokas, when I have been retranslating, I felt very thrilling because there are multiple meanings for single words in this divine language, and I had to choose technical meaning as per the aeronautical engineering knowledge. An assembly of all technical meanings of the words in shloka resulting in a wonderful message which reveals treasures of vast ancient advanced aerospace technology.

Pilot instructions are very interesting and describe something out of the world, as it is too advanced technology difficult to be understood till modern human civilization tries to approach it. Decoding the mysteries is always thrilling and I felt the most exciting experience when I translated shlokas. The re-translations, interpretations, 3d modeling, and analysis I did on Vimana shasthra are performed as per my engineering knowledge I have.

Vimana 3D modeling is another amazing experience, which made my mind expand and think about the unknown ancient ancestors' vimana designs. It has been very hard to do the 3D modeling because all vimanas are almost 100 feet in length, there were many design complications and geometry issues which I had to solve after working a lot on the Sanskrit shlokas and then referring to Vimana diagrams of Vimana shasthra. The CFD analysis was another big task where I had to modify the design dimensions for better analysis results.

Further investigations are being made with the help of scientists who support the research, for discovering in-depth details of procedures and elements used in Vimana Shastras descriptions. Further findings from on-going, deep technical research will be described in the next series of this Book.

3D Printed Vimana models

Kavya Vaddadi in a meet with Giorgio Tsoukalos, famous appearance in Ancient Aliens show in History channel.

Kavya Vaddadi, as an Aircraft Design Engineer in History Channel

Acknowledgment

I am indebted to Mr. Enrico Baccarani for his scholarly section on the Historical Background of Vedic Vimanas. I thank my father Sri. Koteswara Rao Vaddadi and my mother Sri. Annapurna Vaddadi for encouraging me to do research on Vedic Vimanas. I thank a lot to my brother Sri. Karthik vaddadi for helping me with the sources required, Sri. A. Hemanth Kumar Yadav for guiding me Vimanas CFD analysis, Sri. Shashi Kant, CEO of ScheDio CAD Solutions (OPC) Pvt. Ltd., for helping in much deeper analysis of vimanas. I am inspired by my Grandfather Sri. VLN Murthy, who always had a curiosity to learn more and develop more skills. Special thanks to Sri. Raghavan Gopalaswami, a Retired aero-mechanical aircraft engineer and rocket technologist, who is also my main inspiration for his article on Vedic vimanas, which inspired me to do more research on vimana shasthra.

I am also Thankful to Sri. Randall for his great support and his works on antigravity. My Greatest thanks to technical mentors, Supporters, Advisors:
1. Sri. Madhavan Nair - Former Chairman of ISRO
2. Sri. W. Selvamurthy - Former cc R&D DRDO
3. Sri. Prahalad - Former cc R&D - DRDO, VC - DIT
4. Sri. M.S. Prasad - Former Program Director in DRDO
5. Sri. Dr. V. Ashok. Group Director. Aerodynamics and Aerothermal Group. Aero Entity, VSSC, ISRO.

I am also thankful to my family relatives, friends, and SWASTIK team members for encouraging my research.

My Special thanks to Sri. PVN Murthy, founder of Vedic Science Research Institute and Sri. Venkata Ramanan from Bangalore, for making me aware that there will be technical meanings for Sanskrit shlokas describing gods.

Heartfelt thanks to Sri. T. C. Manjunath Chandrashekhar, Sri. VedaRavi Shangar and Sri. Karthikeyan Iyer, who are my special spiritual technical Guides for providing solutions to overcome major obstacles in Vimana research. Another special thanks, to Sri. Eshwar Reddy Cholleti, Founder of 3D Srishti Pvt Ltd, for working on the possibilities of 3D and 4D printing Vedic Vimanas.

A special thanks to his Holiness Sri. Sri. Sri. Ramakrishnananda Saraswathy Swamigal (Founder – VSRF) for proving the power of vedic advanced science and for implementing in ISRO and DRDO.

Another Special Thanks to Dr. APJ Abdul Kalam who understood advanced spiritual dimension science. He once request to see the model of Tripura vimana to a scientist who worked on it material. I am happy that it is going to be fulfilled by me in the shower of blessings of God and all divine people.
3D model designs and 3D printing going to be explained in detail in this book.

I. Introduction: A Journey from Vedic India's 'Vimana' To Modern India's Hyperplane

By Kavya Vaddadi and R. Gopalaswami

"Absence of evidence is not evidence of absence"
Carl Sagan

From 20th Century Rockets, Missiles and Aircrafts to Spaceplanes of the 21st Century. The world has entered the dawn of the 21st Century. Science and technology have enabled man create enter a new form of civilization, more prosperous and comfortable that was known for thousands of years. But while there is unprecedented prosperity in many nations, yet the physical, emotional and intellectual energies of man have not been able to resolve his psychological problems; and great sorrow and suffering continues world over.

A major accomplishment in the early part of the 20th Century was the invention of the rocket in the US (Goddard) and then USSR (Korolev); and winged flight by a heavier-than-air aircraft (Wright brothers). The oxygen-carrying rocket enabled man travel in space where there is no atmospheric oxygen available for combustion of fuel in the propulsion system. The aircraft took advantage of the atmosphere both for propulsive force as well as providing a lift force to keep the aircraft airborne. By the end of the 20th Century, man had mastered travel in both the earth's atmosphere, and in space. Travel across continents became safe and routine for hundreds of passengers at a time; and man travelled, landed on and returned safely from expeditions to the moon.

Even within 60 years after independence from nearly 1000 years of crushing alien invasion, conquest and rule, India has come forth with remarkable achievements in science and technology. Among the most advanced are the accomplishments in aeronautical and space science and technology that are rapidly closing the gap between India and those who had a lead in these technologies for over one century. Still, there are many in India who doubt whether we will ever master these technologies and put it to good uses for enhancing security and prosperity not only for India, but all humanity.

However, even in the Western countries, the limitation of using rockets alone for space travel has been clearly understood. Rockets are cumbersome vehicles, vertically stacked, extremely heavy due to the large amount of oxygen (over 70% of its mass at launch) to be carried on-board. Difficult and complex to handle, prepare and launch, consume too much of fuel, uncomfortable to passengers due to high acceleration levels, still relatively unsafe, and expendable after one launch.

On the other hand, aircraft technologies are much safes and affordable. Even gigantic transport aircraft like the Boeing 747 have magnificent safety records, and are routinely and extensively used by commercial operators. They are comfortable, highly fuel efficient, and can fly non-stop across oceans and continents.

Hence it has been mankind's dream to make access and travel space as safe and affordable as commercial air transportation systems. In other words, for nearly 40 years now the search is to design and build a safe, affordable, reusable space plane by a new form of aerospace vehicle that behaves like an aircraft when in the atmosphere, and a rocket in space!

Definition of Vimana

सू. १.   वेगसाम्या द्विमानो ऽण्डजानामिति ॥

बोधानन्दवृत्तिः —

अण्डजे स्वत्र सूत्रेस्मि न्गृध्राद्याः पक्षिणः स्मृताः ।   आकाशगमने तेषां वेगशक्तिः स्वबेगतः ॥
यस्समर्थो विशेषेण मातुं गणितसंख्यया ।   स विमान इति प्रोक्तो वेगसाम्याच्च शाश्वतः ॥

यद्वा

गृध्रादिपक्षिणां वेगसाम्यं यस्यास्ति वेगतः ।   स विमान इति प्रोक्तो आकाशगमने क्रमात् ॥ इति
इत्थं भावेति शब्दस्या द्विमानार्थ विनिर्णये ॥   ॥ ४५ ॥

लल्लोऽपि —   विसोपमानं गमने येषामस्ति खमण्डले ।
               ते विमाना इति प्रोक्ता यानशास्त्रविशारदैः ॥

नारायणोऽपि —   पृथिव्य प्खन्तरिक्षेषु खगव द्वेगत स्वयम् ।
               यस्समर्थो भवेद्गन्तुं स विमान इति स्मृतः ॥

शंखोऽपि —   स्थानास्थानान्तरं गन्तुं यस्समर्थः खमण्डले ।
               स विमान इति प्रोक्तो यानशास्त्र विशारदैः ॥

विश्वम्भरः —   देशादेशान्तरं तद्वद् द्वीपाद् द्वीपान्तरं तथा ॥
लोका ल्लोकान्तरं चापि योऽम्बरे गन्तु महति । स विमान इति प्रोक्तो खेटशास्त्रविदां वरैः ॥
                                                                                    इत्यादि ।

एवं विमानशब्दार्थ उक्तवा शास्त्रानुसारतः । अथेदानीं तद्रहस्यविचार संप्रकीर्त्यते ॥ ५ ० ॥

"Owing to similarity of speed with birds, it is named Vimaana."

19

Comment: There are different types of Vimanas constructed for various roles and missions, both for atmospheric and space flight. In atmospheric flight, they have the capability to travel from one country to another country or from one island to another island. In space, certain types of Vimanas have the capability for interplanetary missions.

Spaceplanes: A Synergy of Rocket and Aircraft Technologies

Spaceplanes of Ancient India.

Flight in the earth's atmosphere and to space is thought to have originated in the 20th Century. However, that may not be the case. In the Vedic literature of India, recording events that occurred 12,000 to 15,000 years ago, there are many descriptions of flying machines that are generally called **Vimanas.** The Mahabharata speaks of "Two storied celestial chariots with many windows" "They roar off into the sky until they appear like comets." The Mahabharata and various Sanskrit books describe at length these chariots, "Powered by winged lighting...it was a ship that soared into the air, flying to the solar and stellar regions."

Different Representations of Vedic "Vimanas"

RUKMA VIMANA

VERTICAL SECTION

Recently, an Italian scientist Dr. Roberto Pinotti at a World Space Conference reported that India may have had a superior civilization and the flying devices called 'Vimanas' described in ancient Indian texts may underline their possible connections to today's aerospace technology. He held a view that 'Shakuna Vimana' described in the text 'might be defined as a cross between a plane and a rocket of our times and its design might remind one of today's

Space Shuttle.' Quoting from 'Vymanika Shastra' he said the ancient flying devices of India were made from special heat absorbing metals named 'Somaka, Soundalike and Mourthwika.'

Thus, it might appear that mankind's dream of traveling to space; to visit planets and explore the solar system is as old as mankind itself. Why is there no physical evidence of these 'advanced' vehicles, if they were built thousands of years ago?

Vimanas for Space Transportation

It was the use of Vimanas as space transportation systems that might have the clue as to why there is no physical evidence of these ancient aerospace vehicles.

The Atlanteans, known as "Asvins" in the Indian writings, were apparently even more advanced technologically than the Indians. They possessed Vailixi, similar to Vimanas, that were generally "cigar shaped" and had the capability of manoeuvring underwater as well as in the atmosphere or even outer space. Other flight vehicles were saucer shaped, and could apparently travel submerged. It is recorded that between 12000 to 15,000 years ago, nations deploying Vimanas in space with lethal weapons were locked in a global war that destroyed almost all of human life and property on planet earth. Clinching archaeological evidence to this effect has also been found. Thus, the weaponization of space should not be allowed to happen again.

Following are the diagrams of vimanas from Vimana shasthra and the 3D model designs. Going to be explained in detail in the chapter: Structures: 3D modeling of Vimanas.

RUKMA VIMANA

Drawn by T. K. ELLAPPA, Bangalore. 2-12-1923.

Prepared under instruction of Pandit SUBBARAYA SASTRY, of Anekal, Bangalore

3D model design by Kavya

Rukma Vimana

3D model design by Kavya

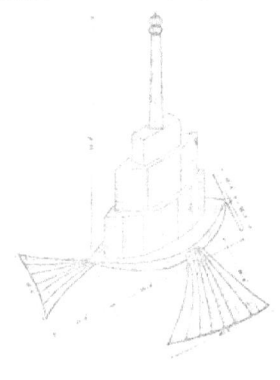
Drawn by T. K. ELLAPPA, Bangalore. 2-12-1923.
Prepared under instruction of Pandit SUBBARAYA SASTRY, of Anekal, Bangalore

Shakuna Vimana

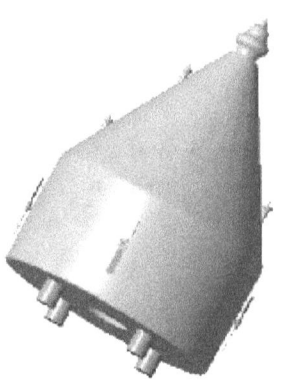
3D model design by Kavya

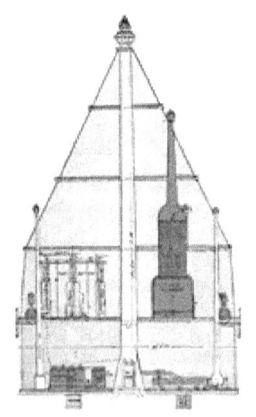
Drawn by T. K. ELLAPPA, Bangalore. 2-12-1923.
VERTICAL SECTION
Prepared under instruction of Pandit SUBBARAYA SASTRY, of Anekal, Bangalore

Sundara Vimana

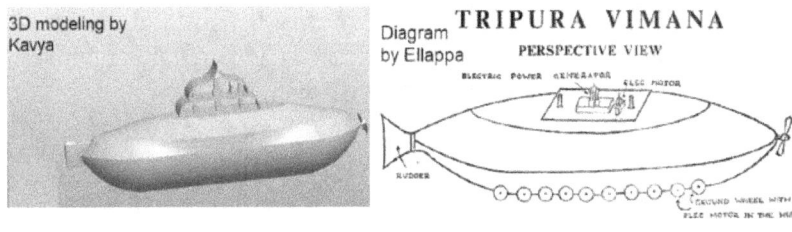

Varieties of Vimanas

Manufacturing Technology

Manufacturing related Re-Translations of Pilot instructions chapter in Vimana shasthra:

1. Mantrika

Sanskrit Pilot instructions Shloka:

१. तत्र मान्त्रिकरहस्यो नाम ॥

मन्त्राधिकारोक्तरीत्या छिन्नमस्ता, भैरवी, वेगिनी, सिद्धान्त्यादिमन्त्रानुष्ठानादुप-
लब्ध सिद्धमार्गोक्त घुटिका, पादुका, दृश्याऽदृश्यादिशक्तिभिः तथा सिद्धाम्बा ओषधी-
श्वर्यादि मन्त्रानुष्ठानैः संप्राप्त ओषधीभिस्तद्द्रावकतैलादिभिश्च सुवनेश्वर्यादिमन्त्रानुष्ठान-
लब्ध मन्त्रशक्तिक्रियाशक्तिभिश्च कलासंयोजनद्वारा अभेद्यत्व अच्छेद्यत्व अदाह्यत्व
अविनाशित्वादि गुणविशिष्टविमानरचनाक्रियारहस्यम् ॥

Re-Translation

Through mantras, with the help of the resonance vibration phenomenon, first we remove obstacles, and achieving abilities of rapid production power, the basic things required for vimana are obtained and then the remaining parts are focused with the energy of magnetism and attraction for materialistic things for manufacturing vimana. The resulting vimana will be strong and cannot be destroyed in any manner.

Interpretation:

Sound waves are connecting to the building blocks of the universe: the energies and properties changing matter. There are various mantras described in Sanskrit shloka, and give a very lengthy ritual kind of information, but when translated into technical terms the above sentence is formed. The Sanskrit words having various meanings and thus when each of these mantras described in Sanskrit shloka when translated into technical terms, the ritual shlokas result in something technical procedure to manufacture a vimana.

Examples of modern Technology:

Matter wave and Quantum mechanics

$$\hat{H}|\psi(t)\rangle = i\hbar\frac{\partial}{\partial t}|\psi(t)\rangle$$

Schrödinger equation

All matter can exhibit wave-like behavior. For example, a beam of electrons can be diffracted just like a beam of light or a water wave. Matter waves are a central part of the theory of quantum mechanics, being an example of wave–particle duality.

The detailed mantras descriptions and translations need to be analyzed with experts and need further investigation. Also, how mantra waves can influence matter need to be studied.

2. Tantrika

Sanskrit Pilot instructions Shloka:

२. तान्त्रिकरहस्यो नाम ॥
महामाया शम्बरादि तान्त्रिकशास्त्रोक्तानुष्ठानमार्गो तच्चच्छक्त्यनुसन्धानरहस्यम् ॥

Re-Translation

By gaining power to remove complications of objects/things and by excelling or improvising their working, and other techniques powers, the vimana will be rewarded with all the technical powers

Interpretation:

Complications of objects or things may indicate structures needing strength, shape needing aerodynamics, etc, and such things are improved by removing complications. This shloka is mentioned in pilot instructions, so this also may indicate guidance for removing complexions related to space flight missions too.

Examples of modern Technology:

It could mean removing, Modern space flight challenges in various Phases:

1. Launch
2. Reaching space
3. Leaving orbit
4. Astrodynamics
5. Transfer energy
6. Reentry
7. Landing
8. Recovery

3. Kritaka:

Sanskrit Pilot instructions Shloka:

३. कृतकरहस्यो नाम ॥
विश्वकर्म, छायापुरुष, मनु, मयादि शास्त्रानुष्ठानद्वारा तत्तच्छक्त्यनुसन्धान-
पूर्वकं तात्कालिकसङ्कल्पानुसारेण विमानरचनाक्रमरहस्यम् ॥

Re-Translation

By the study of natural engineers/architects like universe working laws, similar things to the trees (nature), heart/head roles (monitoring, supplying etc), art or illusions (greediness to material pleasures we should avoid) and others, we can construct Vimanas of various patterns.

Interpretation:

Apart from wars and weapons mentioned earlier, this shloka, It indicates that ancient ancestors referred to the natural creation and observed and followed universal laws to implement vimana technology, so it is clear that they are aware of pollution and space debris.

Examples of modern Technology:

For example, it could mean universal laws like
- Universal law of gravitation
- Law of Vibration
- The Law of Perpetual Transmutation of Energy
- Law of Relativity

Etc.

Comments on Sanskrit Shloka Descriptions:

As described in Vimana shasthra about first Yuga humans, we see similar descriptions of such superhumans in the Ramayana too. Hanuman can grow huge and become tiny, he can fly without machines. An important observation in the mentioned Shlokas is that Dharma or Righteousness were four-footed and men possessed super powers from birth. And this again reminds us the story of Krishna when he was a newly born baby he was able to fight demons. He had strength when he is at that small age itself.

Human evolution, whichever we study in modern science books is not the real fact, but there was a missing link, like a missing puzzle piece we missed it in understanding the evolution of humans especially. The Sanskrit shloka lines mentioned about the human species and era, which has been termed as Yuga in Sanskrit. There were human species which once had supernatural, mystical powers. An era of humans in which everyone had this ability. They can be termed as "Gods"

In the human eras, later on, the ability of super powers started to decrease because of the decrease in positive energies or Dharma or justice. And thus the later humans were termed as "Rakshasas jathi". They were not able to access the memories of the spiritual gene power of ancient ancestors, or they were not able to understand spiritual dimension and not able to use all parts of the brain.

The ancient humans were able to do mind conversations and there was no need for any language. But later on, there was a necessity for creating language. As per the universe and nature based on basic human body, voice sound vibration, frequency etc, they created a language called Veda Bhaasha, which is the language of Rakshasas.

The supercontinent Pangaea had ancient civilizations which were using Veda bhaasha and advanced science and technology. Later on, the land divided due to the great flood and the civilization was torn apart, in which some part was saved in south India, the Sanskrit language is a part of ancient civilizations language.

In modern days near the Himalayas, we see sages who walk on water, and sages who fly. The first era humans had the ability to fly on their own without any machines. They realized that the abilities keep on decreasing for next generations and eras if there is an increase in negative energies in mind. So they wrote documents on how to fly using advanced knowledge about universe and energies.

The sages were given responsibility to educate and pass on the Vedic technology to the next generations and to the humans who are unable to fly by themselves. The later humans lost the ability to superhuman power and the ancient sages started passing on the Vedic technology information or using the help of machines and giving suggestions or writings to their next era humans, For the benefit of humanity. The Sanskrit shloka also contains words like Moral codes, physical material sciences, for travelling in the sky, propagating the art of manufacturing Vimanas, and for attaining wind-speed, all these terms are very interesting and it looks that they are considering moral codes, that is: not harming nature, which is a very good step for world peace.

The shlokas also clearly tell that the manuscripts or documents are written by the first era humans, which again means that the most ancient ancestors, who know advanced technology which doesn't harm the nature. The different types of Vimanas, Maantrika, Taantrika and kritaka are categorized based on human evolution and understanding capability and performance of brain functions of humans to use the Vimanas. In each three categories of Vimanas, there are sub-categories or in each type of Vimanas, there are several Vimanas placed as per the category, based upon functioning of vimana and usage ability by humans. Taantrika Vimanas description is giving more information that vimana uses next generation power or energy extraction which is modified compared to that of Mantrika type of vimanas. The source is described as earth and space.

Kritaka Vimanas all are made of RajaLoha, and similar to performance compared to Taantrika and Maantrika type of Vimanas. This shows that the material used to manufacture was more concentrated maybe because of changes in the atmosphere of the use of Vimanas in different locations of space. Or maybe there was a major event which took place modifying earth's atmosphere drastically.

Spaceplanes of the Modern World

As of 2001, conceptual design of as many as 22 reusable launch vehicle (RLV) concepts were in progress in the US, UK, France, Germany, Japan and India. Eight designs were for "Heavy Lift RLVs" having a capability to deliver large payloads of 10 to 25 tonnes into space. The remaining 14 systems were designed for smaller payloads, less than 5 tonnes in low earth orbit.

There were two basic approaches. Half of the design approaches was based using 20th century rocket propulsion systems. These were two-stage-to-orbit vehicles (TSTO), but once again, very heavy and cumbersome as the early space rockets, reaching orbit in two stages, but with one or both stages returning to earth for reuse. But, these designs have not succeeded, as the costs of building large vehicles were still very high.

Eleven RLV design concepts were based on a combination of air breathing and rocket engines. They fly to orbit directly like an aircraft (in a single stage), hence named single-stage-to-orbit vehicles, (SSTO). None of these is yet flying, but small scale "Technology Demonstrators" like the US X-43 has demonstrated air breathing engines and flight to Mach 10 very recently.

Design Requirements for SSTO Spaceplanes

The basic design requirements for a fully reusable hydrogen fuelled spaceplane, ascending to orbit from a runway take-off and re-entering for a powered landing like any commercial transport aircraft, are as follows:

The hydrogen fuel weight should exceed 56% of the spaceplane's take-off weight.

The time-averaged specific impulse over the flight path from earth-to-orbit should be more than 1200 secs (i.e. 1200 Kgs of propulsive thrust for ever one kg per second of hydrogen fuel flow)

The spaceplane's thrust-to-drag ratio has to be more than 3.5.
The air breathing engines have to be lightweight, with a thrust-to-weight ratio exceeding 14.

The first condition ensures that adequate chemical energy is available in the spaceplane that gets converted to kinetic energy to propel the aircraft-like spaceplane to a height of at least 100 kms and a speed of 8 kms per second. The second and third ensure that the engines operate with an average overall propulsive efficiency of over 40%. The last condition ensures that the payload-to-takeoff weight ratio is maximized.

Promising Contemporary Spaceplane Design Concepts

Out of about 22 design concepts that have been studied, small scale ground and flight tests carried out, three spaceplane design concepts are discussed here as these have the highest promise for successful development

The UK "Skylon" (late 1980's)

The "Skylon" is a heavy lift aerospacevehicle that has a length of 82 meters, a diameter of 6.25 meters and hence a high slenderness ratio of 13.1. It weighs 275 tonnes at take-off, out of which about 11 tonnes or 4% of its take-off weight is useful payload. It carries 218 tonnes of propellant (oxidizer and fuel) at take-off.(151 tonnes oxidizer (liquid oxygen) and 67 tonnes of fuel (liquid hydrogen)

Strictly speaking, "Skylon" is a "rocketplane" or a "winged rocket" and cannot be termed as a "spaceplane". This is because "Skylon" has nearly 79% total propellant fraction at take-off. Hence, at take-off, the vehicle weight consists of about 55% oxidizer (liquid oxygen) and 24% fuel (liquid hydrogen), which is almost identical to that of a conventional space rocket (60% oxidizer, 21% hydrogen fuel). However, there are two distinct differences that make this a promising candidate for a spaceplane:

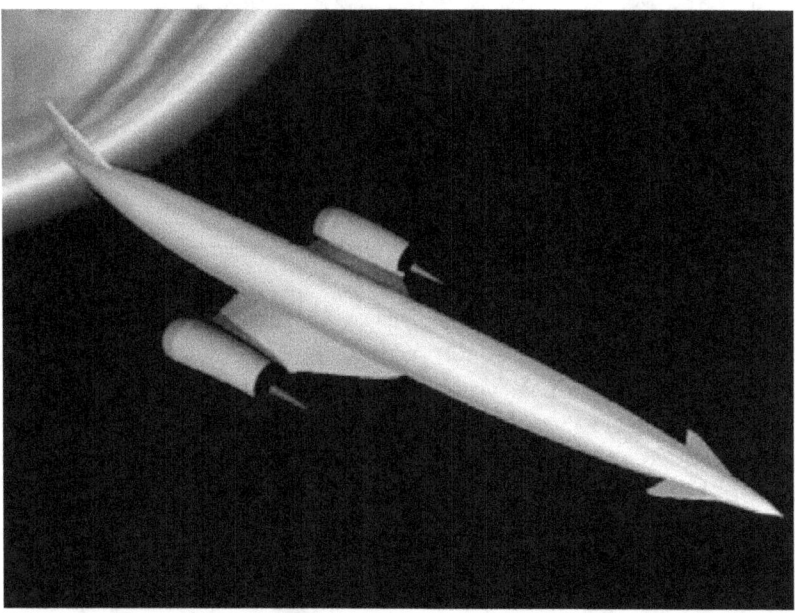

A novel air-breathing liquid rocket engine rocket (also known as a LACE or Liquid Air Collection Engine) the "Sabre" engine operating at an air-to- fuel ratio of 23:1 up to Mach 5. Thereafter, up to orbital speed of Mach 26, the same engine operates as a pure lox-hydrogen liquid rocket engine with

on- board liquid oxygen at a mixture ratio of 6:1. In this way, the oxygen needed to propel the vehicle up to Mach 5 is not carried on board at take-off, thus avoiding carriage of about an additional 218 tonnes of liquid oxygen at take-off [had the take-off engine been a pure lox-hydrogen rocket engine from take- off to orbit].

A winged-body vehicle configuration with podded- engines, that enables the vehicle to glide back and land, like the Space Shuttle. The high thrust-to-drag ratio of such a slender-body rocket configuration, and low structure weight fraction, compensate for its lower hydrogen-fraction at take-off

It is interesting to note that the "Skylon" is cigar- shaped, like the "Vailixi".

The US "Falcon" [Late 1990's, announced August 2003] The USAF has an ongoing programme for a small RLV, the "Falcon". This spaceplane configuration is a classical hypersonic lifting body configuration.

This spaceplane design concept is described as an unmanned hypersonic aircraft "bomber of the future". The technology of SLV is said to lead to a SSTO spacecraft. After take-off, a supersonic turbojet engine is used to reach speeds of Mach 2 or Mach 3, then the scramjet engines take over. At max hypersonic speed, SLV would deploy, either a separate craft to reach space, in which case it would be a TSTO vehicle, or switch from its air-breathing scramjet engine to rocket propulsion to be a SSTO vehicle. The payload would be 1,000-kilogram satellite into sun-synchronous orbits.

It is seen that the "Falcon" resembles the ancient Indian "Shakuna" and "Rukma" Vimanas.

X-43 High Speed Airbreathing Engine (Scramjet) Test Vehicle

Of direct application to the "Falcon" spaceplane programme are the recent successful flight tests carried out on the X-43 Hyper-X" scramjet test vehicles. The tests have demonstrated the efficacy of the supersonic combustion ramjet engine in the flight regime from Mach 8 to Mach 10 at an altitude of about 30 kms. Successful completion of these tests indicate yet one more critical milestone crossed towards direct ascent to near earth orbit for safe, affordable space flight within the next two decades.

The Indian "Hyperplane" or "Avatar" Spaceplane (late 1980's)

The "Hyperplane" / "Avatar" is designed to carry over 60% of its take-off weight as liquid hydrogen. This is made possible by not carrying any liquid oxygen on board at take-off, but collecting the requisite mass of liquid oxygen in high-speed flight. In this way, the spaceplane almost doubles its mass while in hypersonic level flight, while self-refuelling by air collection with simultaneous oxygen liquefaction and on-board storage. A small-scale Flight Technology Demonstrator for "Hyperplane"/ "Avatar" has also been designed.

The main attribute of the "Hyperplane" design concept is its geometric scalability, enabling the design be built for a vehicle as small as 25-tonnes take-off weight (the weight of an advanced fighter aircraft). This is possibly the smallest weight feasible for a reusable SSTO spaceplane, and has a 4% payload ratio, enabling delivery of 1-tonne in parking orbit at Mach 26. Unlike the "Skylon", the "Avatar" can be scaled up to heavy-lift capabilities.

General Comment

The close resemblance between the recent US "Falcon" and Indian "Hyperplane" spaceplane designs to the "Shakuna" and "Rukma" Vimana's, and the UK "Skylon" is cigar-shaped, like the Vimana like "Vailixi". That the "Shakuna", "Rukma" and "Vaillixi" were designed and built 12,000-15,000 years ago indicates that once again after a gap of millennia, mankind has embarked on development of systems and technologies for safe, affordable flight direct to space from a runway take-off.

It is essential that mankind learn from the recorded lessons of the ancient, dangerous past when spaceplane were weaponized and waged from outer space. Mankind must thus ensure, internationally that spaceplanes are not weaponized. These new, revolutionary technologies are to be used for a Second Industrial Revolution for all mankind, and not for domination of the planet by a single nation. Such a Space based industrial Revolution needs to serve space markets in developing countries and south-south cooperation in spaceplane development in partnership with advanced space faring nations would open a new, golden era for all mankind.

No one now in India needs to doubt any longer as to whether we will ever be able master reusable spaceplane technologies and put it to good use for enhancing security and prosperity not only for India, but all humanity. India has done so in the past. As a matter of fact, it is said that Albert Einstein had once remarked

"We owe a lot to Indians, who taught us how to count, without which no worthwhile scientific discovery could have been made."

India's genius will enable it to do it again, in full consciousness that it has to be a globally cooperative mission serving all mankind.

Scientific Overview

Ancient ancestors had 12 strand DNA, hence had more intelligence than modern humans. Sanskrit documents contain advanced science and technology in them, which are documented by ancient ancestors. In the process of giving their valuable information to the next generations of human race, Maharshi Bharadwaja and several other ancient scientists or Rishis provided us Texts like Vimana shasthra. This book also describes modern day rediscoveries and reinventions from Vimana shasthra. Our team SWASTIK (Scientific Works on Advanced Space Technology Investigators for Knowledge) is a group of researchers working on the lost advanced ultimate ancient technology. SWASTIK team works on different types of vimana to make vimana prototypes, their propulsion systems, and modern software works such as 3D modeling design of vimana and aerodynamic, thermal analysis, and ancient materials properties for advanced space radiation are described in this Book.

A part of this work is described in this Chapter where the 2D models described by Subbaraya shasthry and drawn by Ellappa which we now made into 3D models and views are shown. The vimana models are done using the design dimensions of those drawings as well as modern day aerospace and software knowledge. CFD analysis of Rukma Vimana is performed and results are analyzed. The advanced materials and fuels used in vimana, as well as the propulsion which are eco-friendly, are described in details.

Rocket Launch Technology of ancient ancestors is more highly advanced than compared to modern technology. Raja Loha, a high-heat-absorbing alloy used for the bodies of various flying crafts, preparation, and properties of each material in its compositions are also mentioned in the paper and also describes our research works on Rukma vimana, which reveal that it is an advanced interplanetary Vedic space vehicle.

The following points are important events in history describing modern day works on ancient Sanskrit texts:

1) According to ancient astronaut theorists, Buddha had extra-terrestrial contact and visited earth from Vimana, and when Ashoka was educated by Buddha, he had to hide the knowledge and keep it secret to avoid world's destruction by Evil people. Ashoka and a secret society of 9 unknown men are still a mystery. Shivkar Bapuji Talpade was Sanskrit scholar during the British rule. He made attempts to reconstruct ancient vimana using ancient propulsion. Maharaja of Baroda witnessed the event of flying his Marutsakha vimana. Later some of the witness and his vimana were destroyed by evil people. Before World War II,
Hitler and his team worked on Ancient texts which include Ancient Indian Sanskrit texts also. Nuclear weapons, Mercury vortex engine, and Bell shaped flying object named De Glocke have been similar to ancient Indian technology. Few years after the war, blueprints of Antigravity Flying machine were found.

2) NASA recorded the Sacred Hindu OM Sound from Sun's atmosphere. Ancient Sanskrit texts mentioned that sun chants OM, and some texts describe that every atom bonds with another atom with the frequency of sound OM. Sanskrit is declared as the perfect language and can be used in Artificial intelligence.

3) Nikola Tesla was influenced by Swamy Vivekananda and learned Sanskrit terms from him to read the texts. He tried to invent many things mentioned in ancient Indian technology texts and took patents. Some of his inventions are Aerial Transportation vehicle which is similar to Rukma vimana and also Tripura Vimana. This is VTOL aircraft. There are free energy Motors and generators mentioned in Vimana shasthra. According to the studies, Tesla tried to invent Free energy generators.

4) Modern scientists reinventions from Sanskrit texts indicate the value of texts. CSR Prabhu invented following things: Materials such as Raja Loha, Tamogarbhaloha, Pancha Loha, Araara Tamra, and Badhira Loha. Glasses such as Vidyutdarpana, Ravishakti apakarshana Darpana, Ushna Shakti Apakarshana Darpana and Vimana Device, Vakra Prasarana Yantra. Professor Sharon Invented Chumbak Mani and N. G. Dongre invented Dhvantapramapaka Yantra. Ancient Nanotechnology related inventions are done and by Sri Maharshi Research Institute of Vedic Technology. They invented a novel process of preparing nano metal and the products thereof.

Throughout the history there have been attempts to reinvent ancient Indian technology, and now in the 21st-century humans are successful in reinventing. Hence proving the texts as non-mythological and high advanced technological. Our team SWASTIK- (Scientific works on Advanced Space technology Investigators for Knowledge) is a group of researchers working on Vimana technology. Rukma Vimana's Manufacturing, Structures, propulsion, aerodynamics, Space mechanics are described in this book.

There are three types of Vimanas mentioned in Vimana shasthra classified depending upon the Yugas: Maantrika, Taantrika, and Kritaka. All these are described and designed by analyzing the resources available on earth and also depending on the Human intelligence to control the Vimanas. Sanskrit texts date back to several thousands of years ago. Such intelligent advanced technology is known by our ancestors, but how? The Yugas in Vimana shasthra mentioned that the first humans were able to fly themselves without any machine, and the later humans started facing difficulty and had to use mantras, and the next humans had to use machines. Now investigating these facts, it is clear that there are spiritual knowledge consideration and the usage of the brain. Ancient Ancestors use 100% of their brain and can fly without any machines and the usage of brain decreased as generations passed by. There may not be a possibility that the first Humans intelligent and use 100% of the brain by default by birth, But may be influenced by any Extra-terrestrials. The Early man era was described in our history and the Archaeological evidence such as caves describe that the early man was not so intelligent and the planet was often visited by advanced beings in vimanas. There may be a possibility that a planet with advanced technology beings had influenced the apes on earth by altering DNA in order to make a hybrid species called humans. Vishnu Avatars as described in Sanskrit texts are protecting the human race from evil.

We, humans, are poisoning our race by polluting the nature and our brains. There are positive and negative radiations which are always balanced by the destruction through nature. Whatever might be the possibilities for these mentioned things, but one thing is clear that modern humans are being protected and Guided through Sanskrit texts which make humans more intelligent and protect the nature also. Below are some Modern day works on ancient Sanskrit texts: Modern scientists Reinventions from Sanskrit texts indicate the value of texts. CSR Prabhu invented following things: Materials such as Raja Loha, Tamogarbhaloha, Pancha Loha, Araara Tamra, and Badhira Loha. Glasses such as Vidyutdarpana, Ravishakti apakarshana Darpana, Ushna Shakti Apakarshana Darpana and Vimana Device: Vakra Prasarana Yantra. Professor Sharon Invented Chumbak Mani and N. G. Dongre invented Dhvantapramapaka Yantra. Ancient Nano technology related inventions are done and by Sri Maharshi Research Institute of Vedic Technology.

They invented a novel process of preparing Nano metal and the products thereof. Throughout the history, there have been attempts to reinvent ancient Indian technology, and now in the 21st-century humans are successful in reinventing. Hence, proves the texts as non-mythological and highly advanced technology.

II. Aerial Wars, UFO's and Aliens

A description in Vimana shasthra: *"Bharadwaaja implies that the Vimana or airplane constructed according to Vymanika Shastra, may enable men to reach God, and enjoy the benefits of His Divine abode."*

Our Explanation and interpretation:

We can make anything possible. We humans have that power. But the power is hidden. The one who understand this power and able to access this power they will reach god and enjoy all his dimensions of creation. As it is mentioned in vimana shasthra that human body and vimana parts are similarly designed, we can assume that the one who will understand his own body working etc, including all areas of his brain, he can understand the working of vimana mentioned in vimana shasthra. Or the one who understands vimana shasthra working of vimana can understand the creation of god, the human body. If he can implement in form of machines for vimana, that is if he creates vimana similar to god's creation, the human body, then the person will be able to access all dimensions including spiritual, and he will be reaching god. God means the one who created us.

This clearly explains that someone created the human species and is watching us, guarding us, guiding us. In Instructions given for pilot, in Vimana shasthra there are many weapon systems and war-related technical descriptions. Now the Question is why a sage Maharshi Bharadwaja would be so much interested in educating humans about the war and weapons. The answer lies in aerial wars happened with ancient ancestors and paintings describing UFO shaped objects having war with Indian gods. So it clearly again explains to us that there is an alien race trying to destroy the human race, and since sages are against the destruction of god's creation, Hence in the process of trying to protect human race they mention various war weapons systems in vimana shasthra.

TACTICAL AIR DEFENSE

Re-Translations and Interpretations of Vimana shasthra

1. Destruction:

Sanskrit Pilot instructions Shloka:

३२. कर्पणरहस्यो नाम ॥

स्वविमानसंहाराथै परयानपरम्परागमने विमाननाभीमुखस्थ वैश्वानरनालान्तर्गत ज्वालिनीशक्तिप्रज्वलनं कृत्वा सप्ताशीति लिंङ्गप्रमाणोष्णं यथा भवेत्तथा चक्रद्वयकीली- चालनात् शत्रुविमानोपरि वर्तुळाकारेण तच्छक्तिप्रसारणद्वारा शत्रुविमाननाशनक्रिया- रहस्यम् ॥ ८७ ॥

Re-Translation

When enemy planes come together in strength to destroy one's vimana, by setting aflame the fire power in the injurious tube, pipe located at the navel of the plane, and operating the keys of the two wheels, setting temperature to 87 degrees of heat, the burning shakti(power) will envelope/ surround the enemy plane and destroy it.

Devices used:

- Shaktisthaana at the front and right sides.
- At the front of the left side are to be located the Naalapanchakanaala which means something having 5 tubes or pipes.
- Pancha-vaataskandha-naala on the western center.
- Vyshwaanara-naala at the navel center.

Interpretation:

During Aerial wars when the enemy planes attack in a group, then the vimana used for facing them should be equipped or arranged with:
1. Shakthisthaana at the front and right sides of vimana: This is the Location where the energy exists. Setting temperature to 87 degrees of heat, the burning Shakti (power).
2. Naalapanchakanaala, at the front side 5 tubes or pipes: At the front side, 5 cylindrical tubes
3. At the left side should be 5 pipes again termed as Pancha-vaataskandha-naala: 5 cylindrical tubes like missiles.
4. Another pipe at the navel center termed as Vyshwaanara-naala: injurious tube, which also means missile.

The above devices or equipment are used to attack all the enemy planes. By operating these devices, a burning fire will surround the enemy plane and destroy it.

The word naala means tube and maybe the cylindrical missile structure kind of thing, it maybe the location of the missile from which missile shoots onto the enemy planes. Here vimana shasthra mentioned 5 pipes, and it means 5 missiles located hub kind of things which are pipe like cylindrical structures embedded into vimana.

Examples of modern Technology:

An air-to-air missile (AAM) is a missile fired from an aircraft for the purpose of destroying another aircraft. AAMs are typically powered by one or more rocket motors, usually solid fuelled but sometimes liquid fuelled.

Astra Air-to-Air Missile

2. Methods bringing terrific shape:

Sanskrit Pilot instructions Shloka:

१२. विरूपकरणरहस्यो नाम ॥

धूमप्रकरणोक्तप्रकारेण द्वात्रिंशज्जातीयधूमराशिं यन्त्रद्वारा परिकल्प्य तस्मिन् खतरङ्गशस्त्रयुष्णसंजनितप्रकाशं मेळयित्वा पश्चात् विमानशिरोभागस्थ भैरबीतैलसंस्कारित वैद्युदर्पणमुखे पञ्चकचक्रमुखनाळद्वारा पूर्वोक्तप्रकाशशक्तिं सन्धार्य द्वात्रिंशदुत्तरशत-कक्ष्यप्रमाणवेगात् परिभ्राम्यमाणे सति मण्डलाकारेण महाभयप्रदविकाराकारो जायते । विमानद्रूणां तत्प्रदर्शनद्वारा महाभयोत्पादनकार्यरहस्यम् ॥

Re-Translation

As stated in Vapour subject, by producing the 32nd kind of vapour through mechanism and charging it with the solar power or light of heat waves in the sky, and projecting it through the flower-shaped wheel tube to the bhyravee (frightful), oil-smeared Vyroopa (Unique) Darpana (mirror) located at the top of vimana, and whirling with 132nd type of speed (Supersonic or high Mach number) a very fierce and terrifying shape of vimana will emerge, causing utter fright to onlookers.

Devices used:

- Vyroopyadarpana which means different/various shapes/appearance Reflecting mirror

- Padmachakramukha which means wheel appearing like a flower at the shirobhaaga or top head end or crest of the Vimana.

Interpretation:

A method to hide the identification of vimana, by Projecting to the viewers, a terrific shape will make them Scared.

The Word Darpana means not exactly mirror but something which reflects, for example as mentioned in the Vimana parts, the Vimana is similar to the human body, similar to the pressure points, the reaction is seen in different body parts according to the specific pressure point.

Now we can see the reflection of it, so the same way here vimana Darpana meant the projection. 3D projection virtual image techniques methods are described, Vyroopyadarpana is holographic image projector which projects various shapes.

Examples in modern Technology:

Innovation Center Dinosaur Hologram

3. Transforming shape:

Sanskrit Pilot instructions Shloka:

१३. रूपान्तररहस्यो नाम ॥

तैलप्रकरणोक्तप्रकारेण गृध्रजिह्वा, कुम्भिणी, काकजङ्घादितैलसंस्कारितवैरूप्यदर्पणे एकोनविंशज्ञातीयधूमं संयोज्य, तस्मिन् यानस्थकुण्टिनीशक्तिसंयोजनद्वारा विमानद्रष्टॄणां सिंह, व्याघ्र, भल्लूक, सर्प, गिरि, नदी, वृक्षादिविकारेण अन्यथाकल्पितरूपान्तर प्रदर्शनरहस्यम् ॥

Re-Translation

As stated in subject of oils, by preparing griddhrajihwaa, kumbhinee, and kaakajangha oils and rubbing to the distorting mirror in the Vimana with them, applying to it the 19th kind of smoker and charging with the kuntinee shakti (Electricity) in the Vimana, shapes like lion, tiger, rhinoceros, serpent, mountain, river will appear and amaze observers and confuse them.

Interpretation:

This explains creating a smoke/vapor surrounding the vimana using mirror and oils which reflect various animals selected by pilot, the smoke around vimana hides vimana showing only animals

4. Beautiful shape:

Sanskrit Pilot instructions Shloka:

१४. सुरूपरहस्यो नाम ॥

करकप्रकरणोक्तत्रयोदशजातीयकरकशक्तिमाकृष्य हिमोद्गारवायुना संघर्षे पश्चात् विमानदक्षिणकेन्द्रमुखस्थित पुष्पिणी पिञ्जुलादिदर्पणमुखे पूर्वोक्तशक्ति वातप्रसारण-नाळद्वारा संयोज्य, तस्मिन् सुरघाख्यशक्ति सन्धार्य, तद्द्वारा विमानसन्दर्शकानां विविधपुष्पमाल्योपसेबित दिव्याप्सरस्वरूपवत् विकारसंदर्शनक्रियारहस्यम् ॥

Re-Translation

> By attracting the 13 kinds of producing force mentioned in Force production subject, applying snow-surcharged air and protecting it through the air conveying tube to the pushpinee-pinjula (Flowers-stalks) mirrors in the front right side of the Vimana, and focusing on it the suragha (Strong/powerful/robust) beam, a heavenly damsel bedecked with flowers and jewels will appear to onlookers of the Vimana.

Devices used:
- Pushpinee and Pinjulaa which means Flowers and grass type Mirrors are to be on the right side of the center.

Interpretation:

This explains projection of a holographic image of a beautiful lady with flowers and jewelry. This will make the pilots shocked and attract towards the vimana, which would be easy for the vimana to attack them.

5. The secret to sun-like Glow:

Sanskrit Pilot instructions Shloka:

१५. ज्योतिर्भावरहस्यो नाम ॥

अंशुबोधिन्यामुक्तप्रकारेण संज्ञानादिषोडशसूर्यकलासु द्वादशाद्याषोडशान्त कला
प्रभाकर्षणं कृत्वा आकाशचतुर्थपथस्य मयूखकक्ष्यस्थितवायुमण्डले नियोजयेत् ॥
तथैव खतरङ्गशक्तिप्रभामाहृत्य वातमण्डल सप्तमावरणस्थ प्रभाशक्त्या संमेलयेत्
पश्चादेतत् शक्तिद्वयं विमानस्थनालपञ्चकद्वारा विमानगुहागर्भदर्पणयन्त्रतृतीयकोशे
सन्धार्यं तद्द्वारा विमानद्रष्टॄणां बालातपवत् प्रकाशप्रदर्शनरहस्यम् ॥

Re-Translation

As Described in atomic theory, one among out of various and 16 digits of solar glow, by attracting 12th and 16th Digits and by focusing them on bright region of 4th pathway or atmospheric region in the sky, and similarly attracting the force of sun's glow and mingling it with the glow in 7th layer of atmosphere, projecting these forces through 5 tubes in the vimana on to the section of concave mirror, a rich glow like the morning glow of the sun will be produced.

Devices used:

- Guhaagarbha or concave mirror device is to be in the front part of the stomach of the plane.

Interpretation:

It describes a certain portion of the electromagnetic spectrum or wavelength, extracting solar energy from this specific spectrum, by using the device called Guhaagarbha or concave mirror device. This device might be using different projections similar to the holographic projector, and this also maybe reflection principle which reflects solar glow many times in many mirrors which result in the glow like the sun. Here a point to note is that mirrors need to be dealt carefully in order to match the specific reflection with specific arrangements.

6. Facing opposite party Vimana

Sanskrit Pilot instructions Shloka:

१८. विमुखरहस्यो नाम ॥
ऋक्हृदयोक्तप्रकारेण कुबेरविमुखवैश्वानरादि विषचूर्णशक्तीः रौद्रीदर्पणपञ्जर
तृतीयनाळे नियम्य वानरकन्धकीलीचालनद्वारा मूर्छावस्थाप्रदानेन विवर्णकरणं किया
रहस्यम् ॥

Re-Translation

Facing opposite party Vimana: As mentioned in "Rig Hridaya" (Heart of Rig-Veda), by projecting air force of poison powders like Kubera, vimukha and Vyshawaanara, through the third tube of roudree mirror and by operating switch to activate air mechanism, will produce extensive insensibility and coma to the people in enemy Vimana.

Devices used:
- Rowdree mirror.

Interpretation:

It explains the instructions to attack the enemy planes with the help of poison powder which makes the pilots experience extensive insensibility and coma. Rowdree mirror means the projector of poisonous smoke.

Chemical warfare

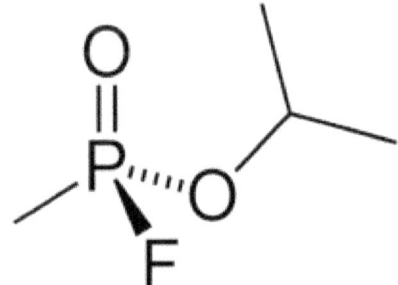

The chemical structure of sarin nerve gas, developed in Germany in 1939.

The Germans may have used poison gas on survivors from the Battle of Kerch, May 1942

7. Stunning, making immobile

Sanskrit Pilot instructions Shloka:

३१.   स्तम्भकरहस्यो नाम ॥
विमानोत्तरपार्श्वेऽस्य सन्धिमुखनाळा दपस्सारधूमं संग्राह्य स्तम्भनयन्त्रद्वारा
तड्डूमप्रसारणाव् परविमानस्थ सर्वजनानां स्तब्धीकरण रहस्यम् ॥

Re-Translation

Stunning, making immobile: By projecting apasmaara (confusion of mind) poison-fume smoke through the tube on the north side on the Vimana, and discharging it with stambhana-yantra (a device which makes immobile), people in enemy planes will be made unconscious.

Devices used:

- Apasmaara or poison gas at the sandhi-naalamukha or junction tube front.
- Sthambhanayantra at the bottom.
- Shabda-kendra-mukha at the left side.
- Shabdaakarshakayantra at the shoulder.

Interpretation:

This pilot instruction again explains poisonous smoke with the tube on the north side, which means apasmaara device is used. Sthambhanayantra is a device which makes people immobile. Shabda-kendra-mukha means a device which is the main producer of sound or maybe a device which produce a sound similar to the speaker. Shabdaakarshakayantra which means a sound attracting device, maybe the people's mind are influenced due to this sound and made them immobile and stunning.

Similar Examples of modern Technology:

Classes of chemical weapon agents

Class of agent	Agent Names	Mode of Action	Signs and Symptoms
Nerve	<ul><li>Cyclo-sarin (G F)</li><li>Sarin (G B)</li><li>Soman (GD)</li><li>Tabun (GA)</li><li>VX</li><li>VR</li><li>Some insectici-des</li><li>Novi-chok ag</li></ul>	Inactivates enzyme acetylcholin esterase, preventing the breakdown of the neurotransmitter acetylcholine in the victim's synapses and causing both muscarinic and nicotinic effects	<ul><li>Miosis (pinpoint pupils)</li><li>Blurred/dim vision</li><li>Headache</li><li>Nausea, vomiting, diarrhea</li><li>Copious secretions/sweating</li><li>Muscle twitching/fasciculations</li><li>Dyspnea</li><li>Seizures</li><li>Loss of consciousness</li></ul>

Category	Agents	Mechanism	Symptoms
Asphyxiant/Blood	• Most Arsines • Cyanogen chloride • Hydrogen cyanide	• **Arsine**: Causes intravascular hemolysis that may lead to renal failure. • **Cyanogen chloride/hydrogen cyanide**: Cyanide directly prevents cells from using oxygen. The cells then use anaerobic respiration, creating excess lactic acid and meta...	• Possible cherry-red skin • Possible cyanosis • Confusion • Nausea • Patients may gasp for air • Seizures prior to death • Metabolic acidosis
Vesicant/Blister	• Sulfur mustard (HD, H) • Nitrogen mustard (HN-1, HN-2, HN-3) • Lewisite (L) • Pho...	Agents are acid-forming compounds that damages skin and respiratory system, resulting burns and respiratory problems.	• Severe skin, eye and mucosal pain and irritation • Skin erythema with large fluid blisters that heal slowly and may become infected • Tearing, conjunctivitis, corneal damage • Mild respiratory distress to marked airway damage
Choking/Pulmonary	• Chlorine • Hydrogen chloride • Nitrogen oxides • Phosgene	Similar mechanism to *blister agents* in that the compounds are acids or acid-forming, but action is more pronounced in respiratory system, flooding it and resulting in suffocation; survivors often suffer chronic breathing problems.	• Airway irritation • Eye and skin irritation • Dyspnea, cough • Sore throat • Chest tightness • Wheezing • Bronchospasm
Lachrymatory agent	• Tear gas • Pepper spray	Causes severe stinging of the eyes and temporary blindness.	Powerful eye irritation

Incapacitating	• Agent 15 (BZ)	Causes atropine-like inhibition of acetylcholine in subject. Causes peripheral nervous system effects that are the opposite of those seen in nerve agent poisoning.	• May appear as mass druginto-xication with erratic behaviors, shared realistic and distinct hallucinations, disrobing and confusion • Hyperthermia • Ataxia (lack of coordination) • Mydriasis (dilated pupils)

Cytotoxic proteins	Non-living biological proteins, such as:
	• Ricin
	• Abrin
		Inhibit protein synthesis	• Latent period of 4-8 hours, followed by flu-like signs and symptoms
			• Progress within 18-24 hours to:
			  • Inhalation: nausea, cough, dyspnea, pulmonary edema
			  • Ingestion: Gastrointestinal hemorrhage with emesis and bloody diarrhea; even-

8. Invisibility

Sanskrit Pilot instructions Shloka:

९. अदृश्यरहस्यो नाम ॥

शक्तितन्त्रोक्तरीत्या सूर्यस्थेषादृष्ट प्राङ्मुख पृष्ठकेन्द्रस्थ वैणरथ्य विकरणादि-शक्तिभिः आकाशतरङ्गस्थ शक्तिप्रवाहमाकृष्य वातमण्डलस्थ बलाहाविकरणादिशक्तिपञ्चके नियोज्य तद्द्वारा श्वेताघवन्मण्डलाकारं कृत्वा तदावरणात् विमानादृश्यकरणरहस्यम् ॥

Re-Translation

Invisibility: According to the study of energy extraction (electricity through wires), by means of the modification of elements in air useful for flying chariots and modification of other forces/energies from the centre of sun, extracting the force from outside earth passing through space and sky, and mingle it with the dense thick air and modifying its energies by extracting in the aerial globe, produces thereby a White cover which will make the vimana invisible.

Interpretation:

It speaks about hiding the identity of vimana, by making it invisible through covering it with white sphere kind of projection. It explains energy extraction and modification of elements in air for energy. An interesting thing mentioned is sun's energy extraction, which may be similar to solar panels or maybe more advanced. As modern humans start approaching better technology, we are able to understand ancient texts in a better way. Because the advancement in modern technology is not up to the high range of ancient's and is lagging in many areas.

9. Presence

Sanskrit Pilot instructions Shloka:

९. अपरोक्षरहस्यो नाम ॥
शक्तितन्त्रोक्तरोहिणीविद्युत्प्रसारणेन विमानाभिमुखस्थवस्तूनां प्रत्यक्षनिदर्शन-
क्रियारहस्यम् ॥

Re-Translation

> Presence: According to the subject of energy techniques, by the projection of the brightest beam of light, things in front of vimana are made visible.

Interpretation:

The words Brightest Beam of light projection could mean object detection scanning using energy techniques, which makes the objects in front of vimana to be visible.

10. Destruction

Sanskrit Pilot instructions Shloka:

१७. प्रळयरहस्यो नाम ॥

ऐन्द्रजालिकप्रळयपटलोक्तरीत्या यानपुरोभागकेन्द्रस्थ उपसंहारयन्त्रनाळात् सप्तम
जातिवद्धूममाकृष्य षड्गर्भविवेकोक्त मेघधूमेऽन्तर्धाय तद्धूमं विद्युत्संसर्गात् पञ्चस्कन्ध
वातनाळमुखेषु प्रसार्य तद्द्वारा सर्वपदार्थानां प्रळयवत् नाशक्रियाकरणरहस्यम् ॥

Re-Translation

Destruction – as described in book of magical destruction, attracting the 5 kinds of smoke through the tube of pressurising machine in front part of vimana and merge it in vaporized form (cloud smoke) mentioned in the book shadgarbhaviveka, and pushing it by electric force through five-limbed aerial tube, will destroy everything as in a cataclysm.

Interpretation:

It is very interesting to know that ancient ancestors had the book of magical destruction. Pressurizing 5 kinds of smoke is another interesting feature, the smokes could be explosive chemical related and thus resulting in destruction after pushing out through five-limbed aerial tube.

11. Radiance

Sanskrit Pilot instructions Shloka:

१९.ताररहस्यो नाम ॥

वात जलसूर्यकिरणप्रभाशक्तीनां दशसप्तषोडशांशान् खतरङ्गशक्त्या संयोज्य
तच्छक्तिं नागमुखदर्पणद्वारा विमानमुखकेन्द्रशक्तिनाळमुखप्रसारणात् सर्वेषां नक्षत्र
मण्डलवत् प्रदर्शनक्रियारहस्यम् ॥

Re-Translation

Radiance: By mixing with unearthly (other planets or stars) force, 10 units/fraction of air strength, 7 units/fraction of water force, and 16 units/fraction of solar glow, and projecting it by means of the mirror facing star, through the frontal tube of the Vimana, the appearance of a star-like bright sky is created.

Interpretation:

Making vimana glow like Radiance of a star is explained in this pilot instruction. Mirror facing the star could be Projector facing from a star or facing to Star. There could be devices in ancient times which can project or reflect anything which it faces, and thus vimana in this shloka is explained to have an appearance like a bright star in the sky.

12. Ultra-sound art of confusing
Sanskrit Pilot instructions Shloka:

२०. महाशब्दविमोहनरहस्यो नाम ॥

विमानस्थसप्तनाळवायुमेकीकृत्य शब्दकेन्द्रमुखे अन्तर्धाय पश्चात् तत्कीलीं प्रचालयेत् । तद्वेगात् शब्दप्रकाशकोक्तरीत्या द्विषट्विमानकलासहूट्टनशब्दवत् महान् शब्दो जायते । तच्छब्दस्मरणात् सर्वेषां हृदयकम्पनं भवति । किण्वत्रयप्रमाणकम्पने यदा भवति तदा स्मृतिविस्मरणं भवति । तद्द्वारा परेषां विमोहनक्रियारहस्यम् ॥

Re-Translation

The ultra-sound art of confusing: By concentrating the air force in the seven tubes of the Vimana and operating the switch, as stated in the book "Shabdaprakaashikaa" will activate a huge loud thunder-like noise, which will stun people, and makes them broken with fear and become insensible.

Interpretation:

This describes another special book called "Shabdaprakaashikaa". Air force concentrating through seven tubes is made to activate huge loud thunder like noise. Here, in this case, the tubes should be special which are made of special material which produces too much sound even with the air force. Or there could be another loudspeaker kind of device and also it may be making use of sonic boom (sound associated with the shock waves created by an object traveling through the air faster than the speed of sound). The main focus of this instruction is to make people fear and to make them become insensible by making huge loud thunder-like noise.

13. Snake-like movement
Sanskrit Pilot instructions Shloka:

२२. सर्पगमनरहस्यो नाम ॥

दण्डवक्त्रादि सप्तविध आतरिश्चार्किकिरणशक्तीराकृष्य यानमुखस्थ वक्त्रप्रसारण-केन्द्रमुखे नियोज्य पश्चाचदाहृत्य शब्दयुद्गमनाले प्रवेशयेत् । ततः तत्कीलीचालनात् विमानस्य सर्पवत् गमनक्रियारहस्यम् ॥

Re-Translation

> Snake-like movement: By attracting Elephant-like force and other seven forces of air, and joining with solar rays, passing it through the zig-zagging center of the Vimana, and by operating the switch related to it, the Vimana will have a zig-zagging motion like a snake.

Interpretation:

A device called "Vakra Prasarana Yantra" is mentioned by modern scientist. This device is said to make vimana move like a snake. The shloka here may refer to pilot instruction explaining the same device at zig-zagging center, which make vimana move like a snake or zig-zag motion.

14. Wavering/unsteady

Sanskrit Pilot instructions Shloka:

२३. चापलरहस्यो नाम ॥
शत्रुविमानसंदर्शनकाले विमानमध्यकेन्द्रस्थ शक्तिपञ्जरकीलीचालनेन एक-
छोटिकायच्छिन्नकाले सप्ताशीत्युत्तरचतुस्सहस्रमहातरङ्गवेगो जायते । तत्प्रसारणात्
शत्रुविमानकम्पनक्रियारहस्यम् ॥

Re-Translation

> Wavering/unsteady: On sighting an enemy plane, by operating the switch in the force center in the middle section of the vimana, atmospheric wave with speed 4087 revolutions an hour is generated from vimana and will shake up the enemy plane.

Interpretation:

This shloka refers to creation of atmospheric wave from vimana to shake the enemy plane, by creating to it an aerodynamic flutter through Turbulence.

Atmospheric wave:

An atmospheric wave is a periodic disturbance in the fields of atmospheric variables (like surface pressure or geopotential height, temperature, or wind velocity) which may either propagate (traveling wave) or not (standing wave).

15. Facing all sides

Sanskrit Pilot instructions Shloka:

२४. सर्वतोमुखरहस्यो नाम ॥
खगथे खविमानविनाशनार्थं परविमानशतै: आवृते सति तदा खविमानशिर:-
केन्द्रकीलीचालनादनेकविमानवत् सर्वतोमुखसञ्चारक्रियारहस्यम् ॥

Re-Translation

Facing all sides: When enemy planes in group formation come to attack one's vimana, by operating switch which controls the upper end or crown of the vimana, make it revolve with alertness and face all sides.

Interpretation:

Making vimana easy to observe, defend or counter-attack enemy Vimanas by facing all sides. It could also mean the attacking procedure of projecting missiles through all locations in the circumference of vimana.

16. Foreign/alien vimana sound observing

Sanskrit Pilot instructions Shloka:

२५. परशब्दग्राहकरहस्यो नाम ॥
सौदामिनीकलोक्तप्रकारेण विमानस्थ शब्दग्राहकयन्त्रद्वारा आकाशप्रथममण्डल
परिधिमारभ्य सप्तममण्डलपर्यन्तं परविमानस्थ जनसंभाषणादिसर्वशब्दाकर्षणरहस्यम् ॥

Re-Translation

Foreign/alien vimana sound observing: As explained in "Sowdaamineekalaa" or the Art of listening, by means of the sound capturing Device with the science of electronics in the vimana, to hear the talks and sounds in enemy vimana flying in the sky.

Interpretation:

sound capturing Device Maybe something which hacks black box (flight data recorder) of alien planes in the sky.

Examples of modern Technology:

News Article:
This company can 'hack' and completely take over enemy drones for the US military

17. Visual view attraction

Sanskrit Pilot instructions Shloka:

२६.   रूपाकर्षणरहस्यो नाम ॥
विमानस्थ रूपाकर्षणयन्त्रद्वारा परविमानस्थित वस्तुरूपाकर्षण रहस्यम् ॥

Re-Translation

Visual view attraction: By means of the photographic device in the vimana, to obtain a television view of things going on inside the enemy plane.'

Interpretation:

This could be a tiny device which shoots the enemy plane, which makes way into the inside of the plane to capture what's happening inside it.

18. Action observing

Sanskrit Pilot instructions Shloka:

२७. क्रियाग्रहणरहस्यो नाम ॥

विमानाधःकीलीचालनाव् शुद्धपटप्रसारणं भवति । ईशान्यकोणस्थद्रावकत्रं शक्तिसंयोजनं कृत्वा तच्छक्ति सप्तमवर्गसूर्यकिरणेषु सन्धार्य तत्किरणान् त्रिशीर्षदर्पण नाळदण्डेऽन्तर्धाय पूर्वोक्तशुद्धपटं दर्पणाभिमुखीकरणं कृत्वा तन्मुखात्पूर्वोक्तशक्ति प्रसारणपूर्वकार्धकीलीचालनद्वारा विमानाधोभागस्थित पृथिव्यन्तरिक्षेषु यद्यत्क्रिया रहस्यानि अन्यैः क्रियन्ते तत्स्वरूपप्रतिबिम्बः शुद्धपटे मूर्तवच्चित्रितो भवति । तद्द्वार क्रियाग्रहणरहस्यम् ॥

Re-Translation

Action observing: By operating key which controls bottom of vimana a white cloth is made to appear for display. By electrifying three acids in the north-east part of the Vimana, and exposing them to the 7 kinds of solar rays, and passing the resultant force into the tube of the Thrisheersha mirror or Three-Headed mirror and making the cloth screen face the mirror, and switching on the upper key, all the activities going on down below on the ground, will be projected on the screen.

Interpretation:

A device which is made to capture what's happening below on the ground, and to project in the vimana on cloth which is used as a screen. Maybe used for unplanned or emergency landings in unknown areas on the ground.

19. Sky like appearance

Sanskrit Pilot instructions Shloka:

२९. आकाशरहस्यो नाम ॥

आकाशतन्त्रोक्तरीत्या कृष्णाभ्रवारिणा पिचुकन्दमूलभूनागद्रावकाभ्यां यानाबर णाभ्रपट्टिकामालिप्य तस्मिन् वायुपथकिरणशक्तिसंयोजनद्वारा विमानमाकाशाकार त्वप्रदर्शनरहस्यम् ॥

Re-Translation

Sky like appearance: according to Sky-techniques or sky-devices, by mixing black mica solution with neem and bhoonaagaor earth-snake decoctions and smearing the solution on the outer body of the Vimana made of mica plates, and exposing to solar rays, the plane will disappear and look like the sky and become invisible appearing like the sky.

Interpretation:

The chemicals or materials described in this shloka maybe producing a sky blue color solution which when applied on vimana makes It look like sky making vimana disappear or un-noticed.

Here, in this case, there should be a special device which can make this solution spread all over the vimana, which could be possible if there are any tiny openings all over the skin of vimana connecting to the produced chemical solution inside vimana.

20. Cloud like appearance

Sanskrit Pilot instructions Shloka:

२०. जलदरूपरहस्यो नाम ॥
करकाम्ल बिल्वतैल शुल्बलवण धूमसार ग्रन्थिकरस सर्षपपिष्ट मीनावरणद्रवाण
शास्त्रोक्तप्रकारेण भागांशसंमेलनं कृत्वा, मुक्ताफलशुक्तिकालवणसारे संयोज्य सम्मि-

Re-Translation

Cloud like appearance:

Mixing pomegranate juice, bilva (wood apple) or bael oil, copper-salt, kitchen smoke, granthika (resin of Guggul tree) or gugul (Balm) liquid, mustard powder, and fish scale decoctions, and adding sea-shell and rock-salt powder, and collecting smoke of the same solution and exposing it to solar heat enveloping the cover, the Vimana will have the appearance of a cloud.

Interpretation:

This again describes the similar style of making vimana un-noticed, by projecting it similar to cloud. The materials used may produce clouds like gray or white color chemical solution which will be surrounding vimana.

Or the solution will form Vapour all around vimana which makes it look like a cloud and the vimana surrounding in cloud-like appearance to escape an enemy attack or alien people, and also to be un-noticed.

21. Absence

Sanskrit Pilot instructions Shloka:

परोक्षरहस्यो नाम ॥

मेघशक्तिप्रकरणोक्त शरन्मेघावरणपट्केषु द्वितीयावरणपथे विमानमन्तर्धाय विमानस्थ शक्त्याकर्षण दर्पणमुखा सन्मेघशक्तिमाहृत्य पश्चा द्विमानपरिवेषचक्रमुखे निवेशयेत् । तेन स्तम्भनशक्तिप्रसारणं भवति । पश्चाद्द्वारा लोकस्तम्भनक्रिया- रहस्यम् ॥

Re-Translation

1. Absence: According to the subject energy extraction from clouds, by entering the layers of atmosphere where the second of the summer clouds exist, and attracting the power from them with the energy extraction mirror or force attraction mirror in the vimana and applying the energy to the area of vimana, a paralysing force is generated, making the enemy Vimanas cannot perform any action.

Devices used:

- Shaktyaakarshana or energy attracting mirror.
- Parivesha mechanism above the hood of the Vimana.

Interpretation:

Energy Extraction from clouds subject is very interesting to know and this shloka indicates that ancient ancestors have deep knowledge about layers of atmosphere and energies which can be extracted from each layer. The word mirror again indicates projector kind of device because the mirror cannot extract forces or energies but it can only reflect. It is again interesting to know that paralyzing force generated. This could be a force or energy which makes the enemy aircraft unable to control its own mechanical parts etc and so it will become paralyzed. Further investigation is needed for deep understanding. This again speaks about tactical air defense, by making enemy vimana not to perform any action. Devices used are Shaktyakarshana for energy extraction and Parivesha mechanism is mentioned above the hood of vimana.

22. Darkness trap

Sanskrit Pilot instructions Shloka:

२६. तमोमयरहस्यो नाम ॥

दर्पणप्रकरणोक्त तमोशक्त्यपकर्षणदर्पणद्वारा तमश्शक्तिमाहृत्य विमानपञ्जर वायव्यकेन्द्रस्थ तमोयन्त्रमुखात् तमोविद्युति सन्धाय तत्कीलीचालनात् मध्याह्नकालेपि अमारात्रिवत् तमोविकारप्रदर्शनरहस्यम् ॥

Re-Translation

Darkness trap: Thamo-Yantra (Darkness force capturing device). As described in mirror subject, by means of the dark force mirror, the dark force is passed after captured through dark force capturing device called thamoyantra which is located at front-left part of vimana (north-west) and by operating its switch, will produce darkness like the new moon night, even when there is overhead sun at the noon.

Devices used:

- Thamoyantra at the north western side.

Interpretation:

This describes the procedure of spreading darkness which makes viewers on the ground and in air, unable to see suddenly and trap in the darkness. Mirror subject might be something related to projections and reflections. Darkness force might be capturing from space and projecting it or reflecting through the device mentioned. All these may be related to the modern definition of vacuum or dark energy: In essence, it suggests that space itself produces energy, which is "pushing" the universe outward. ... This "repulsive" force could begin to explain the acceleration of the universe. In other words, it might be the dark energy. Today, physicists explain the cosmological constant as the vacuum energy of space.

23. Solar flare Escaping Secrets

Sanskrit Pilot instructions Shloka:

५. गूढरहस्यो नाम ॥

वायुतत्त्वप्रकरणोक्तरीत्या वातस्तम्भाष्टम परिधिरेखापथस्थ यासा, वियासा, प्रयासादि वातशक्तिभिः सूर्यकिरणान्नर्मत तपश्शक्तिमाकृष्य तत्संयोजनद्वारा विमाना-च्छादनरहस्यम् ॥

Re-Translation

As explained in the book of atmospheric layer properties, by extracting powers from air travel region, escape velocity travel region and space travel region, in the outer atmospheric layer covering the earth, to attract the solar power from exosphere and use it to hide vimana from the enemy.

Devices used:
- Solar power attractor at the top of the Vimana.

Interpretation:

Energies extraction from various regions of space and on earth atmosphere also known by ancient ancestors. It's strange thing that these energies are used to hide vimana from the enemy. This could be taken in another way that the vimana might be hiding by traveling through space-time dimensions. Or this could also be indicating that the vimana escape quickly from the enemy by using these energies. Solar power attractor at the top of vimana makes sense because it is the only best location for attracting solar energy and also the vimanas in Ellappa's diagrams also show the same location of Solar power attractor.

III. SPACE AND FLIGHT MECHANICS

Landing Gears

Vimana had ground wheels which were used when moved on the ground and in water also. Modern landing gear mechanisms are similar to this description of folding links and hinged joints, sliding covers, rack and pinion arrangements etc. It appears that the landing gears were retracted when vimana goes into the water.

Re-Translations from Vimana shasthra

Flight mechanics related Re-Translations of Pilot instructions chapter in Vimana shasthra with Vimana Devices used for it:

1. Defense/offense

Sanskrit Pilot instructions Shloka:

२१. लघ्नरहस्यों नाम ॥
वायुतत्वप्रकरणोक्तप्रकारेण वातमण्डलपरिधिरेखासु विमानसञ्चारकाले यदा सूर्यगोलवाढवायुत्वकिरणज्वालापवाहः विमानाभिमुखो भवति तेन विमानः प्रज्वलितो भवति । तत्सञ्जिवारणार्थं विमानस्थविद्युतशक्तिमेकीकृत्य, विमानस्थप्राणकुण्डली-स्थाने सञ्चाव्य पश्चात् कीरीपालनेन विमानोद्दीपनमात्रं कुम्भारहस्तवत् रेखान्तर-लघ्नक्रियारहस्यम् ॥

Re-Translation

Defense/offense: As stated in Air properties subject, when the Vimana is crossing from higher atmospheric layers into another, the Vimana experiences very close flares glow of the sun and catches fire. In order to prevent that, the electric force and air force in the Vimana should be combined and centered in the center of mass/gravity of Vimana, and by operating switch, Vimana will move into safety region.

Interpretation:

This shloka might be indicating solar flares region, van Allen belts etc. To prevent this, there are methods explained such as electric force with air force inside vimana, and also the centre of mass/gravity is mentioned, a switch is described which makes vimana move into safety region. The switch could be related to direction control systems.

2. Intermediate Space

Sanskrit Pilot instructions Shloka:

४. अन्तराळरहस्यो नाम ॥
आकाश परिधिमण्डल शक्ति सन्धिस्थानेषु विमानप्रवेशो यदा भवति, तदा उभयशक्तिसंमर्दनेन चूर्णितो भवति । अतः विमानस्य तत्सन्धिप्रवेशसूचनात्तदन्तरालेषु विमानस्तम्भनक्रियारहस्यम् ॥

Re-Translation

Atmospheric electricity and ionospheric regions can damage the vimana if it is not driven carefully. But by observing the warning system installed in vimana, then such dangers could be overcome and driven with care.

Interpretation:

This shloka similar to previous one, explaining mechanisms in space and earth's atmosphere, to make vimana travel safely. Atmospheric electricity as per modern definition: Atmospheric electricity is the study of electrical charges in the Earth's atmosphere (or that of another planet). The movement of charge between the Earth's surface, the atmosphere, and the ionosphere is known as the global atmospheric electrical circuit.

3. Visibility

Sanskrit Pilot instructions Shloka:

६. दृश्यरहस्यो नाम ॥
आकाशमण्डले विद्युद्गात किरणशक्त्योः परस्परसंमेलनात् सञ्जात बिम्बकृत्छक्तेः विमानपीठपुरोभागस्थ विश्वक्रियादर्पणबिले प्रतिफलनं कृत्वा पश्चात्प्रकाशसंनिवेशन-द्वारा मायाविमानप्रदर्शनरहस्यम् ॥

Re-Translation

Visibility: By extracting electricity through wind force by collision on vimana while traveling in the atmosphere, a glow is created and made to reflect on universal operation mirror located at the front part of vimana. By manipulating it, a cloaked vimana image is formed.

Devices used:

- Vishwakriyaadarpana or mirror of outside views.

Interpretation:

The line "extracting electricity through wind force by collision on vimana while traveling" Could indicate similar to modern Magneto hydrodynamic generator. This electricity is used for projection through Projector which creates a Holographic 3D image in the sky or cloaking. The device used is Vishwakriyaadarpana, which could be understood as universal action operation or projection. Further investigation needed.

Examples of modern Technology:
Cloaking device

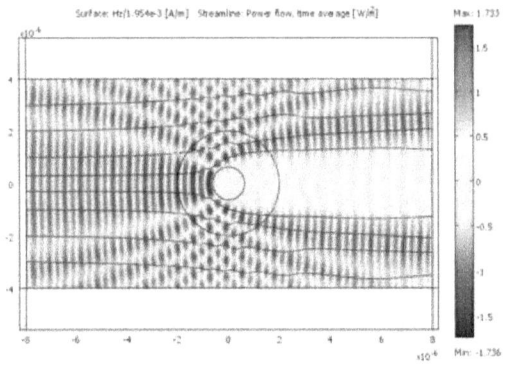

Simulation of how a cloaking device would work.
Cloaking device deactivated: Light is reflected and absorbed by the object, causing it to be visible

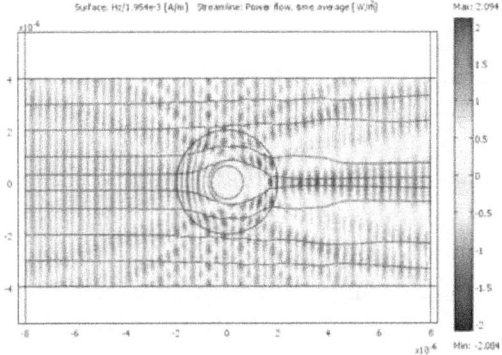

Simulation of how a cloaking device would work. Cloaking device active: Light is deflected around the object, causing it to be invisible

The Airborne Holographic Projector [Source: Air University]

4. Contraction

Sanskrit Pilot instructions Shloka:

१०. सङ्कोचनरहस्यो नाम ॥
यन्त्राङ्गोपसंहाराधिकारोक्तरीत्या अन्तरिक्षे अतिवेगात् पलायमानानां विस्तृत-
खेट्यानानां अपायसम्भवे विमानस्थसप्तमकीलीचालनद्वारा तत्तदङ्गोपसंहारक्रियारहस्यम् ॥

Re-Translation

Contraction: As mentioned in the machine parts manual section, when the vimana is flying at speed with fully extended wings, and there is danger ahead, so by operating the 7th switch in vimana, the wings can be made to fold.

Devices used:

- Angopasamhaara yantra is a folding up device used for contraction of vimana part

Interpretation:

Danger could be that in higher altitudes, due to supersonic speed an expanded wing causes drag and also could catch fire due to the friction of air and wings. It also explains Foldable wings which could be having Sweep back wing mechanisms also. Angopasamhaara yantra could be the device or links responsible for folding the expanded wings of vimana.

5. Expansion

Sanskrit Pilot instructions Shloka:

११.   विस्तृतरहस्यो नाम ॥

आकाशतन्त्रोक्तप्रकारेण आकाशतृतीयपञ्चम परिधिमण्डळस्थानीय मूलवात-परिधिकेन्द्रस्य विमानानां वाल्मीकिगणितोक्त विमानप्रस्ताररेखाविन्यासमनुसृत्य विमानस्थ एकादशरेखामुखस्थानीय कीलीचालनद्वारा तात्कालिकोपयुक्त विमानविस्तृतक्रिया-करणरहस्यम् ॥

Re-Translation

Expansion: According to Ether technology when the vimana is in the first and third regions of the sky where there is air in abundance, by operating the switch in the 11th section of the plane, the wings are expanded suitably as described in "vaalmeekiganita".

Devices used:

- Vistritakriyaa is an expanding location in the middle of a section of vimana

Interpretation:

This explains the lower and higher atmospheric regions and air abundance region in which the vimana can fly with expanded wings. Ether technology is mentioned because in this shloka, energy source also they may want to describe.

Or there could be another reason where Ether means Aakash or space, the space technology terms also can be interpreted into this shloka. Needs further investigation. Vimana part used is Vistritakriyaa, which is responsible for the expansion of wings in vimana.

6. Direction display

Sanskrit Pilot instructions Shloka:

२८. दिक्प्रदर्शनरहस्यो नाम ॥
विमानमुखकेन्द्रकीलीचालनेन दिशांपतियन्त्रनाळपत्रद्वारा परयानागमनदिक्प्र
दर्शनरहस्यम् ॥

Re-Translation

Direction display: by operating key which controls front of the vimana, the direction monitor device will show the direction from which enemy plane is approaching.

Devices used:

- Dishaampatiyantra at the left front.

Interpretation:

This shloka describes device similar to modern cockpit devices, for direction display monitoring and displaying. Mostly this could be detecting enemy planes location and indicating in the display device. A device termed as Dishaampatiyantra and located at the left front of vimana.

Other Flight mechanics related Vimana Parts described in Vimana shasthra are:

- Vakraprasaarana at the side of Vimanaadhaara. This is a device which enables vimana to move like snake

- Shirahkeelaka at the head of the Vimana. Main hinge located at top of vimana

- Pata-prasaarana at the bottom center. Hydraulic system for landing gears

- Vaataskandha keelaka at the bottom center. It is a hinge at the bottom of vimana

Vimana Flight and space mechanics:

There are 500 principles of flight mentioned in Vimana shasthra. Rukma, Sundara, Tripura, Shakuna are main vimanas focused in the research, and all might lead to be antigravity with different principles.

Following is special investigation work on Rukma vimana.

Rukma Vimana

The wing-like structures of Rukma vimana is located such a way that it can be swept back and also rotated to act as a fin for vimana during VTOL.

Numbering the vimana wings starting from positive X axis, clockwise manner numbering as described in the following figure.

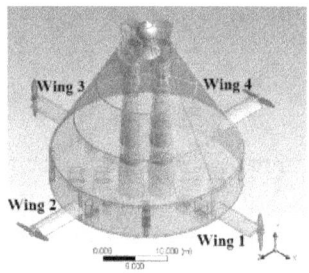

Vimana wing numbers for direction control description

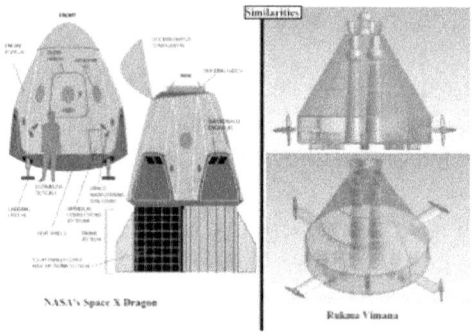

Rukma vimana and Space X dragon similarities

Table: Vimana maneuvers

Direction of movement	Wing number as shown in figure	Wing Rotation	Sweep back
+x	2 and 4	No rotation	-x direction
+y	1, 2, 3, 4	90⁰	-y direction
+z	1 and 3	No rotation	-z direction

-x	2 and 4	No rotation	+x direction
-y	1, 2, 3, 4	90⁰	+y direction
-z	1 and 3	No rotation	+z direction

The positive Y direction movement of vimana is VTOL. 4 Wings rotate 90⁰ and act like fins of a rocket by its sweep back positioning.

IV. Structures and Flight: 3D modeling Vedic Vimanas

3D Modeling by Kavya Vaddadi
and

Structural Analysis by Shashi Kant,
CEO, ScheDio CAD Solutions (OPC) Pvt. Ltd.

VIMANA 3D MODELS:-

A Part of our works on Reinvention of ancient Indian Vimanas are shown below. Referring to the Drawings done by Ellappa under instructions of Subbaraya shasthry who used the third eye to visualize the Vimanas. Similar procedure was done by Maharshi Bharadwaja to transmit the knowledge to subbaraya Shasthry. The third eye of the human brain is mysterious and unknown to us even in modern advanced science. Investigating more on it could lead us to understand how sages used the advanced part third eye of the brain.

VIMANA ANALYSIS:-

➢ Vimana Meshing-

The Vimana is divided into small control volumes which provide elements, nodes, cells, edge faces, boundary faces, in order to apply the conditions further inserted in the analysis, such as material, boundary condition like force, gravity, loads etc, and the equations used to calculate the structural analysis.

➢ Self Weight Analysis-

Self Weight of vimana is defined as the force that gravitation exerts upon a Vimana, equal to the mass of the Vimana times the local acceleration of gravity: commonly taken, in a region of constant gravitational acceleration, as a measure of its mass.

➢ Impact Analysis-

Impact Analysis define as in mechanics, an impact is a high force or shock applied over a short time period when two or more bodies collide on vimana. Such a force or acceleration usually has a greater effect than a
lower force applied over a proportionally longer period.

RUKMA VIMANA :-

RUKMA VIMANA Reverse Engineering

As described in vimana shasthra, 3D modeling is done. Base plate 50 feet radius, 10 feet high. Which would be the ground floor of vimana, remaining part of the vimana three floors in a conical shape. This describes aerodynamic shape similar to NASA space x dragon plane in modern technology. The Rukma vimana has total 8 propellers, which make us feel that it is a subsonic aircraft but the electromagnets used in this vimana make us think about antigravity propulsion too. In fact there are many areas to explore such as vacuum propellers, a cap covering vessel shaped containers of propellers similar to space x dragon plane, and much more to be described in next series of this book. This vimana has wing-like structures which makes it possible to rotate along its central axis. The vimana is obviously a VTOL plane. Similar to the transportation vehicle attempted by Nikola tesla. Rukma vimana also has descriptions such as motors and generators. Mica pillars are limited length till the first floor of vimana starts. 4 fan pipes or ducts for propellers are mentioned.

VTOL: As per my investigation on Vimana shasthra diagrams, and as per my interpretation of modern technology, two different positions of side Propellers vimana are noticed.

1. Rukma vimana with propellers in vertical position
2. Rukma vimana with propellers in horizontal position

There could be different purposes of this positions. Such as VTOL and sideways movement of vimana. Another interesting point to be noted is that the working of specific propellers when others are in rest motion will give the result to a maneuver of vimana which is non-existing in modern aeronautics.

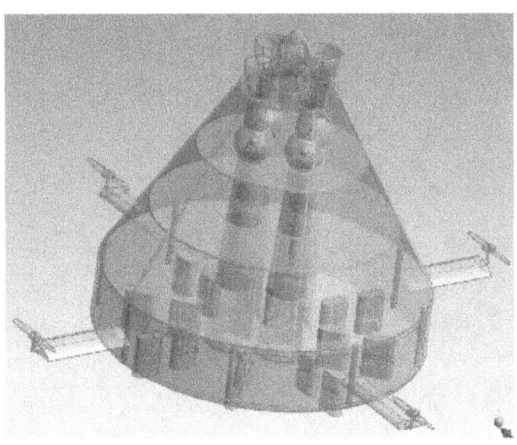

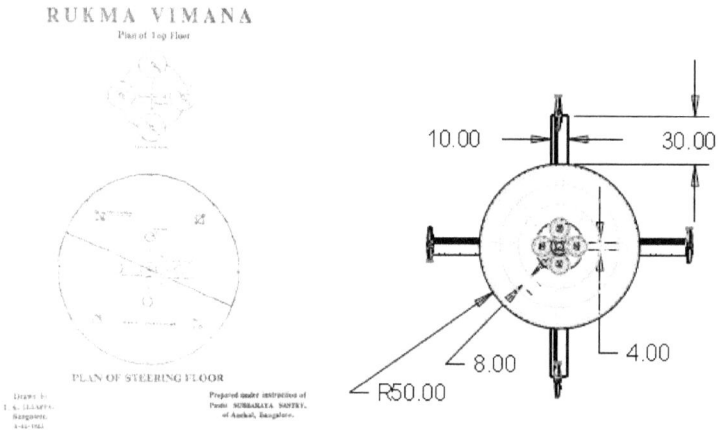

Left side: Ellappa Drawings from vimana shasthra ; Right side: Top view of 3D model of Rukma vimana

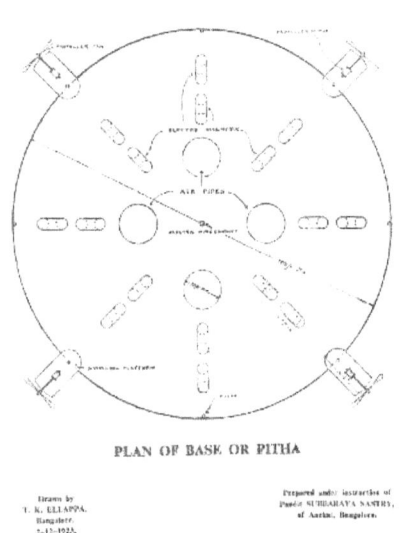

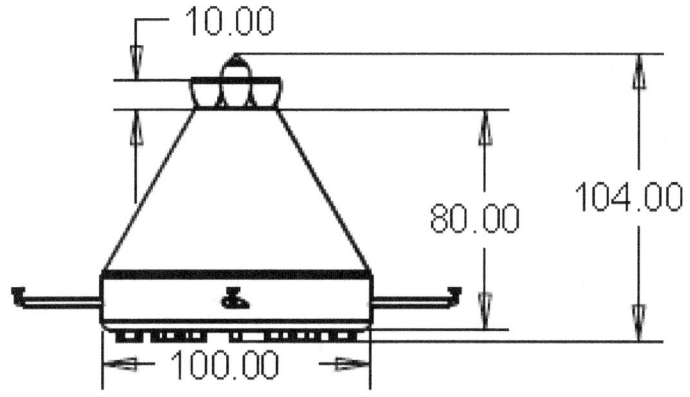

Dimensions of Rukma vimana 3D model

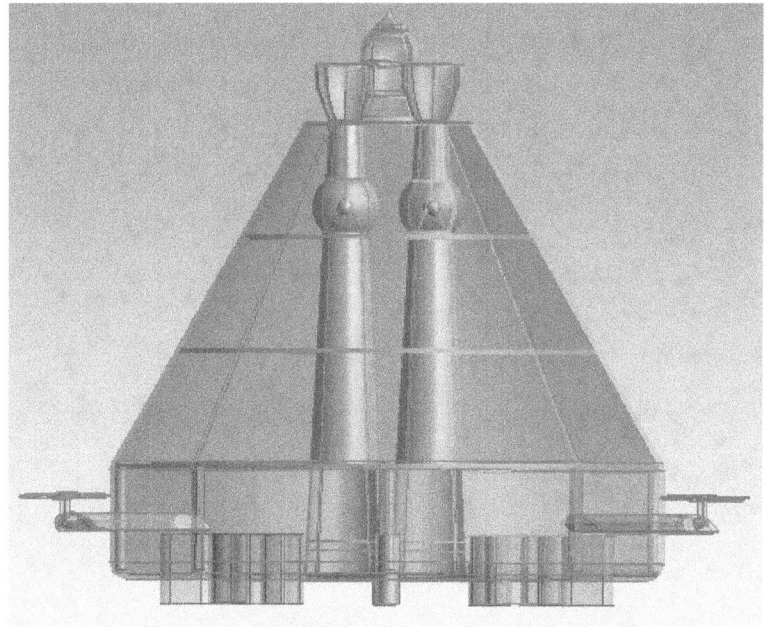

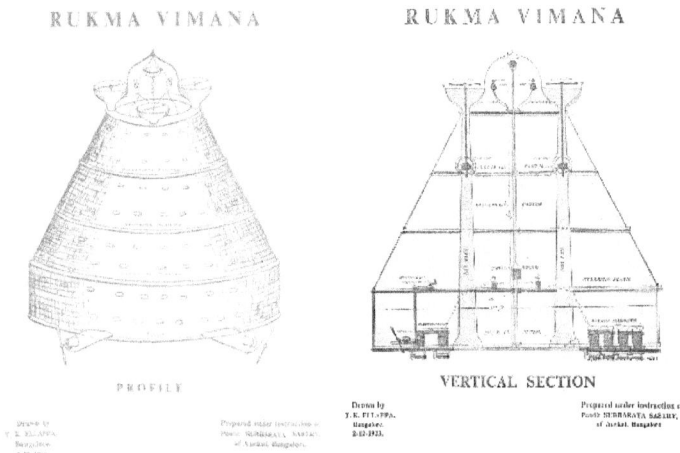

Drawings and interpretation by Ellappa

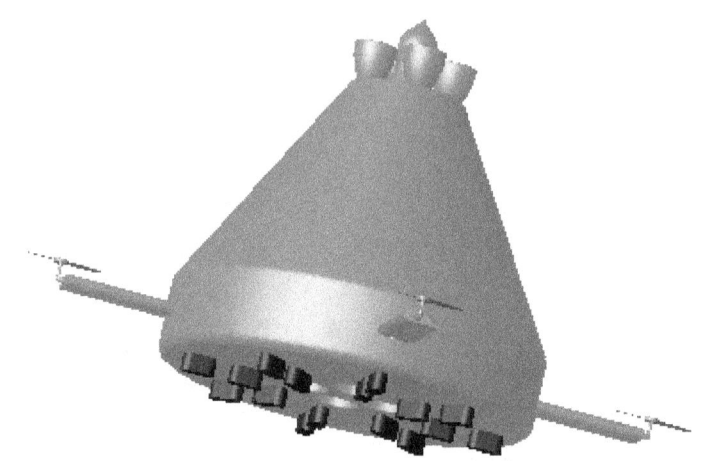

3D model

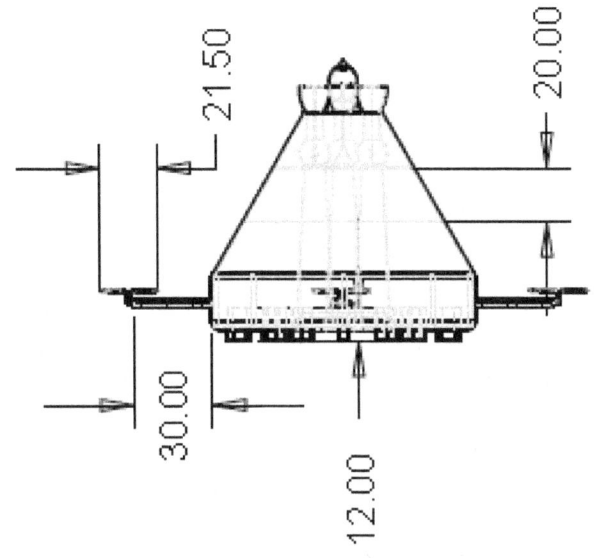

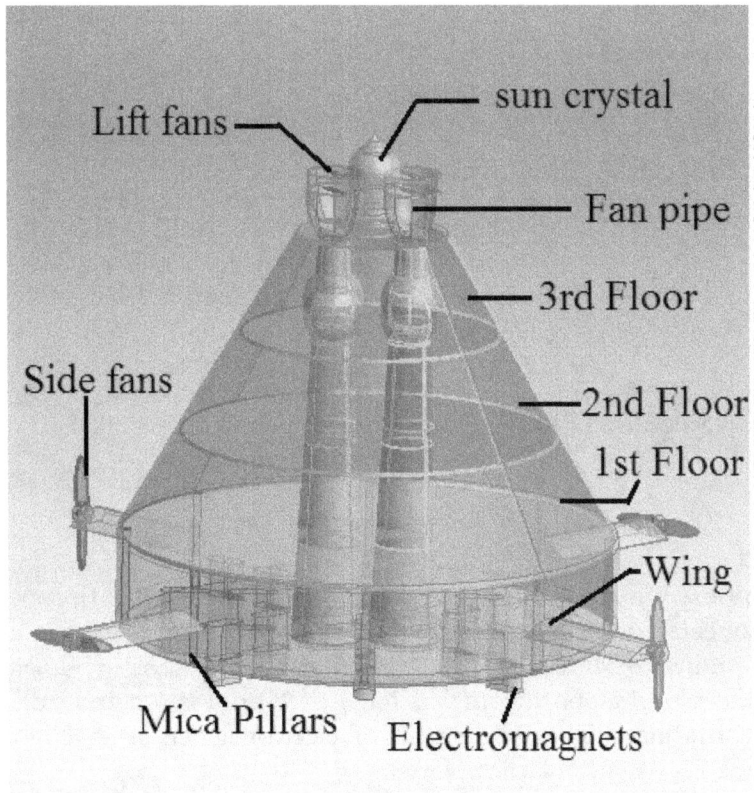

Various parts of vimana

The 3D model of Rukma vimana is developed based upon descriptions of Vimana in Sanskrit texts and also by referring to Drawings made by Yellappa guided by Pandit Subbaraya shasthry.

For 3D modeling, Rukma vimana is divided into following parts:

1) **Rukma Vimana main body:**
   All the floors of vimana are 20 feet high, ground floor containing landing gears, electromagnets, and wings. Whereas remaining floors are passenger cabins. According to Vimana texts, the third floor is used as a cockpit for pilots. There is goblet-shaped cup-like structures at the top of vimana in order to fix propellers or lift fans inside them to suck the air from the top. Sun crystal is Dome shaped on the top of vimana third floor placed in the centre of all the goblet shaped cups. There are four Wings of vimana around the circumference of the ground floor. There are slots provided for locating the Electromagnets in assembly.

2) **Propellers:** There are 8 propellers for Rukma vimana

   Table: Vimana Propellers Details

Propellers	Location	Purpose
4 Lift Fans	Above the third floor of vimana, in goblet-shaped cup-like structures	Sucks air from top of vimana for lift
4 Side Fans	At the end of 4 Wings extension with movable mechanisms	for VTOL and direction control of vimana

3) **Fan pipes:** Wires run through fan pipes from 4 lift fans or propellers and are connected to the motor at the third floor of Vimana.
4) **Mica pillars:** 8 Mica pillars are located at ground floor of vimana, pillars are 20 feet long support from which the third floor is constructed.
5) **Electromagnets:** There are 8 pairs of electromagnets arranged 45 degrees from the central axis of each pair.

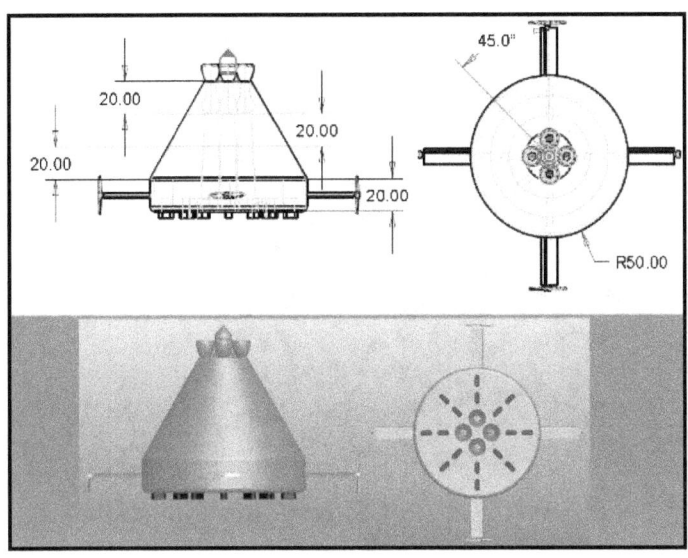

Assembled Rukma Vimana and Dimensions

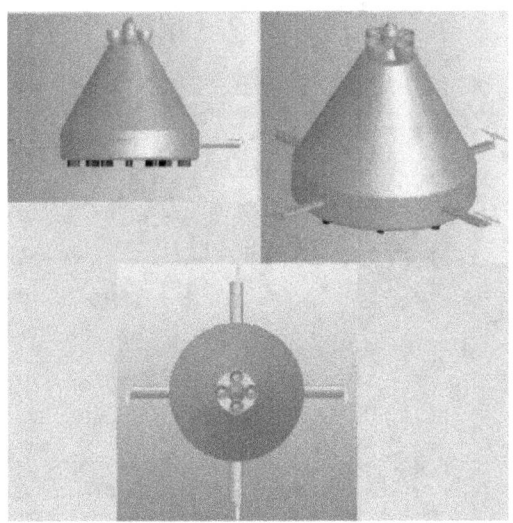

Rukma Vimana 3D model views

Structural analysis Rukma Vimana:-

Design Modifications: In Rukma vimana model, the propellers are removed to avoid complications in the further analysis procedures.

Mesh view with propellers:

Mesh view without propellers

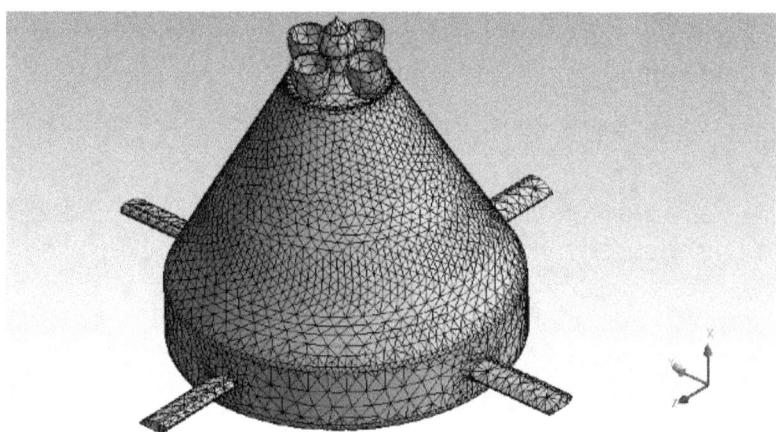

Self weight analysis:

Total Deformation-

After analysis of total deformation due to the Self weight of vimana, the results appear as shown in the below image, displaying and demonstrating the total deformation of rukma vimana due to self weight load, where red colour shows maximum Deformation and blue colour showing minimum deformation. So we can observe and study that the maximum Deformation occurs at the tip of the wing.

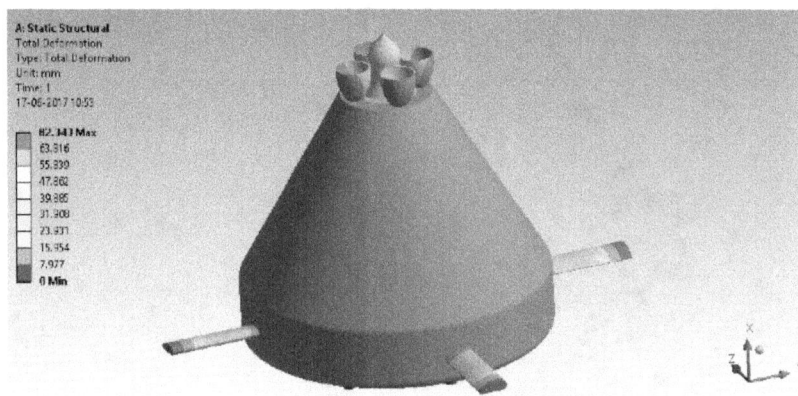

Equivalent stress:-

After analysis of Equivalent Von mises stress due to the Self weight of vimana, the results appear as shown in the below image, displaying and demonstrating the Von mises stress of rukma vimana due to self weight load, where red colour shows maximum stress and blue colour showing minimum stress. So we can observe and study that the maximum stress occurs at the starting of the wing.

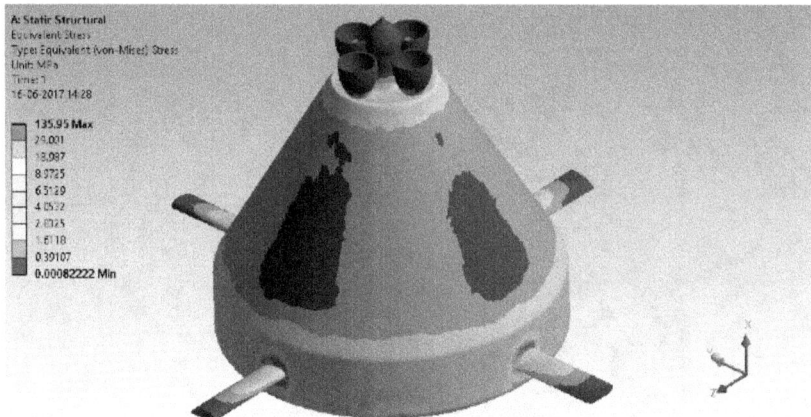

Impact analysis:

Total Deformation-

After analysis of total deformation due to the impact load of vimana, the results appear as shown in the below image, displaying and demonstrating the total deformation of rukma vimana due to impact load, where red colour shows maximum Deformation and blue colour showing minimum deformation. So we can observe and study that the maximum Deformation occurs at the tip of the wing.

Equivalent Stress-

After analysis of Equivalent Von mises stress due to the impact load on vimana, the results appear as shown in the below image, displaying and demonstrating the Von mises stress of rukma vimana due to impact load, where red colour shows maximum stress and blue colour showing minimum stress. So we can observe and study that the maximum stress occurs at the starting of one of the wing where impact load starts acting.

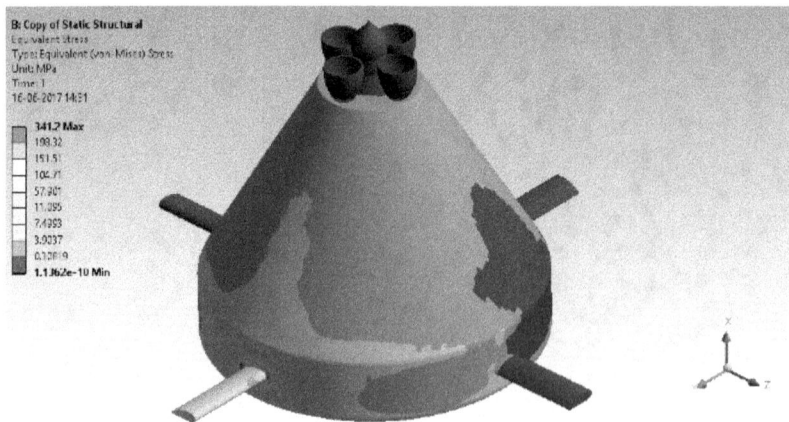

SHAKUNA VIMANA

Shakuna means bird, Shakuna vimana looks like large 100 feet ornithopter like vimana, using flapping wing mechanisms and rudder to lift off. This vimana description has too special features like extending wings to infinite length. As per my investigation, this might be using various layers of wings (sheets kind of structures) overlapped over one another expanding them as per the location and maneuver of the vimana.

Different types of devices or yantras are mentioned in specific locations of this vimana (These devices basic investigations are described in later chapters of this book). Vimana Working: The Propeller gives thrust force to move forward and the wings lift off the vimana. 3D model is done referred to drawings from vimana shasthra, but there is also a dome around the floors and the design and its CFD analysis, and detailed devices working is going to be shown in the next series of this book.

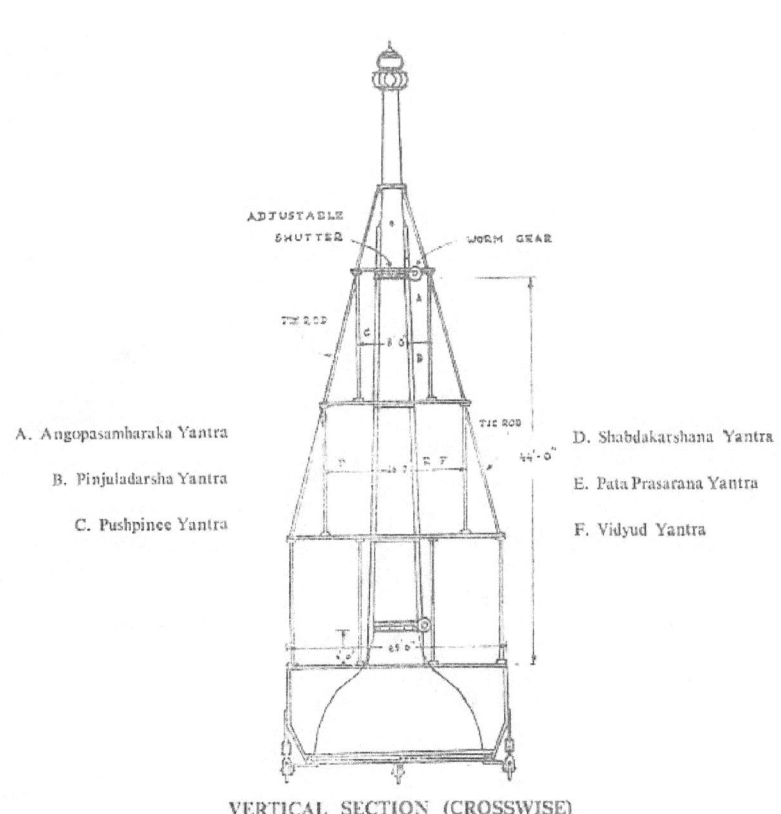

A. Angopasamharaka Yantra

B. Pinjuladarsha Yantra

C. Pushpinee Yantra

D. Shabdakarshana Yantra

E. Pata Prasarana Yantra

F. Vidyud Yantra

VERTICAL SECTION (CROSSWISE)

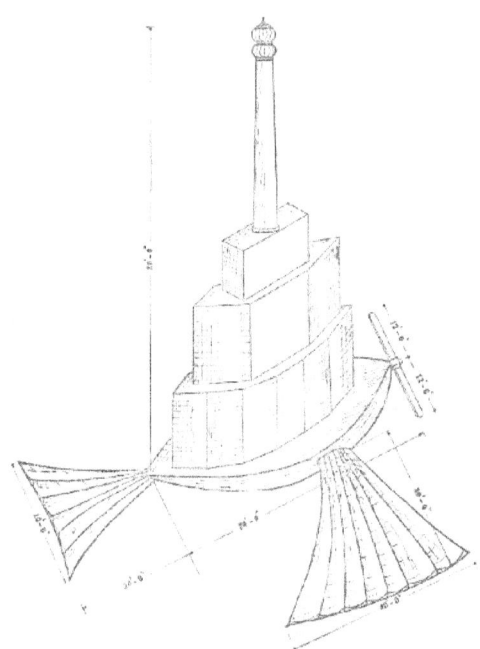

PERSPECTIVE VIEW

3D model

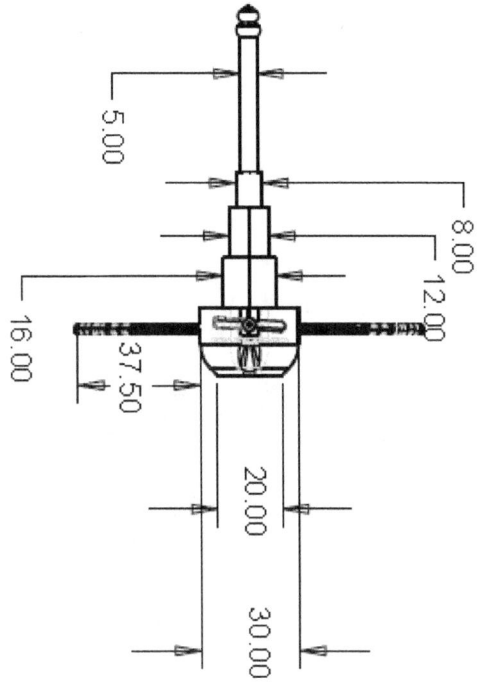

VERTICAL SECTION (CROSSWISE)

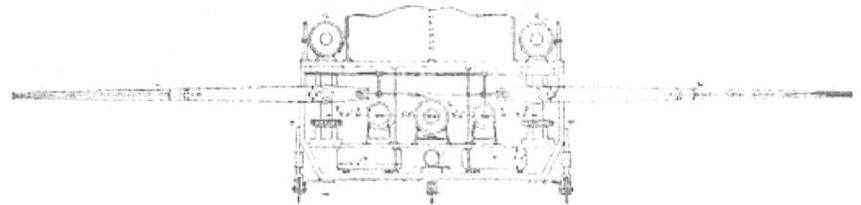

LEGEND: G. Steam Boiler H. Engine J. Oil Tank K. Air Vessel
L. Air Heater P. Pivot T. Turning Wheel W. Wing

VERTICAL SECTION AT THE WING JOINT

3D model

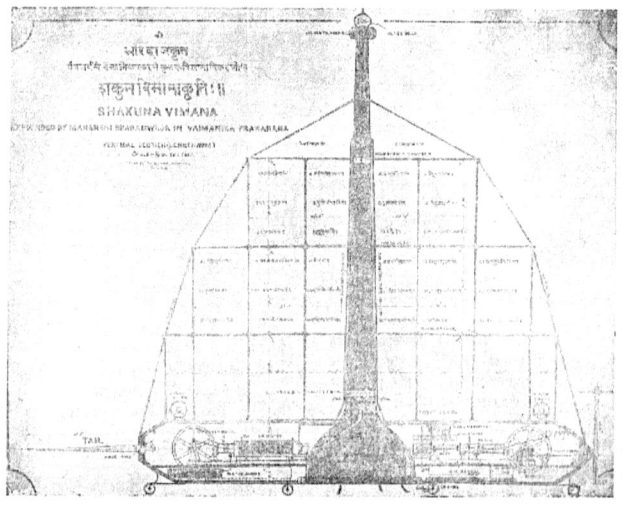

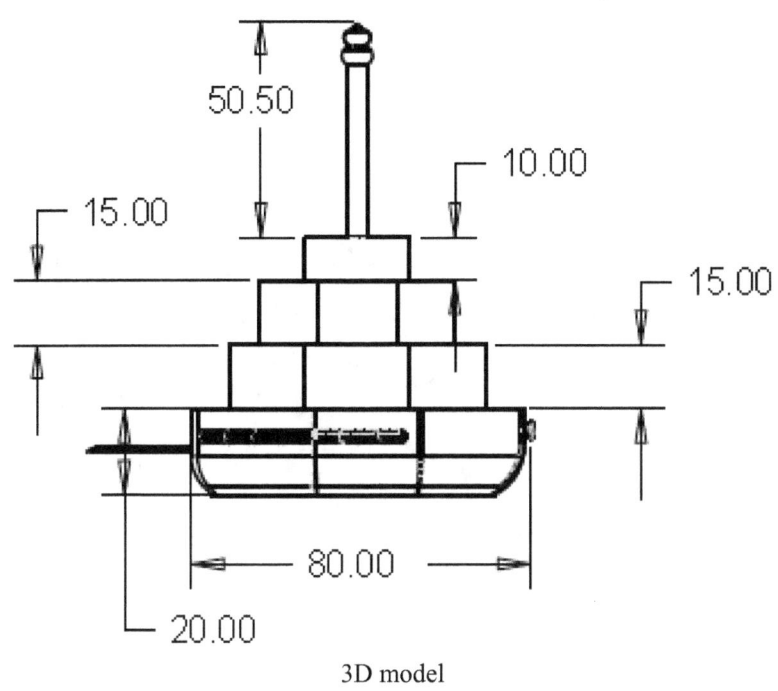

3D model

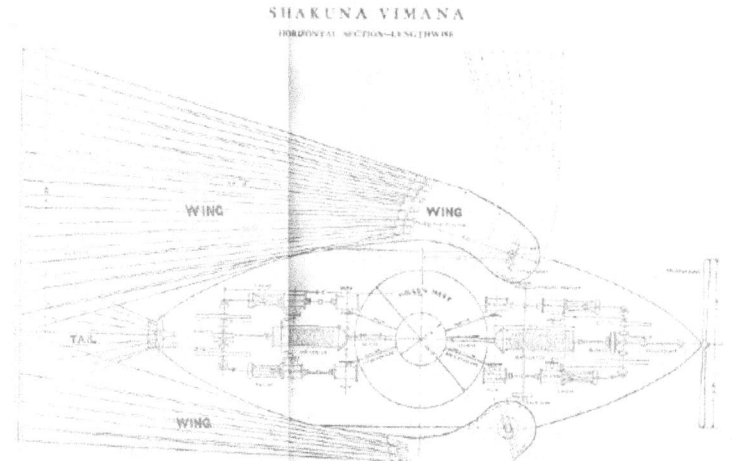

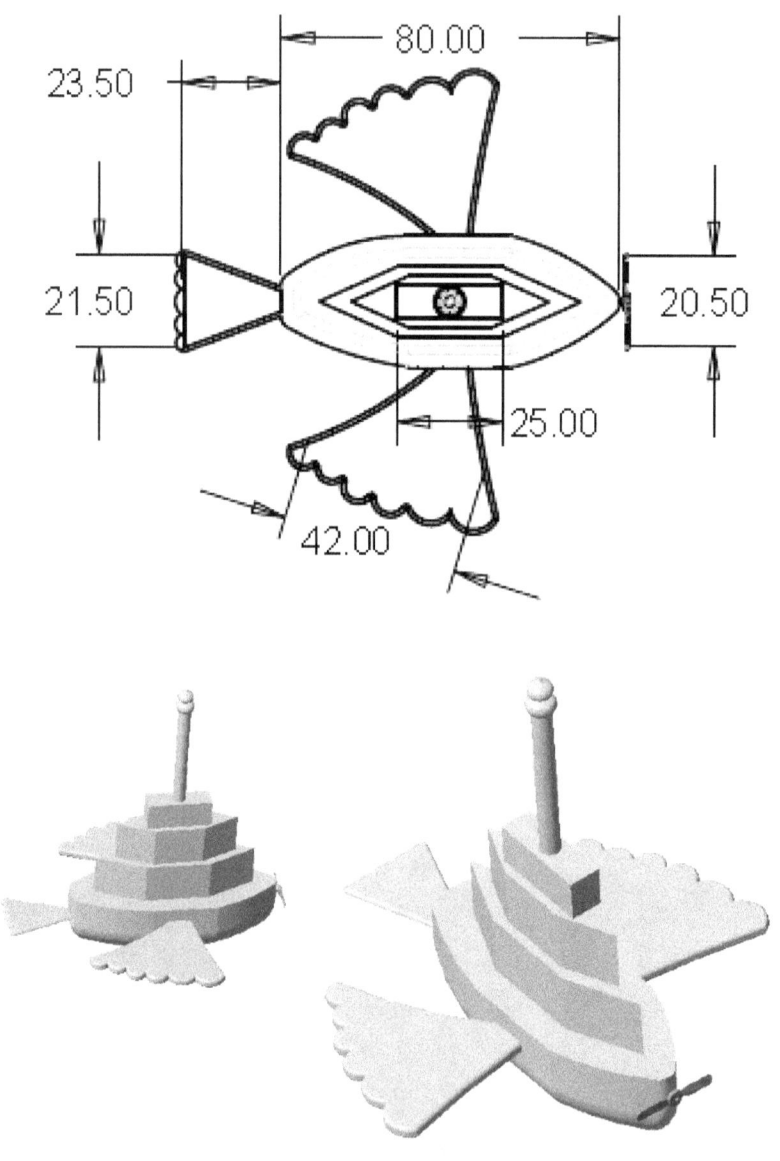

3D model

Shakuna Vimana 3D model views

Structural Analysis Shakuna Vimana:-

Design Modifications: Shakuna vimana wings are made into simpler design to avoid complications in further analysis procedures.

Self weight Analysis:

Total Deformation-
After analysis of total deformation due to the Self weight of vimana, the results appear as shown in the below image, displaying and demonstrating the total deformation of Shakuna vimana due to self weight load, where red colour shows maximum Deformation and blue colour showing minimum deformation. So we can observe and study that the maximum Deformation occurs at the tip of the wing.

Equivalent stress-

After analysis of Equivalent Von mises stress due to the Self weight of vimana, the results appear as shown in the below image, displaying and demonstrating the Von mises stress of Shakuna vimana due to self weight load, where red colour shows maximum stress and blue colour showing minimum stress. So we can observe and study that the maximum stress occurs at the starting of the wing.

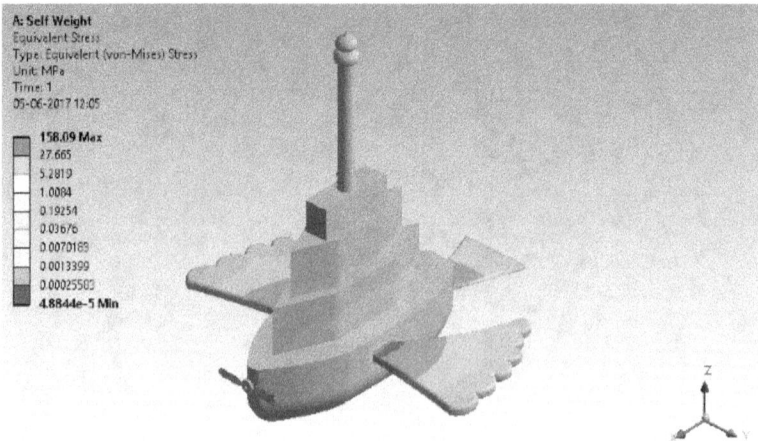

Impact analysis:

Total Deformation-

After analysis of total deformation due to the impact load of vimana, the results appear as shown in the below image, displaying and demonstrating the total deformation of shakuna vimana due to impact load, where red colour shows maximum Deformation and blue colour showing minimum deformation. So we can observe and study that Maximum Deformation occurs at the tip of the nose part where impact load starts acting

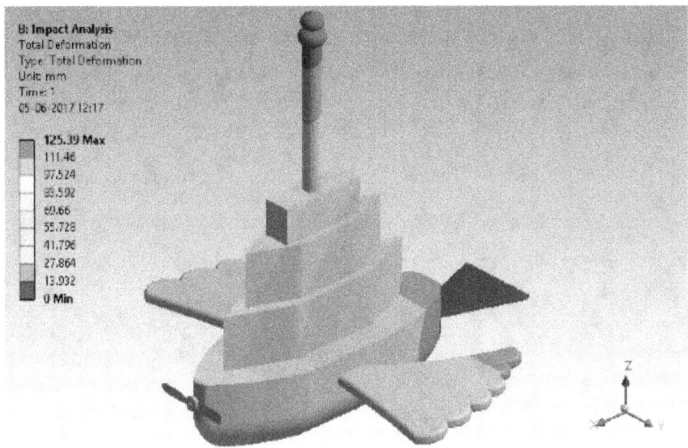

Equivalent stress-
After analysis of Equivalent Von mises stress due to the impact load on vimana, the results appear as shown in the below image, displaying and demonstrating the Von mises stress of shakuna vimana due to impact load, where red colour shows maximum stress and blue colour showing minimum stress. So we can observe and study that Maximum stress occurs at the starting of one of the starting of wings and rudder, where the impact load effects more.

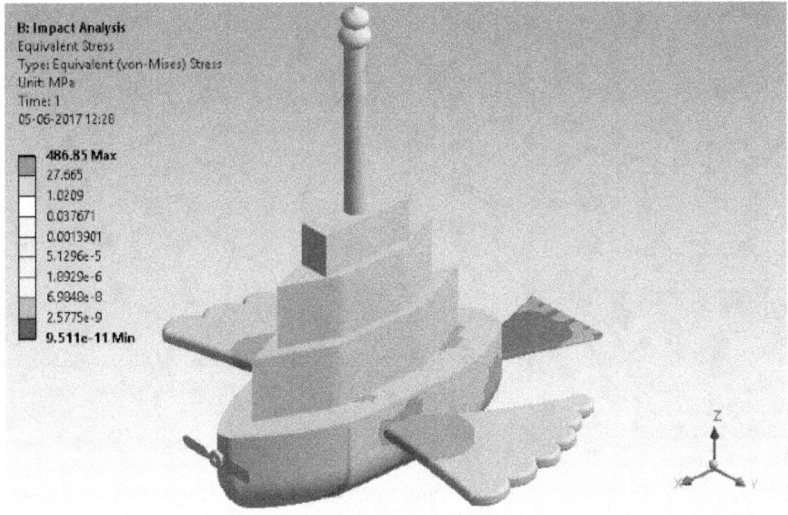

SUNDARA VIMANA

Sundara vimana looks similar to modern hyperspace plane, but with propellers around its circumference. This is strange to see but there might be reasons such as utilizing vimana for multipurpose, that is on earth and as interplanetary Vedic space vehicles too, this could be the exact reason for such configuration.

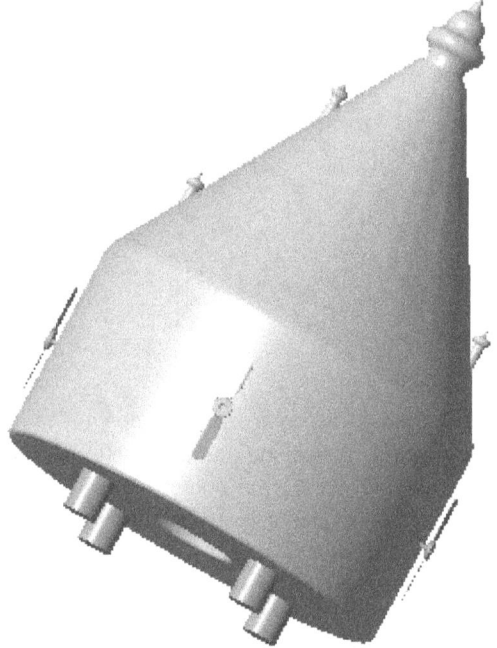

3D model

FLOORS

Drawing by Ellappa

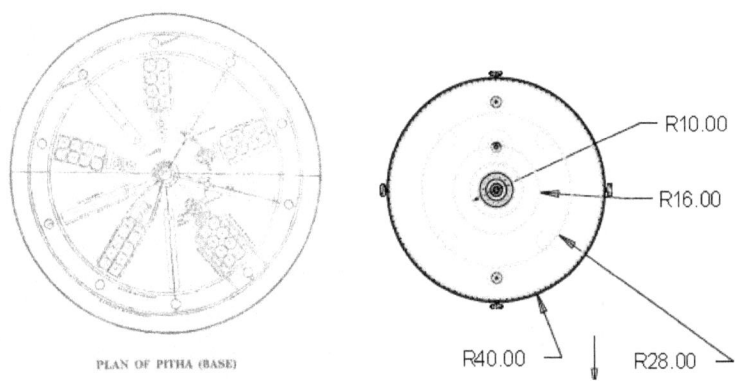

PLAN OF PITHA (BASE)

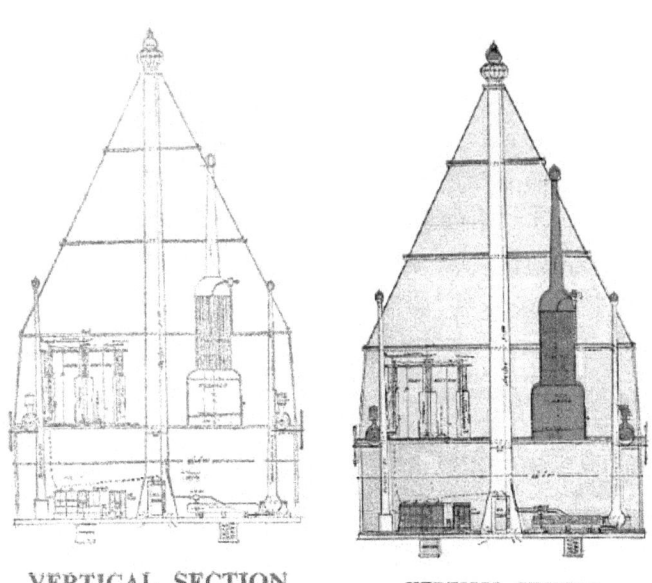

VERTICAL SECTION   VERTICAL SECTION

Drawing by Ellappa

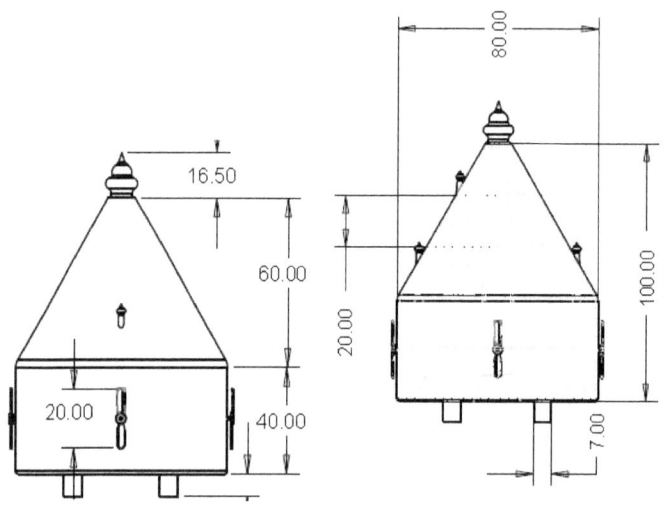

3D model

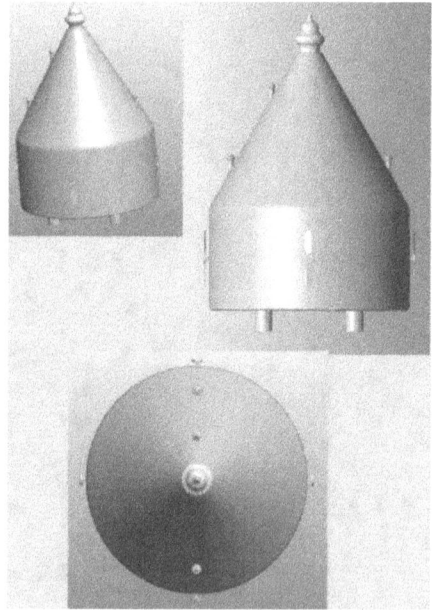

Sundara Vimana 3D model views

Structural Analysis Sundara Vimana:-

Sundara Vimana Mesh view-

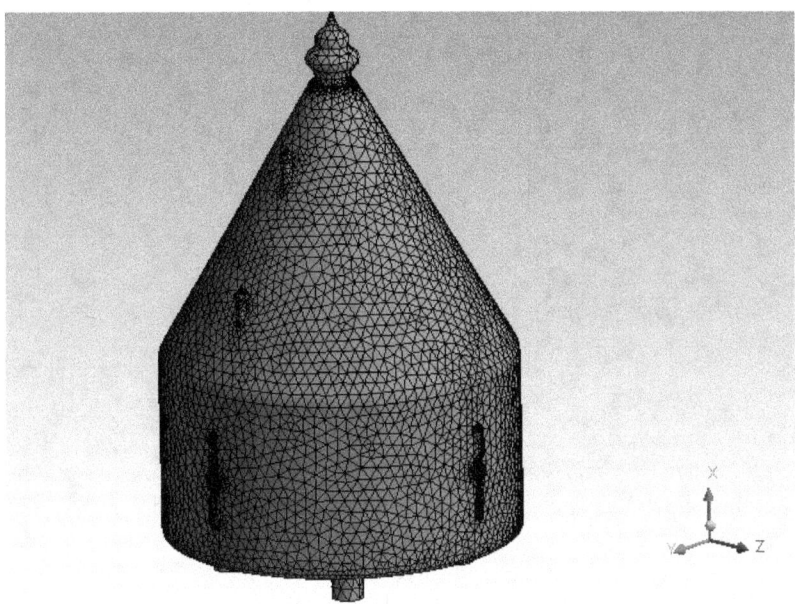

Self weight analysis:

Total Deformation-
After analysis of total deformation due to the Self weight of vimana, the results appear as shown in the below image, displaying and demonstrating the total deformation of Sundara vimana due to self weight load, where red colour shows maximum Deformation and blue colour showing minimum deformation. So we can observe and study that the Maximum Deformation occurs at the propeller in the circumference of vimana dome.

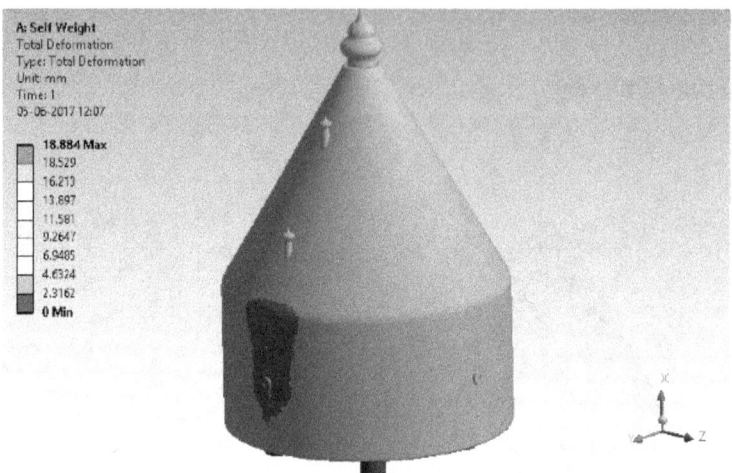

Equivalent stress-

After analysis of Equivalent Von mises stress due to the Self weight of vimana, the results appear as shown in the below image, displaying and demonstrating the Von mises stress of Sundara vimana due to self weight load, where red colour shows maximum stress and blue colour showing minimum stress. So we can observe and study that the maximum stress occurs at the starting of lower floor, the circumference of vimana dome.

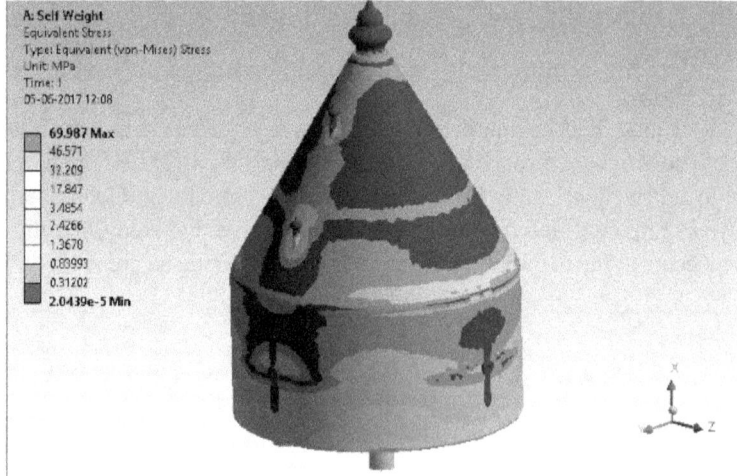

Impact analysis:

Total Deformation-

After analysis of total deformation due to the impact load of vimana, the results appear as shown in the below image, displaying and demonstrating the total deformation of Sundara vimana due to impact load, where red colour shows maximum Deformation and blue colour showing minimum deformation. So we can observe and study that the maximum stress occurs at the circumference of vimana dome at the bottom where impact load starts effecting.

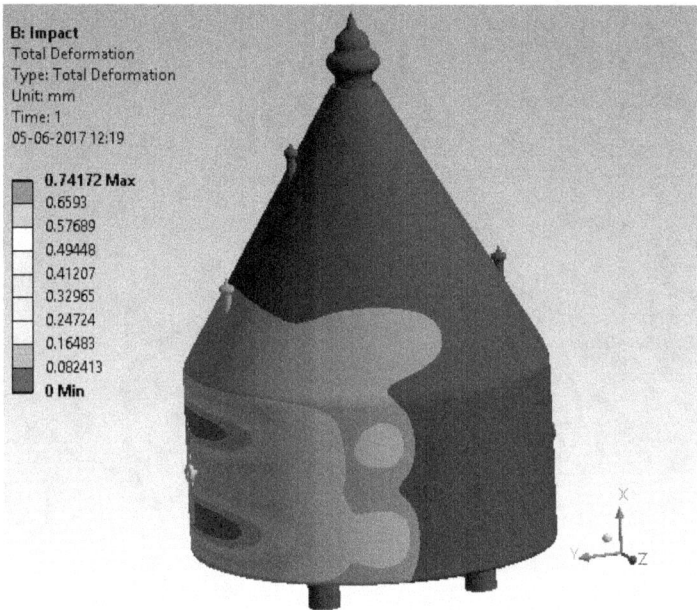

Equivalent Stress-

After analysis of Equivalent Von mises stress due to the impact load on vimana, the results appear as shown in the below image, displaying and demonstrating the Von mises stress of shakuna vimana due to impact load, where red colour shows maximum stress and blue colour showing minimum stress. So we can observe and study that the maximum Deformation occurs at the circumference of vimana dome from the point where the impact load starts.

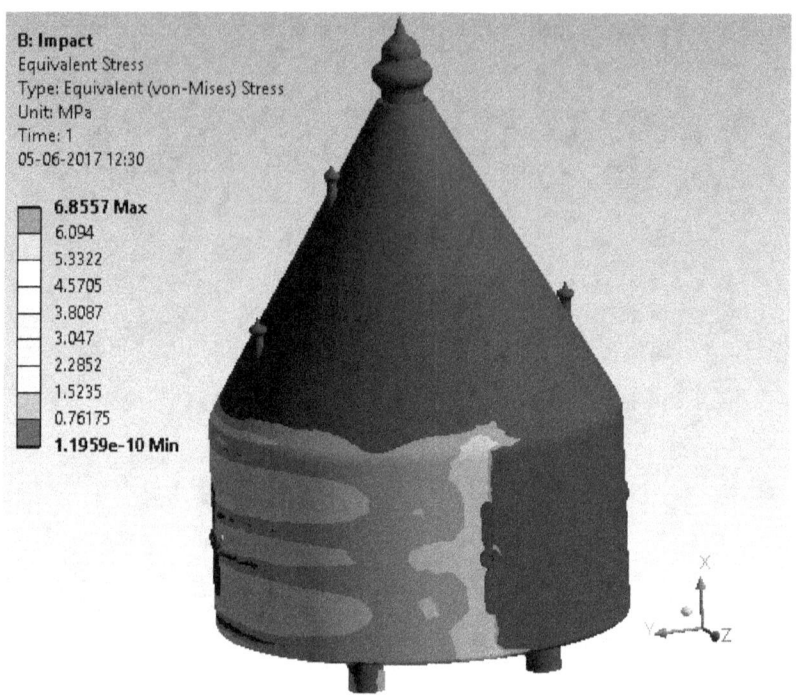

TRIPURA VIMANA

Tripura Vimana is a three-way usable vehicle, which can go under/on water, on land and in air. This vimana also have special features such as foldable links, transforming the vimana, enabling it to be used in various mediums such as water air and on land. It has combined properties of Submarine, ship, aircraft and spacecraft. This vimana also uses special devices to transform vehicle to be used in various medium. There is a dome around the electric power generator interpreted and designed a 3D model of it as shown below. Designs are going to be improvised along with the inner devices designs and workings and shown in the next series of this book.

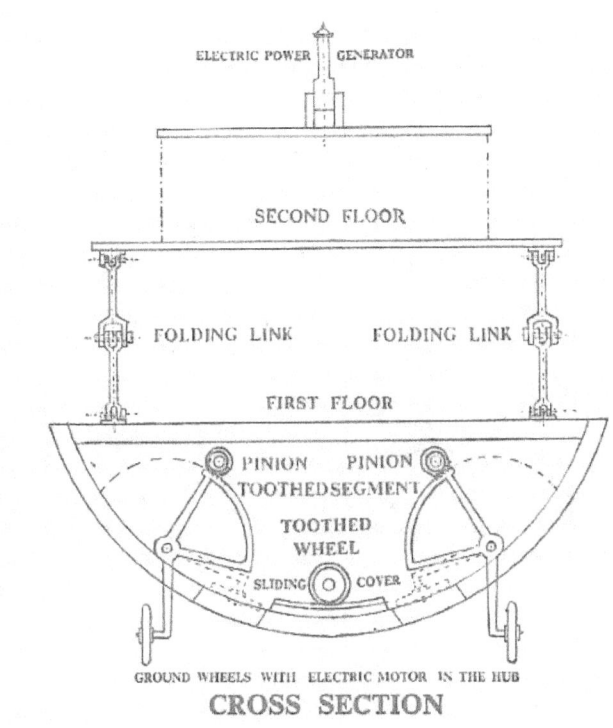

CROSS SECTION

TRIPURA VIMANA

PERSPECTIVE VIEW

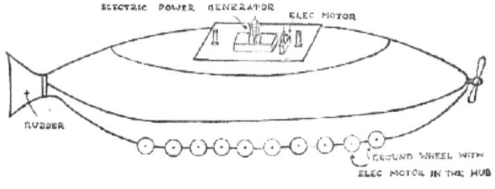

VERTICAL SECTION

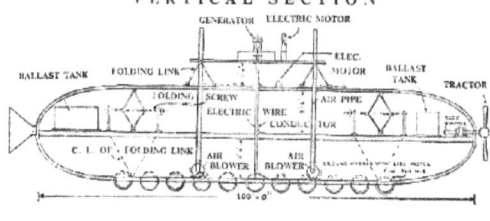

PLAN

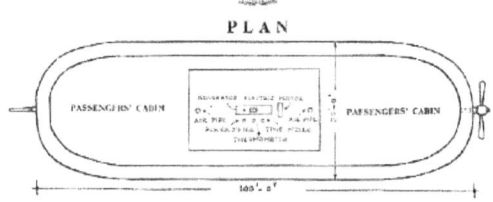

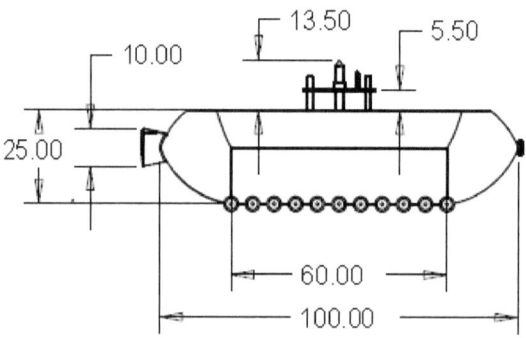

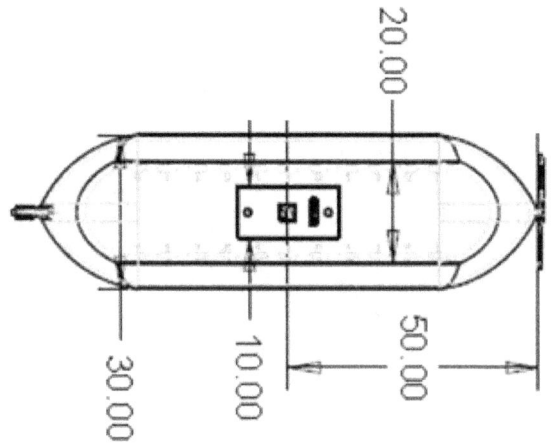

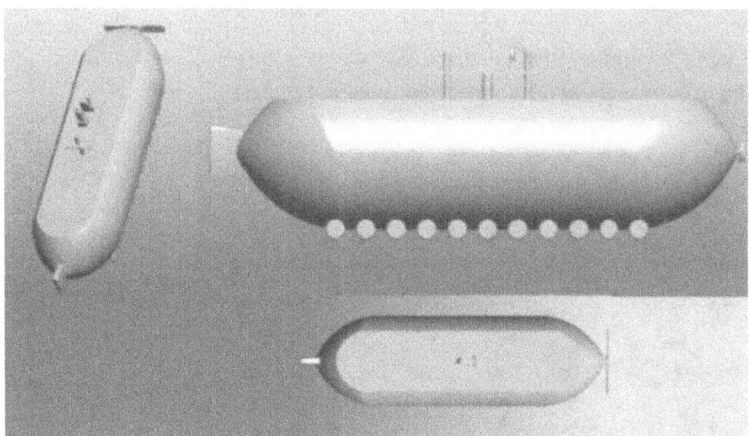

Tripura Vimana 3D model views

Structural analysis of Tripura Vimana:-

Design Modifications: Tripura vimana has been shown in multiple forms due to its advanced features of overlapping floors. Considering the features, two types of tripura vimana are designed

Tripura Vimana Mesh view-

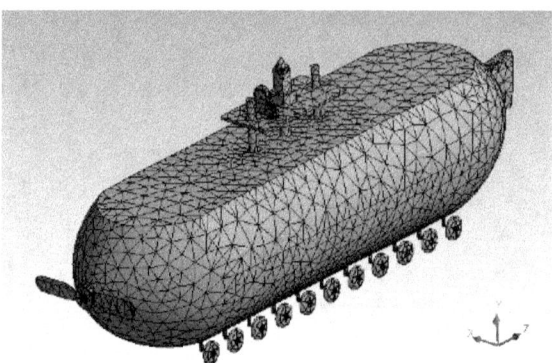

Self weight analysis:

Total Deformation-
After analysis of total deformation due to the Self weight of vimana, the results appear as shown in the below image, displaying and demonstrating the total deformation of Tripura vimana due to self weight load, where red colour shows maximum Deformation and blue colour showing minimum deformation. So we can observe and study that Maximum Deformation occurs at the links connecting to landing gears

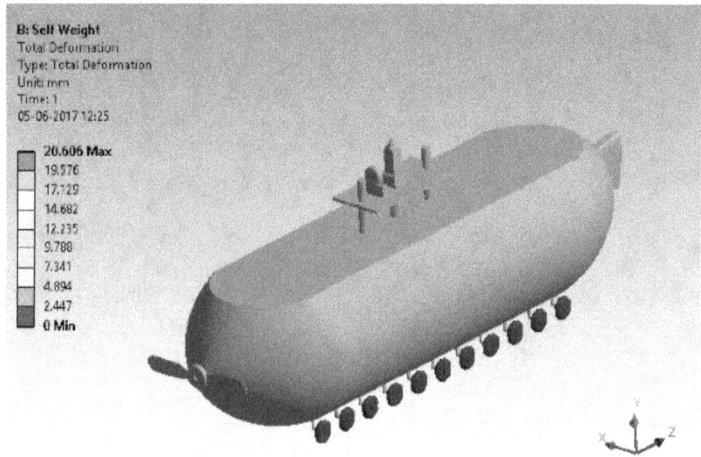

Equivalent stress-
After analysis of Equivalent Von mises stress due to the Self weight of vimana, the results appear as shown in the below image, displaying and demonstrating the Von mises stress of Tripura vimana due to self weight load, where red colour shows maximum stress and blue colour showing minimum stress. So we can observe and study that the maximum stress occurs at the links connecting to landing gears

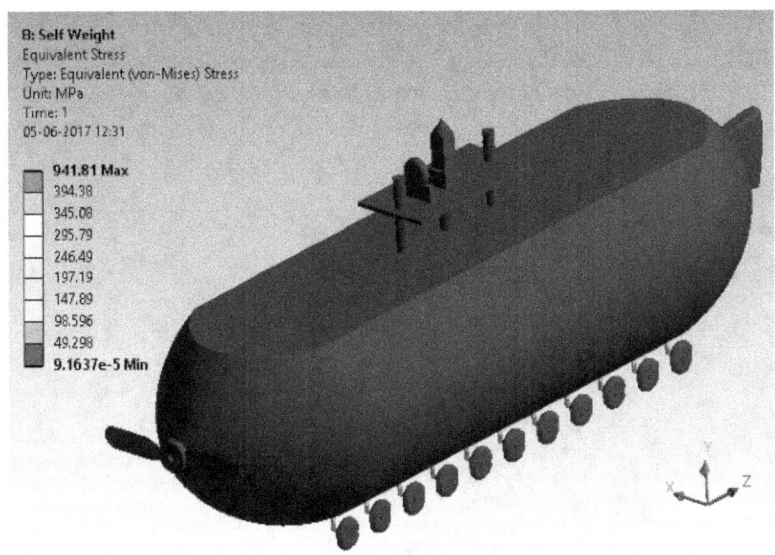

Impact analysis:

Total Deformation-

After analysis of total deformation due to the impact load of vimana, the results appear as shown in the below image, displaying and demonstrating the total deformation of tripura vimana due to impact load, where red colour shows maximum Deformation and blue colour showing minimum deformation. So we can observe and study that the maximum stress occurs at the nose part of vimana dome at the Propeller.

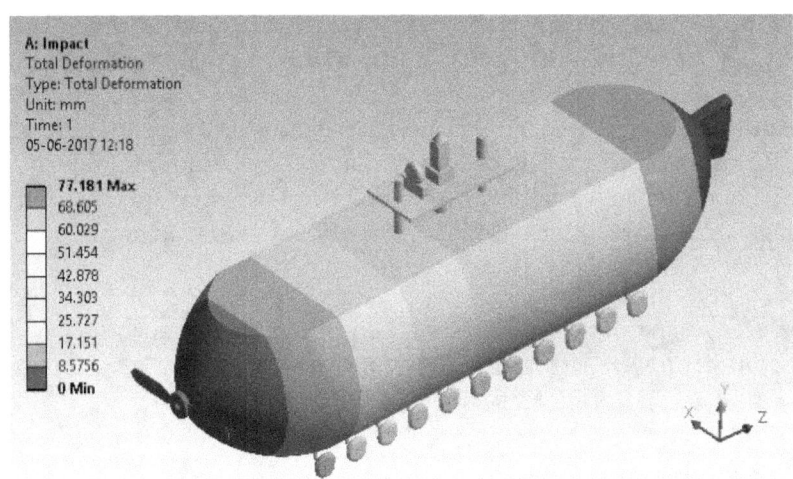

Equivalent Stress-

After analysis of Equivalent Von mises stress due to the impact load on vimana, the results appear as shown in the below image, displaying and demonstrating the Von mises stress of shakuna vimana due to impact load, where red colour shows maximum stress and blue colour showing minimum stress. So we can observe and study that the maximum Deformation occurs at the Rudder of vimana dome due to the effect of impact load.

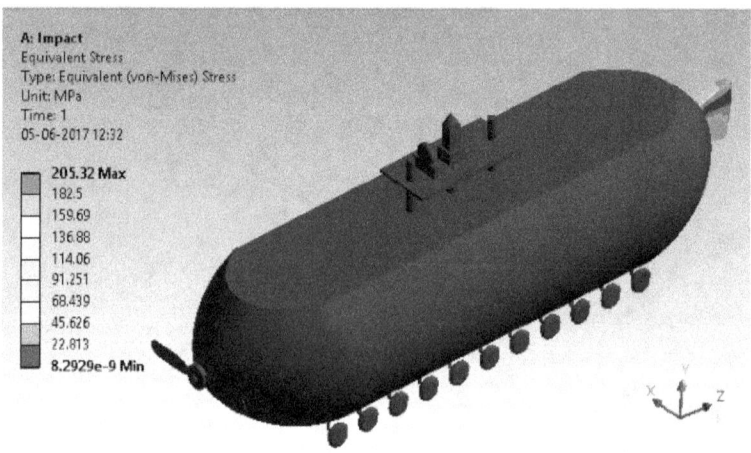

V. AERODYNAMICS

By Kavya Vaddadi and
A. Hemanth Kumar Yadav

Aerial Routes

Sanskrit shloka descriptions clearly say that vimana could travel from land to earth, to much beyond in the space regions, for which the pilot should drive carefully. As it is not planned mission like modern, ancient vimana used to explore entire universe single time/interplanetary missions etc. Ancient vimana used to explore entire universe single time/interplanetary missions etc
Computational Fluid Dynamics Analysis is carried out for Vimana model with the same input values in reference to space shuttle such as Gauge pressure, Mach number, and Velocity

CFD Analysis of Vimana done with reference to Space Shuttle analysis (This Work is done considering modern technology comparisons also)

Input Values:

Material	Air
Density	Ideal Gas
Mach number	0.6
Guage pressure	0
Turbulent intensity	10
Hydraulic diameter	10
Inlet velocity	208.2526 m/s

RUKMA VIMANA

After CFD analysis of the VTOL Rukma Vimana, we can clearly observe in the results, that there is more pressure on the top of vimana and also at the beginning of wings. The pressure values vary depending upon the speed with which the vimana flies.

Pressure contour on Rukma Vimana model:

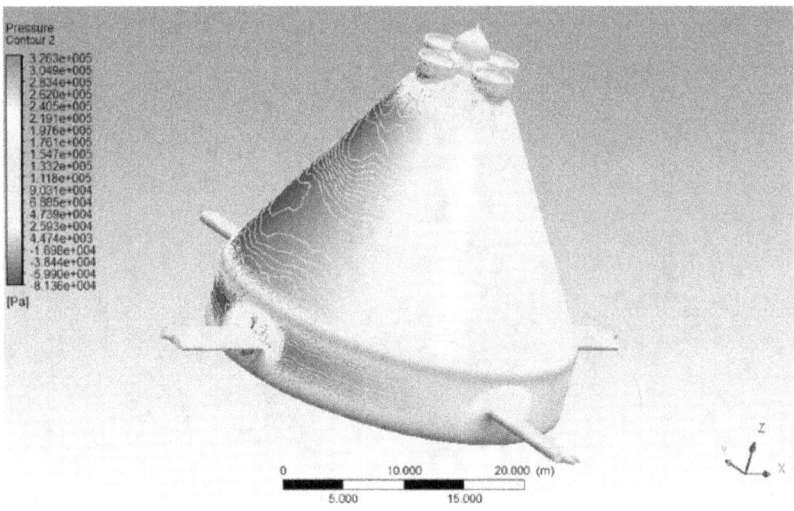

Pressure contour on a plane at the cross-section of vimana:

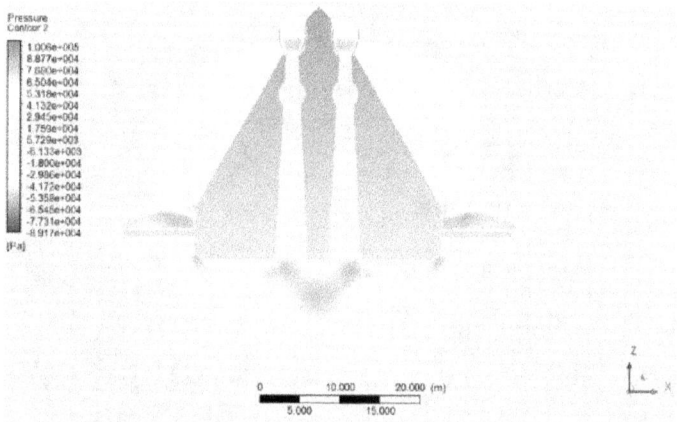

Velocity Streamlines of Rukma Vimana:

The air flow on Rukma Vimana is found to be smooth and hence the vimana is aerodynamic, moreover the vimana has advanced features such as foldable wing mechanisms, a dome on the top propellers when doing space travel. The CFD analysis has been done on the Vimana designed in one of its form as shown in Vimana shasthra diagrams. Further deep analysis on various forms of this vimana will be presented in the next version of this book.

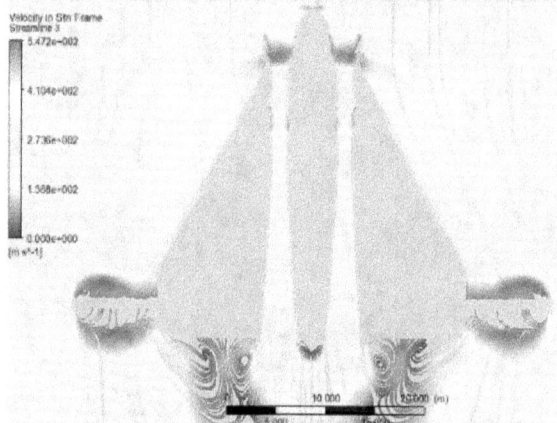

Analysis Results:

The analysis results on Rukma vimana is found to be similar to the space vehicles design in modern days. The Rukma vimana is similar to NASA Space X Dragon plane and also exhibit similar results when done analysis on it.

SUNDARA VIMANA

After CFD analysis of the VTOL Sundara Vimana, we can clearly observe in the results, that there is more pressure on the top of vimana and also around the cone shaped dome of vimana. The pressure values vary depending upon the speed with which the vimana flies.

Pressure contour on Sundara Vimana model:

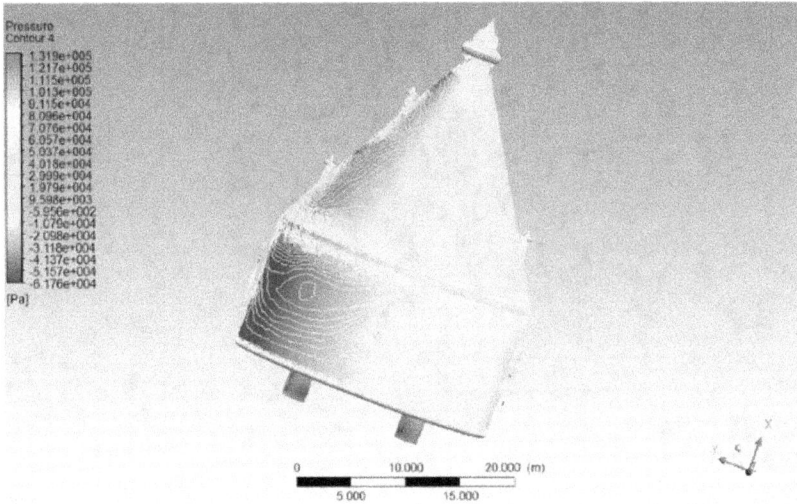

Pressure contour on a plane at the cross-section of vimana:

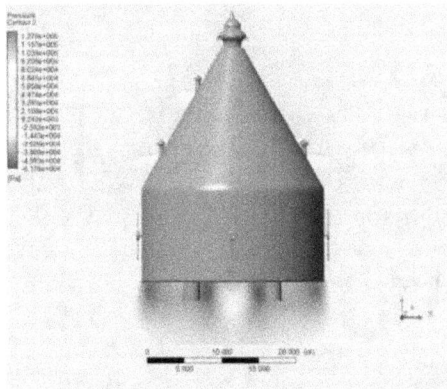

Velocity Streamlines of Sundara Vimana:

The air flow on Sundara Vimana is found to be smooth and hence the vimana is aerodynamic for doing space travel. The CFD analysis has been done on the Vimana designed in one of its form as shown in Vimana shasthra diagrams.

Further deep analysis on various forms of this vimana will be presented in the next version of this book.

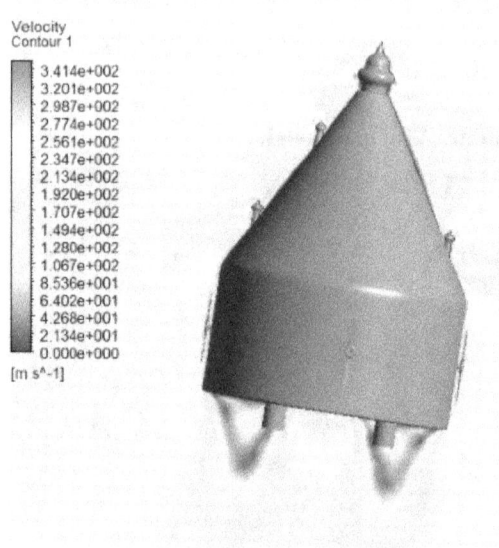

Analysis Results:

The analysis results on Sundara vimana is found to be similar to the space vehicles design in modern days. The Sundara vimana is similar to modern space shuttle with Rocket Boosters, and also exhibit similar results when done analysis on it.

SHAKUNA VIMANA

After CFD analysis of the Sundara Vimana, we can clearly observe in the results, that there is more pressure on the front part, throughout the length of vimana and also at the wings. The pressure values vary depending upon the speed with which the vimana flies.

Pressure contour on Shakuna Vimana model:

Pressure contour on a plane at the cross-section of vimana:

Velocity Streamlines of Shakuna Vimana:

The air flow on Shakuna Vimana is found to be smooth and hence the vimana is aerodynamic, moreover the vimana has advanced features such as foldable wing mechanisms, a dome on the top floors and also for doing space travel. The CFD analysis has been done on the Vimana designed in one of its form as shown in Vimana shasthra diagrams. Further deep analysis on various forms of this vimana will be presented in the next version of this book.

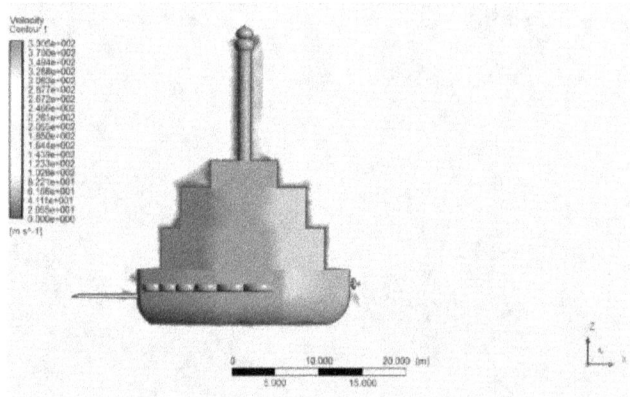

Analysis Results:

The analysis on Shakuna vimana is found to be similar to the advanced aerospace vehicles design in modern days. The Shakuna vimana is similar to a modern Ornithopter but with much advanced features and also exhibit similar results when done analysis on it.

TRIPURA VIMANA

After CFD analysis of Tripura Vimana, we can clearly observe in the results, that there is more pressure on the front part, of vimana and also at the vimana motor. The pressure values vary depending upon the speed with which the vimana flies.

Pressure contour on Tripura Vimana model:

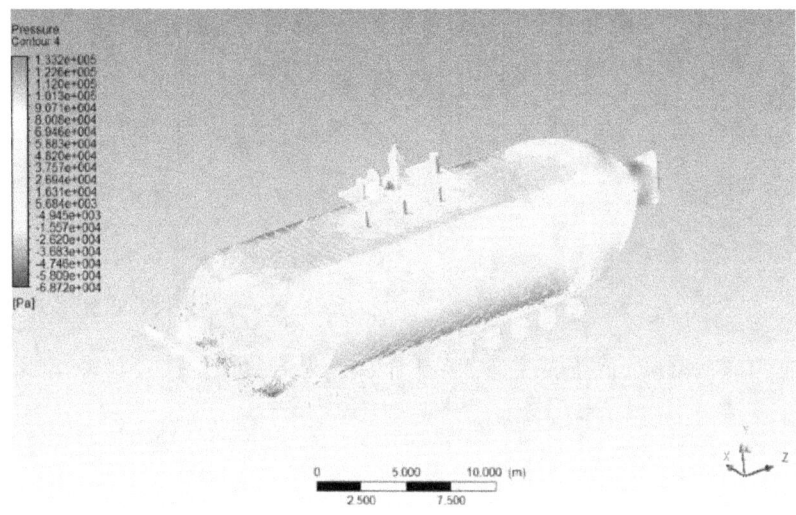

Pressure contour on a plane at the cross-section of vimana:

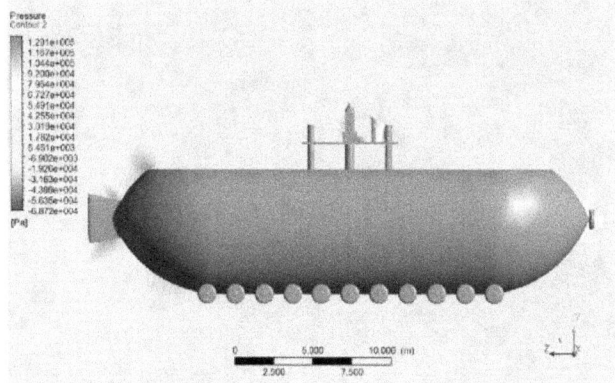

Velocity Streamlines of Tripura Vimana:

The air flow on Tripura Vimana is found to be smooth and hence the vimana is aerodynamic, moreover the vimana has advanced features such as retractable landing gear mechanisms, overlapping floor mechanisms, and a dome on the top floor covering the propulsion devices, and the vimana is multipurpose craft, travelling on land, in/on water, in air and also for doing space travel. The CFD analysis has been done on the Vimana designed in one of its form as shown in Vimana shasthra diagrams. Further deep analysis on various forms of this vimana will be presented in the next version of this book.

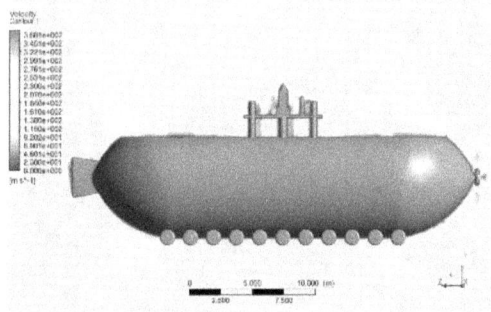

Analysis Results:

The analysis Results on Tripura vimana is found to be similar to the advanced aerospace vehicles design in modern days. The Tripura vimana is similar to modern airship but with advanced features and flexible materials, and also exhibit similar results when done analysis on it.

VI. Ancient Aerospace Materials

Puratana Aakasha-Yantrika Nirmana Sadhanavasthu (Ancient Aero-mechanical manufacturing materials)

Manufacturing Materials of Vimana

Rukma, Sundara, Shakuna vimanas are made of RajaLoha. And Tripura vimana is made of Trinetra Loha.

Testing material properties on Rukma vimana model. RajaLoha is used to make Rukma vimana, RajaLoha means king of all metals. The term RajaLoha was used for the resultant material obtained by combining various metals and herbs. In order to protect from heat and radiations, the alloy was used for the bodies of various flying crafts in ancient times.

Raja Loha material compositions and descriptions:

- Ammonium chloride: NH_4Cl
- Bengal gram:

Table: Bengal Gram Composition

Constitute	Composition
1) Calorific value	350 (cal./100 g)
2) Crude protein	26.2 percent
3) Fat	1.2 percent
4) Carbohydrate	56.6 percent
5) Calcium (mg)	185 (mg./100 g)
6) Iron (mg)	8.7 (mg./100 g)
7) Phosphorus (mg)	345 (mg./100 g)
8) Vitamin (mg)	
a) B_1	0.42 (mg./100 g)
b) B_2	0.37 (mg./100 g)
c) Niacin	2.0 (mg./100 g)

- Lodhra plant: It is used in RajaLoha to protect the humans inside the vimana from diseases caused by Space Radiations.
- Benzoin: $C_{14}H_{12}O_2$
- Lead: [Xe] $4f^{14}\ 5d^{10}\ 6s^2\ 6p^2$
- Sea-foam: Sea foam, ocean foam, beach foam, or spume is a type of foam created by the agitation of seawater, particularly when it contains higher concentrations of dissolved organic matter (including proteins, lignins, and lipids) derived from sources such as the offshore breakdown of algal blooms.
- Iron pyrites: The mineral pyrite, or iron pyrite, also known as fool's gold, is an iron sulfide with the chemical formula FeS_2
- Iron: chemical element with symbol Fe (from Latin: Ferrum) and atomic number 26. Electron configuration [Ar] $3d^6\ 4s^2$
- Mercury: Electron configuration [Xe] $4f^{14}\ 5d^{10}\ 6s^2$
- Natron, $NaHCO_3$
- Salt-petre, KNO_3
- Borax: $Na_2B_4O_7 \cdot 10H_2O$
- Mica: Chemically, micas can be given the general formula $X_2Y_{4-6}Z_8O_{20}(OH,F)_4$ in which

X is K, Na, or Ca or less commonly Ba, Rb, or Cs;

Y is Al, Mg, or Fe or less commonly Mn, Cr, Ti, Li, etc.;

Z is chiefly Si or Al but also may include Fe^{3+} or Ti.

Structurally, micas can be classed as dioctahedral (Y = 4) and trioctahedral (Y = 6). If the X ion is K or Na, the mica is a common mica, whereas if the X ion is Ca, the mica is classed as a brittle mica.

- Silver: Electron configuration [Kr] $4d^{10}\ 5s^1$
- Aconite: $C_{34}H_{47}NO_{11}$
- 5 sweets: curd, milk, ghee, sugar, honey mixture give result in the golden color thick paste.

VIMANA MATERIALS: RAJALOHA INVESTIGATION

The importance of raja loha: Hatakasya alloy -> high heat absorbing alloy used for bodies of various flying crafts.

Properties of Mica and its uses:
- It gives shiny and glittery appearance

- It has high thermal resistance allows it to be used as an insulation various electronics
- It is invariably used for fillers extenders along with providing smoother uniformity

Properties of Mercury and its uses:
- The chief source of mercury is cinnabar (HgS)
- The metal was named after the messenger of gods in roman mythology.
- Heavy silver white in color liquid.
- It is a liquid at ordinary temperatures and expands and contracts evenly when heated and cooled.
- Mercury and its vapor conduct electricity, its vapor is also a source of heat for power usage.
- Mercury amplifies sound waves and does not lose timber in quality.
- High frequency sound waves produce bubbles in the liquid mercury when the frequency of bubbles grow to match that of the sound waves the bubbles implode releasing a sudden burst of heat.
- A mercury -filled flywheel can be used for stabilization and propulsion in discoid aircraft/spacecraft.
- Mercury atom offers the most stable gyro device in nature and has the additional advantages of saving space and weight. This is particularly valuable on long distance flights where all space and weight must be very carefully calculated and conserved.

Properties of Iron Pyrites (FeS_2):
- In 16th and 17th-century pyrites are used as a source of ignition in early guns
- It is used as semiconductors within crystal detectors. In the crystal detectors, these pyrites behave as rectifier (it turns A.C current to D.C current) in crystal radios
- If pyrite is cut and polished to form jewelry. It can broadly have classed as a gemstone.

Properties of Borax ($Na_2B_4O_7$), Boric Acid (H_3BO_3) and its uses:
- It is used as neutron absorber in nuclear reactors
- It is used as flame retardant
- It is used for the manufacture of fibreglass, household glass and the glass used in LCD displays.

Properties of Saltpeter (KNO_3) -> Potassium Nitrate:
- It is used as rocket propellants
- It is used in fire works

- It is soluble in water, hygroscopic and it absorbs 0.03% water in 80% relative humidity
- It works as an aluminum cleaner
- In heat treatment of metals as a medium temperature molten salt both usually in combination with sodium nitrate
- Its oxidizing quality, water solubility and low cost make it a short term rust inhibitor
- In molten stage form with the solar energy, it can act as a thermal storage medium in power generation systems.

After observing the composition we can understand that ancient manufacturing process was also eco-friendly and because of the extinct plants/trees in a modern day, we will have to replace them with other substitutes which can be similar to it. In order to understand the Vimana materials and similarities with modern available materials, our SWASTIK team did Thermal analysis on a 3D model of Rukma vimana. Each time different material is used for the vimana, such as Copper, silver, gold, mica, titanium, tungsten and ceramics. Input values taken are: Heat Flux = 500 W/m², Convection = 500 W/m².°C, Radiation = 1, Temperature = 1000 °C. Convection and Radiation are applied to all faces of vimana but Heat flux is applied to the shell part of three floors of vimana where the passengers and pilots are seated.

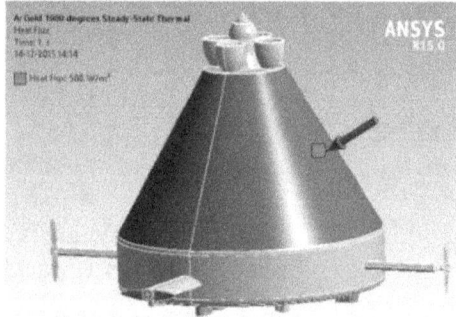

Rukma vimana Meshing and Heat flux applied on passenger cabin surface [7]

Rukma Vimana is one of the Kritaka Vimanas and it is similar to a rocket. Rukma Vimana had long vertical ducts with fans on the top to suck air from the top and send it down the ducts, generating a lift in the process. Utilization of electrical energy to operate Electro-Mechanical arrangement to enable vimana to lift off and accelerate Directional control is through conventional rudder system provided at the base.

Rukma vimana was made of Raja Loha and is deciphered by modern scientists. But it is not completely manufactured and It is observed as a high-heat-absorbing alloy used for the bodies of various flying crafts. Based on the Vimana shasthra texts, Raja loha composition, and in our SWASTIK teamwork, we have found several important properties of Raja Loha as described below.

The ancient manufacturing process was also eco-friendly and because of the extinct plants/trees in the modern day, we will have to replace them with other substitutes which can be similar to it. In order to understand the Vimana materials and similarities with modern available materials, our SWASTIK team did Thermal analysis on a 3D model of Rukma vimana. Each time different material is used for the vimana, such as Copper, silver, gold, mica, titanium, tungsten and ceramics. Input values taken are:

Heat Flux = 500 W/m², Convection = 500 W/m².°C, Radiation = 1, Silver input temp = 961.8 °C, Gold input temp = 1064 °C and Temperature = 1000 °C, for other materials. Convection and Radiation are applied to all faces of vimana but Heat flux is applied to the shell part of three floors of vimana where the passengers and pilots are seated.

Table . Thermal analysis results

Metals	Copper	Copper Alloy	Silver Input temp = 961.8 °C	Gold Input temp = 1064 °C	Mica	Titanium Alloy	Tungsten	Ceramics
Max Temperature	1001	1001	962.37	1064.6	1001.1	1001	1001	1001.1
Max Heat Flux	540.41	521.23	470.13	428.7	2.8175	120.03	372.48	34.146
Max Directional Heat flux	430.89	431.6	355.76	329.2	2.1121	78.249	243.67	23.418

The mixture used to make Raja Loha has mica which gives less heat flux compared to ceramics. The proportions of silver, mica, lead, mercury and other materials would result in raja loha which would have properties similar to that of NASA space shuttle heat shield tiles. When these mixtures are taken in exact proportions mentioned in texts we get the perfect raja loha and it can be used for Modern space vehicles.

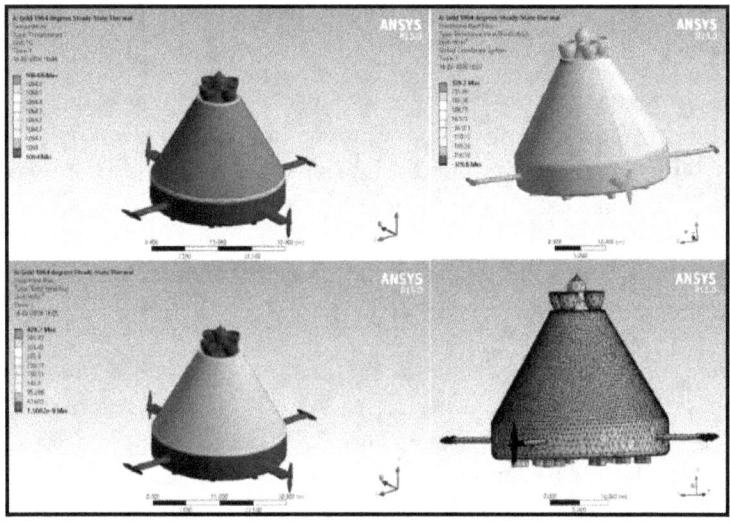

Fig (a). Thermal analysis results of Rukma vimana using material as gold - Input Temperature 1064 °C

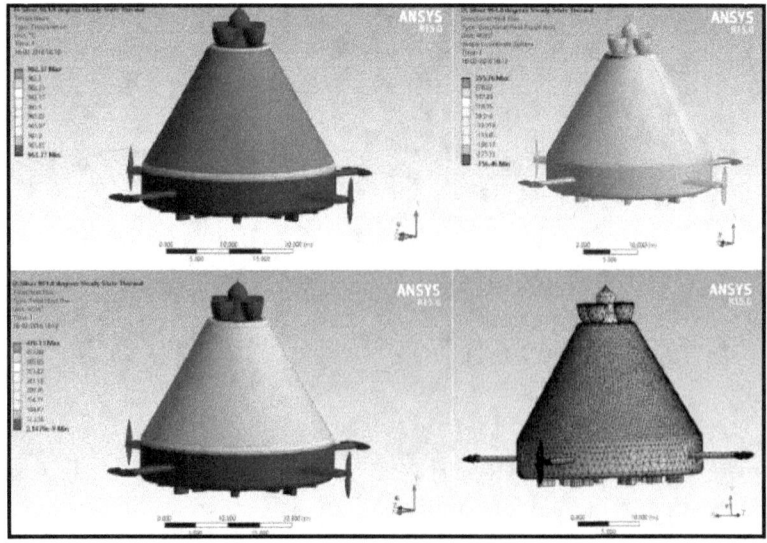

Fig. (b). Thermal analysis results of Rukma vimana using material as Silver - Input Temperature 961.8°C

Propulsion materials

Sun crystal

The energy source of Vimana is the sun, solar energy is absorbed by Sun crystal. Yellappa Drawings show the description of electrodes and acids. Inside the dome of sun crystal, there are 4 containers of acids and electrodes. This equipment all together produce electricity and distribute to generators and motors of Vimana. Inside the dome of sun crystal, there are 4 containers of chemicals which are electrolytes, and cathode anode reactors. The sunlight enters the sun crystal dome, which can absorb high energy from solar rays and the chemical reactions with the help of electrodes process gives result in electricity which can be stored and supplied all over the vimana. The entire equipment of sun crystal is a device that is able to absorb energy from the sun and store electrical energy in the form of chemical energy and convert that energy into electricity. This working of sun crystal is deciphered by our team SWASTIK, and available data shows that sun crystal is advanced than solar panels used today.

The following data are the description for Sun crystal composition:

"Ravi Shakti Apakarshana darpana (glass)": A special glass concentrating (visible) light energy in sunlight

Status: Already produced and study of optical properties is not yet done.

"Ushna Shakti Apakarshana darpana (glass)": A special glass for concentrating the heat energy in sun light

Status: Fully deciphered and to be produced in the laboratory.

Surya Shaktya Pakarshana yantra or collector of solar energy or Solar heat extracting Yantra:

In order to relieve the excessive cold of the winter months, the soorya shaktyapakarshana yantra should be installed on the vimaana.

Says Yantra Sarvasva,

"In order to protect from the cold of the 4 winter months, the solar heat storing machine is now explained. The 27th kind of mirror capable of capturing solar heat is to be used in its making."

It is said in Darpana prakarana: Sphatika or alum, manjula or madder root, sea-foam, sarja salt or nation, sand, mercury, garada or aconite, kishora or wild liquorice, gandhaka or sulphur brimstone, karbura or yellow orpiment, praanakshaara or ammonium chloride, in the proportion of 12, 1, 5, 1, 13, 12, 8, 10, 27, 4, 3, 7, 8, 5, 1. 5, 8, 3, 9, 2, purified, to be filled in antarmukha crucible, placing it in shuka-mukha furnace, and boded. Then pour it into antarmukha yantra or vessel and turn the churning key. When cooled in the mould a fine, light, strong, golden. colored, solar heat collecting glass will be formed.

Sun Crystal of Rukma vimana

Electromagnets

In Vimana shasthra Rediscovered by ADA, it has been written that Rukma vimana is subsonic aircraft on earth. But SWASTIK team discovered something Extraordinary about the vimana materials. Rukma vimana is antigravity interplanetary vedic space plane. The electromagnets used in the vimana is responsible for levitation of vimana. Magnetic materials and systems are able to attract or press each other apart or together with a force dependent on the magnetic field and the area of the magnets, For example, the simplest example of lift would be a simple dipole magnet positioned in the magnetic fields of another dipole magnet, oriented with like poles facing each other, so that the force between magnets repels the two magnets. Essentially all types of magnets have been used to generate lift for magnetic levitation; permanent magnets, electromagnets, ferromagnetism, diamagnetism, superconducting magnets and magnetism due to induced currents in conductors.

To calculate the amount of lift, a magnetic pressure can be defined.

For example, the magnetic pressure of a magnetic field on a superconductor can be calculated by:

$$P_{mag} = B^2 / 2\mu_0$$

where P_{mag} is the force per unit area in pascals, B is the magnetic field just above the superconductor in teslas, and $\mu_0 = 4\pi \times 10^{-7}$ N·A^{-2} is the permeability of the vacuum.

An in-depth understanding of the responses of materials to electromagnetic waves may even enable us to design and fabricate materials with properties not found in nature.

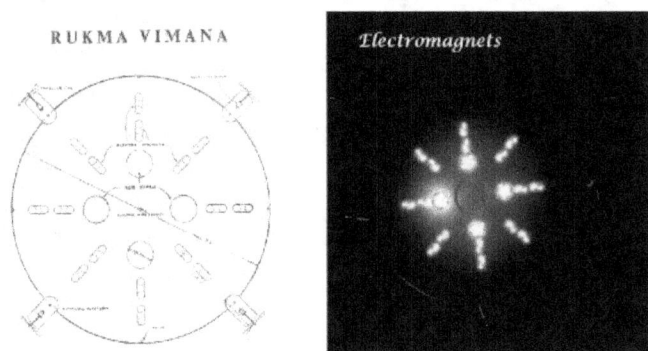

Electromagnets Rukma vimana

Mercury for Ion Engine: Ion propulsion technology development at Glenn began when Dr. Harold Kaufman, now retired from NASA, designed and built the first broad-beam electron-bombardment ion engine in 1959. It used mercury as fuel but is otherwise similar to the engine flying today on DS1. The laboratory tests of variations of the original ion engine were promising enough for Glenn to begin suborbital flight tests in the early 1960's. By 1964, an ion engine launched on the Space Electric Rocket Test I (SERT I) operated for all of its planned 31 minutes before returning to Earth. In 1970, two modified ion engines were launched on SERT II; one operated for nearly three months and the other for more than five. Both engines suffered grid shorts, believed to have been being caused debris from thruster grid wear, before the planned end of the mission. After an attitude control maneuver cleared its grid of the short in 1974, one of the engines was started and was operated on and off for six more years. The information learned from these genuine space success stories was used to refine and improve the technology that today flies on communications satellites and, of course, on DS1.

Problems faced by mercury ion engine: Early ion engines used mercury or cesium instead of xenon as propellants. (Glenn researchers had worked on cesium ion engine technology in the mid-1950's.) But both proved to be difficult to work with. At room temperature, mercury is a liquid and cesium is a solid, making them easy to store. But both had to be heated to turn them into gasses. Then there was the clean-up. After exiting the ion engine, some mercury or cesium atoms would condense onto the ground test hardware, causing numerous clean-up difficulties. In the 1970's, NASA managers decided that if ion propulsion research was to continue, it would have to be environmentally clean and less hazardous. Glenn researchers soon turned to xenon as a cleaner, simpler fuel for ion engines, with many of the same characteristics as mercury. One of the first xenon ion-engine-like devices ever flown was a Hughes Research Laboratories design launched in 1979 on the Air Force Geophysics Laboratory's Spacecraft Charging at High Altitude (SCATHA) satellite. It was used, not to propel the spacecraft, but to change its electrical charge. Researchers then studied the effects of the "charging" on spacecraft system performance. In 1997, Hughes launched the first commercial use of a xenon ion engine on the communications satellite PanAmSat 5. This ion engine is used for station keeping that is, keeping the satellite in its proper orbit and orientation with respect to Earth. But according to the study, if pure mercury is used, it could give better efficiency.

Table . Flight Thrusters efficiency

Flight Thrusters	Beam diameter cm	Propellant	Specific Impulse, s	V_D, V	η_u, %
SERT I	10	Mercury	5000	46	80
SERT II	15	Mercury	4200	37	80
XIPS-13	13	Xenon	2565	25-28	90

Mercury in Indian Vimanas

Vimanas of Ancient India used mercury as fuel in their ion engines. The mercury used was a purified one. Mercury has been known to Indians 11000 years ago. Dharnidhar Samhita gives the 16 steps to purify Mercury and make a SOLID Shiva Lingam out of it. For the one hundred and ninetieth "richa" (verse) of the Rig Veda and the aeronautical treatise of Bharadwaja mention that flying machines came into full operation when the power of the sun`s rays, mercury and another chemical called "Naksha rassa" were blended together. This energy was, it seems, stored in something like an accumulator or storage batteries. The Vedas refer to eight different engines in the plane and Bharadwaja adds that they are worked by electricity.

Shiva Linga made of mercury

Rasa Shasthra

In Ayurvedic medicine, the traditional medical lore of Hinduism, rasa shastra is a process by which various metals and other substances, including mercury, are purified and combined with herbs in an attempt to treat illnesses. Methods The methods of rasa shastra are contained in a number of Ayurvedic texts, including the Charaka Samhita and Susruta Samhita. An important feature is the use of metals, including several that are considered to be toxic in evidence-based medicine. In addition to mercury, gold, silver, iron, copper, tin, lead, zinc and bell metal are used. In addition to these metals, salts and other substances such as coral, seashells, and feathers are also used. The usual means used to administer these substances is by preparations called bhasma, Sanskrit for "ash". Calcination, which is described in the literature of the art as shodhana, "purification", is the process used to prepare this bhasma for administration. Sublimation and the preparation of a mercury sulfide are also in use in the preparation of its materia medica. A variety of methods is used to achieve this. One involves the heating of thin sheets of metal and then immersing them in oil (taila), extract (takra), cow urine (gomutra) and other substances. Others are calcined in crucibles heated with fires of cow dung (puttam). Ayurvedic practitioners believe that this process of purification removes undesirable qualities and enhances their therapeutic power.

Toxicity

Modern medicine finds that mercury is inherently toxic and that its toxicity is not due to the presence of impurities. While mercury does have anti-microbial properties and formerly was widely used in Western medicine, its toxicity does not warrant the risk of using it as a health product in most circumstances. The Centers for Disease Control and Prevention have also reported a number of cases of lead poisoning associated with Ayurvedic medicine. Other incidents of heavy metal poisoning have been attributed to the use of rasa Shastra compounds in the United States, and arsenic has also been found in some of the preparations, which have been marketed in the United States under trade names such as "AyurRelief", "GlucoRite", "Acne nil", "Energize", "Cold Aid", and "Lean Plus". Ayurvedic practitioners claim that these reports of toxicity are due to failure to follow traditional practices in the mass production of these preparations for sale, but modern science finds that not only mercury but also lead is inherently toxic.

Solution

In NASA after exiting the mercury ion engine, some mercury or cesium atoms would condense onto the ground test hardware, causing numerous clean-up difficulties. In the 1970's, NASA managers decided that if ion propulsion research was to continue, it would have to be environmentally clean and less hazardous. Glenn researchers soon turned to xenon as a cleaner, simpler fuel for ion engines, with many of the same characteristics as mercury. But using Mercury in an ion engine has given more specific impulse when compared to xenon. And using the purified mercury will boost up the efficiency of an ion engine. The solution to the problem faced by NASA mercury ion engine can be found in the ancient science of purifying metals known as Rasa Shastra. We can improve the efficiency of ion engines, by using mercury purified by Rasa Shastra. If mercury is purified and used in ion engines then compared to xenon ion engines, Mercury ion engines would be a better choice for space exploration as it will give a better specific impulse according to the experiments and more over there will be much better efficiency even with larger beam diameter used in the ion engines. Upon considering all the experiments done in the modern world, there is a clear path laid in front of us, describing that the ancient writings are not a myth anymore, they are not science fiction anymore. So, by further studying on Rasa Shasthra and implementing the knowledge for experiments on mercury, we can use mercury in Ion Engine, as we are able to get more efficiency. This can make space exploration much simpler. By deeper understanding and practical experiments on the book Yantra Sarvaswa and Rasa shasthra, many more advanced technology can be Rediscovered and Reinvented, which will modify the world into the much better way.

Nanomaterials - Nanotechnology in Ancient India

Ayurvedic Bhasma - A Nano preparation: Bhasma used in Ayurveda for the treatment of various disease for the past several centuries is the oldest form of nanotechnology. Bhasma is ancient but ultra-modern nanomedicine prepared from metal after the scientific process to the raw material into the therapeutically active form. This is done through the classical process by repeated incineration and grinding with some herbal juice and other specified drug. Due to its small size, basic character gets changed. It is mainly due to change in electrical, thermal, inorganic, optical, chemical and biological behavior. Swarna Bhasma is a therapeutic form of gold metal of Nano size particle. When evacuated by various tool and techniques like AFM (atomic force microscope), SEM (scanning electron microscope), it was found that size of the particle was 56 nm. Analysis by FT-IR and XRD shows that pure Au in Zero valence state.

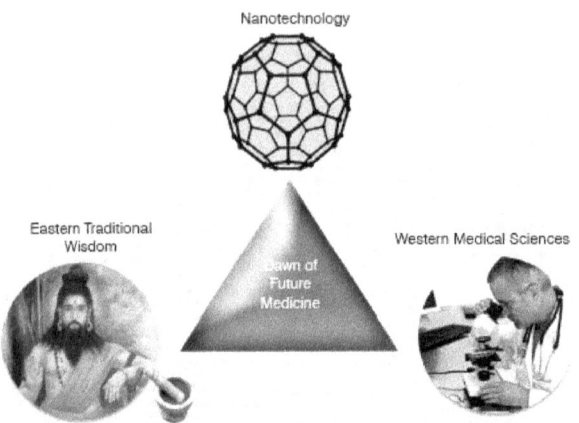

Ancient advanced nanotechnology

Comment

Upon considering all the experiments done in present day, we can understand that we are lacking in technology when compared with our ancient ancestors. It is clear that Vedic texts are not mythology, but historical documents on advanced technology. So, by further studying on Rasa Shasthra and implementing the knowledge for experiments on mercury, we can use purified mercury in Ion Engine, as we are able to get more efficiency. This can make space exploration much simpler. Among deciphered Vimana materials mentioned, the ancient materials can be used during manufacturing of space vehicles for radiation resisting and deciphered glass can be used for spacecraft to extract energy from the sun in outer space. The analysis done by our SWASTIK research team hence proves that there are high chances of space radiation resisting which can be better for future space missions. Also, further research on our findings of Sun crystal and electromagnets for propulsion will entirely change the existing complications in technology and makes the space exploration much easier.

VII. PROPULSION

The Power

Following Devices used in Power Generation of Vimana as mentioned in Vimana shashtra Power chapter:

- Vidyuddwaadashaka at the north-east side.
- Praanakundala at the moola of the Vimana.
- Shaktyudgama at the navel of the Vimana.
- Shaktipanjara in the central portion.
- Pattikaabhraka at the center of the hood of the Vimana.
- The Kuntinee-shakti mechanism is to be in the neck of the Vimana.

Interpretation:

Praana Kundala is the Power/Energy storage at the moola of the vimana, similar to human Moola chakra and spirit, the vimana have energy at the bottom of Vimana. Shakthi Udgama is energy production mentioned at the navel of the vimana. Similar to human production of energy at the navel connected to babies inside.

Shakthi Panjara is energy trapping location mentioned at the central portion. Here we can see that Energy extracted, stored and used something similar to battery little.

Electric Generator

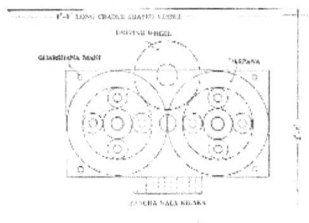

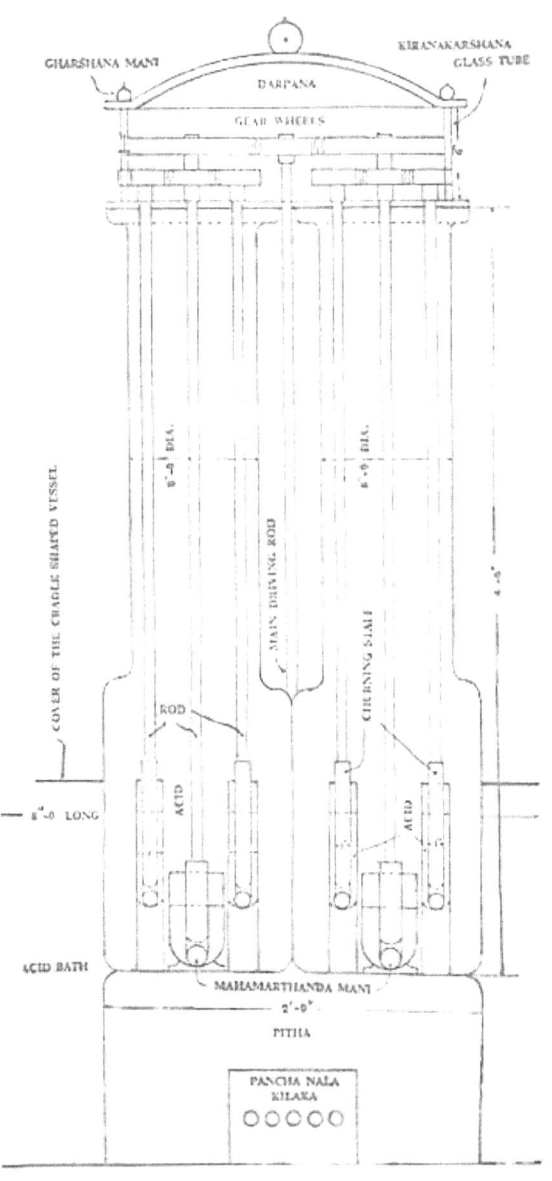

SECTIONAL ELEVATION

Interpretation:

It could be a Mercury vortex generator used for antigravity propulsion of vimana. Initial lines describe acids, and maybe it was battery fuel kind of thing. A mirror glass and crystal are mentioned which absorb solar energy. This means that something similar to solar panels. Acids and solar power absorbing glass are used to extract Electricity and passed to 5 different points. Further investigation is needed.

The Electric Motor

ELECTRIC MOTOR

- AMSHUPA DARPANA
- AMSHUMITRA MANI
- GEAR WHEELS
- SIMHIKA SHANKHA
- BHAMUKA GRAHINI MANI
- DRIVING WHEEL
- FINE WIRE CAGE
- WIRES IN GLASS TUBE

ELEVATION

PLAN

Drawn by
T. K. ELLAPPA,
Bangalore.
2-12-1923.

Prepared under instruction of
Pandit SUBBARAYA SASTRY,
of Anekal, Bangalore

Interpretation:

Simhika shankha is a battery using chemicals or acids and crystals. 5 rods might indicate the cathode-anode type of rods. Toothed wheels obviously indicate gears revolving. And rubbing might indicate creating a Spark or fire. The solar powered stored in the Battery is used to send to various other motors of vimana with the help of crystals called "bhaamukhagraahineemani".

Ganapa-yantra

Vighneshwara

The Ganapa Yantra is a machine similar Obstacle Destroyer. Gears are connected and made to rotate.

Interpretation:

It could be in vimana to remove obstacles like gravity force. Further investigation needed. Sapta-shashthishankha could be 760 units fillable/capable container shell or conch called simhikaa. Covering made of kravyaada metal meant fully covering Airtight metal. Jeevaavaka acid may be living grassy plant acid. Bhaamukhagraamukha manis could be shining sparkling Crystals
Umbrellas meant saucer-shaped structures, this might be electrolysis process. Cathode anode and chemical solution need to be identified by further investigation.

Description is looking like preparation of solar panels to extract solar power. Which is converted into mechanical energy rotating the gears described earlier. The later lines describe that output is transmitted through wires which again means current electricity. It mentions thermometer, so maybe there is heat energy also coming in output. The further investigation needs to be done to understand much about this.

Electric Dynamo

It describes that there are 32 devices for generating electricity such as by

1. Friction
2. Heating
3. Waterfall
4. Combination
5. Solar rays etc

Interpretation:

In modern technology, we have less than 32 devices for generating electricity. Waterfall can be understood by the modern example of hydro-power and solar rays mentioned we can understand as solar panel technology. Out of the mentioned electricity generations, the combination of methods is treated as a better process for generating electricity and suitable for vimanas because as per my analysis and knowledge the combined forces of Solar power and heat, and friction and heat, water force etc. are observed in natural faced by air transport.

It is very great to know that Sage Agastya explains power generation in shakthitanta. If it is a book, if it is available, then we may get much-advanced power generating inventions by referring to it.

Below are 3D modeling pictures of the parts used in electric dynamo, designed as per the dimensions described. For visualization I did this 3D modeling, and we can see similar things in unearthed archaeological evidence and ancient batteries too used similarly shaped structures etc.

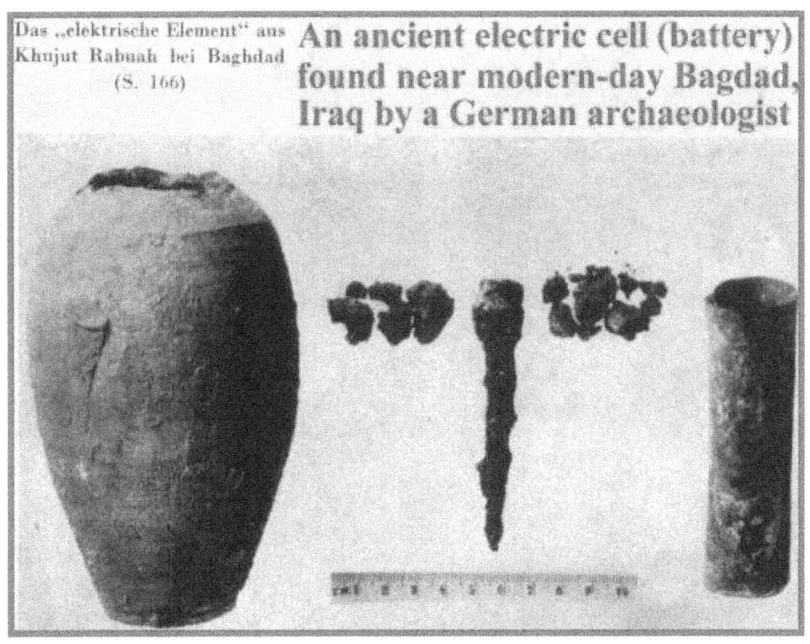

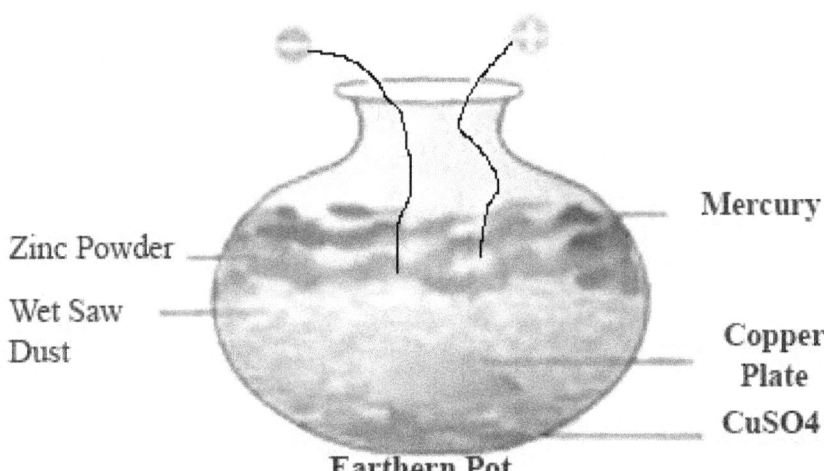

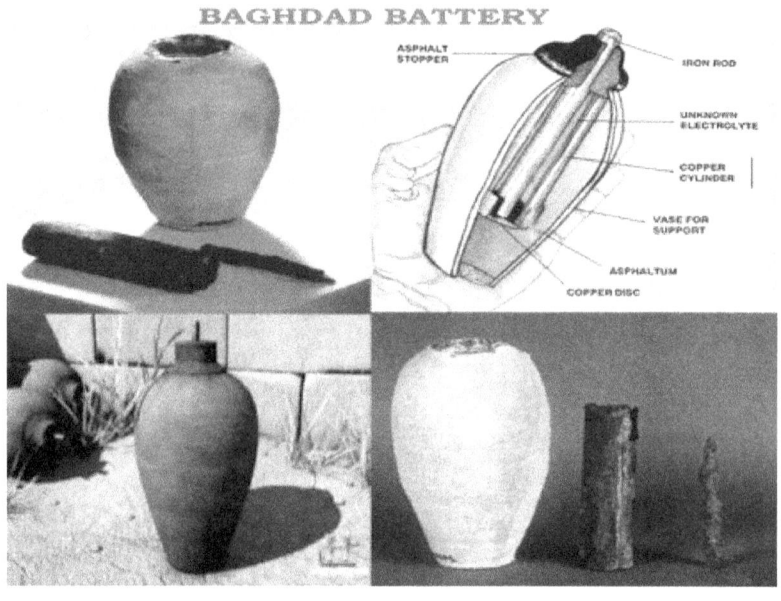

3D modeling by kavya

35 feet in diameter metal base plate with 5 spots

Vessels 4 feet wide, 2 feet high, shaped like a pot

Cylindrical pipe 1 foot wide and 1 foot high. Top of the cylinder should be 4 feet wide and round

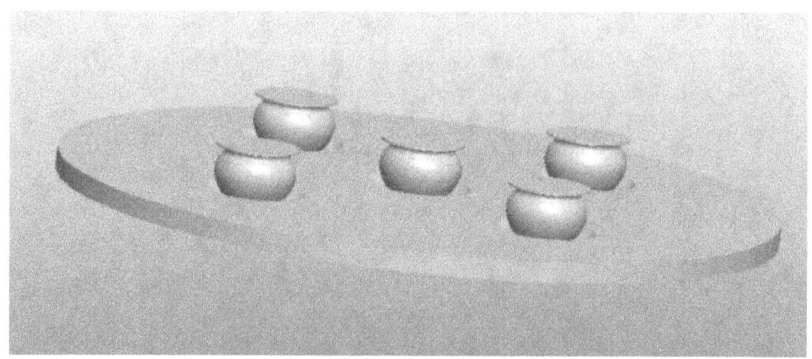

Assembled view all parts

Interpretation:

The chemicals, materials, and acids used are explained in detail. Usage of animal's urine etc cannot be neglected because Ayurveda uses modern cow urine for human body vimana, the same way ancient for vimana energy production they might be knowing something which we modern humans don't know. Hence all need to be tested and verified in the lab. Results may not come as described in the texts if we do not follow proper procedure and also if we do not follow correct material as per correct translation

The mechanism, procedure, the process is also explained clearly but not able to understand and it looks complicated or sophisticated description with lengthy sentences. Need to separate the big sentences and should be gone through in detail and need to be investigated.

Electricity production is explained and location of the machine in vimana is also mentioned in detail.

Rukma Vimana Propulsion

1) Rukma vimana extracts electricity from the sun through sun crystal located at the top of vimana. Sun crystal absorbs more electricity from the sun when compared with modern-day solar panels. Hence it was more efficient and advanced than solar panels known to us. The procedure of electricity extraction from the sun, has been deciphered by our SWASTIK team. The 4 electrodes, electrolytes, acids located inside the sun crystal perform electrolysis process and supply to the motors of vimana through fan pipes in which electric wires are contained.
2) There were no space vehicles at the time when Yellappa drew the vimana diagrams guided by Subbaraya shasthry. There are electromagnets men-

tioned in the Figures of vimana, which are earlier than today's NASA news that says "Electromagnetic drive propulsion is possible"

Nikola Tesla made attempts to create free energy generators and antigravity VTOL vehicles with the inspiration and reference of Vimana texts. Swamy Vivekananda helped Tesla to understand Sanskrit texts and its technical terms.

3) Tesla's works describe that excess electricity to an object will lift which is antigravity propulsion or levitation. The purpose of electromagnets in Vimana may indicate such type of propulsion. Tesla also worked on free energy concepts with the help of ancient Sanskrit texts. Vimana has motors and generators which are mentioned in vimana shasthra. Tesla's works on free energy generators and motors might give a clue on how the ancient ancestors tapped free electricity from the ether. Free energy concepts have been neglected and avoided by many countries but recent invention from India describes free energy generator which gives more than 200% efficiency by taking power from the vacuum.

Our works on Sun Crystal have shown us good results and given confidence that it is much more advanced than modern solar panels. Ancient advanced solar power generation is a research platform working on methods of generation of solar power using ancient technology with advance knowledge.

Propulsion part of vimana is not decoded completely but after more research, modern humans may be able to reach it.

Modern Spherical solar cell similar to Rukma vimana sun crystal

Modern Spinning solar cell similar to sun crystal

Advanced Antigravity Technology

By Randall E Mielke, Ph.D.
Chief Technical Officer (CTO),
Blue Ocean Sciences, LLC

The ability to fly through the sky, like a bird, has inspired generation after generation; to examine how a man might be able to fly into the sky with the birds. The ability to get off the ground is relatively easy; put enough energy into a specific mass, causing it to accelerate enough to 'jump' into the sky above the earth. The jumping can occur through the use of muscles, in the case of man, that uses the energy needed to cause them to move in a specific direction and resulting in both of our feet leaving the earth, but only for a few seconds at most, until the gravitational forces exceed the upward motion, and we fall back. The energy is still being put into the muscle systems, but the voids or the spaces between the air molecules, and our inability to impact the air itself in order to cause further lift by the use of man's muscle system means that we fail to maintain flight. The initial understanding of motion in the planetary and physical sense was conceived in Aristotelian Physics (Aristotle 384-322 BCE) where earth and water were the heaviest of the four elements (Air, Fire, Water, & Earth) and predicting which will fall farther than the others when intermixed (1). Earth is heavier than water as demonstrated when a rock is put into water, it will generally sink to the bottom of the container (2). The following chapter will not go through every step in our understanding of how gravitational fields or their mathematics were generated, but to advance to the mathematics relating directly to the antigravity technologies and the ultimate building of devices enabling man to sail upon the sky.

    Man's ability to sail upon the sky relates directly to the gravitational field present on earth, or upon any other planet or huge body. This

gravitational force is the result of the interaction of two objects, at a distance apart from each other, with a gravitational constant causing a force between the two masses, and has been considered in Newton's Law of Universal Gravitation, as originally written in 1687 by Sir Isaac Newton in Principia (3).
Equation 1.

$$F = G \frac{m_1 m_2}{r^2}$$

The gravitation Force (F) is equal to the Gravitational Constant (G) multiplied by the relative mass of two objects (m1 and m2) divided by the square of half of the distance (r) between the objects. Newton was not the first to consider the concept, as early as 1945 Ismaël Bullialdus described that a force, felt by the sun acting upon planetary objects, would follow the inverse-square law (4). When we consider a man standing on earth and the force being applied to him is only relative to the center of distance (r) between the objects, due to the infinitely larger mass of the earth and the centrifugal force of the earth rotating on its axis (5). The gravity on earth is 9.80665 m / s2 (32.1740 ft / s2) at 45° latitude that was initially adopted by the International Committee on Weights and Measures in 1901 (6,7). It also should be noted that gravity does vary from the equator and the poles respectively, 9.780 m / s2 and 9.832 m / s2, and also is affected by the altitude of the actual object. Since the gravitation attraction is constant, Newton's Law of Universal Gravitation can be simplified to:

Equation 2.

$$F = mg$$

The gravitational force (F) is equal to the mass (m) multiplied by the constant vector (g) and which, upon the earth, the average magnitude considered is approximately equal to 9.81 m / s². While adhering to these same gravity assumptions, the potential energy of a human body over small distances above the earth can be expressed by the following equation:
Equation 3.

$$E_p = mgh$$

The potential energy (E_p) is equal to an object of mass (m), multiplied by the acceleration due to gravity (g), and a small distance (h) above the earth. This potential energy is the energy required to maintain an object suspended above the earth. With this calculation, we could theoretically know the amount of energy required to maintain an object hovering above the ground and could be considered the energy required for antigravity to occur. Since we can calculate the amount of energy needed for a launch, we then need to ask the question:
Where can we generate enough energy over an extended period of time in order to counter the effect of earth's gravitation forces, and the

centrifugal force of an object on the earth to finally escape from this planet and venture beyond?

Currently, we are using fossil fuels for aircraft with engines having the power to life aircraft into the sky; noting that combustible fuel is easy to transport but considered a non-renewable product. The Institute for Energy Research has shown that the United States transportation requires 95% usage of fossil fuels and 66.9% of electrical powers are generated by fossil fuels [8]. The amount of energy generated from one specific jet, using A-1 fuel is 43.15 MJ / kg (12.00 kWh / kg) with an energy density of 34.7 MJ / L (9.6 kWh / L) that generating 124,536 BTU / gallon, which is slightly less than using normal diesel fuels [9].

When you are considering how most of our present spaceships are propelled into space by the use solid or liquid fuels it is important to understand that rocket fuel is extremely different from the fossil fuel used for propelling airplanes. For fossil fuel using airplanes, the combustion of a fuel occurs within the engines that then use the production of the gasses created to cause the movement of the mechanical parts and this is highly inefficient, due to loss of power through friction from the air. Friction is not an issue for rocket engines, due to how the combustion chamber and the resultant mixing of the oxygen and propellant, transfers the exhaust gasses via a restriction of those gasses through a device known as a De Laval nozzle [10]. The force generated is described using the formula below:

Equation 4.

$$F = pA$$

Where the pressure change (p) is equal to the pressure in the combustion chamber ($P_{combustion}$) minus the pressure outside the exhaust ($P_{ambient}$) multiplied by the area (A) of the exhaust chamber.

Equation 5.

$$F = (P_{combustion} - P_{ambient}) A$$

These exhaust particulates can reach an astounding 5 km / s traveling speed and, as could be expected, the rocket fuel that accelerates the exhaust particulates at the fastest rate is the best at getting a spacecraft into space and outside of the earth's gravitational field. By using the kinetic theory and Ideal gas law the mass of the particulates can evaluate by the velocity and temperature of the generated gasses:

Equation 6.

$$\frac{1}{3} m\overline{c^2} = kT$$

Where the mass (m) of the particles is multiplied by the "c squared bar" or the mean of the squared velocity of the particles is equal to the Boltzmann constant (k) multiplied by the temperature (T). By rearranging to evaluate the root mean square velocity of the gas particles:

Equation 7.

$$c_{R.M.S.} = \sqrt{\frac{3kT}{m}}$$

The relationship between temperature and particle mass can be shown to be inversely proportional, where higher temperatures and smaller molecular mass particulates will result in the movement of the gas particulates at a faster rate. This rate of the gasses escaping the exhaust manifold is the velocity ($V_{exhaust}$) of exhaust jet that can be evaluated through the ratio of specific heat (λ), which is generally between 1.1 and 1.7.

Equation 8.

$$V_{exhaust} = \sqrt{\frac{2\gamma kT_{Combustion}}{(\gamma-1)m}\left[1 - \left(\frac{P_{combustion}}{P_{exhaust}}\right)^{\frac{\gamma-1}{\gamma}}\right]}$$

The above velocity equation uses the pressure in the combustion chamber ($p_{combustion}$) and the ambient pressure ($p_{ambient}$) along with the specific heat (λ), the temperature of combustion ($T_{combustion}$), and Boltzmann constant (k) to be able to evaluate the highest velocity from different generated molecular mass particles. It was observed earlier and similarly with this equation that higher combustion temperatures generate greater velocities of the exhausting jet. Since solid rocket fuels do burn at greater temperatures, but conversely generate large mass particulates in the exhaust, this results in a slower velocity of the exhaust jets, for solid fuel when compared using liquid rocket fuels as the propellant.

Liquid rocket fuel will use liquid oxygen along with a liquid propellant resulting in combustion, with various by-products, depending upon the propellant. Using liquid hydrogen as the propellant does generate the higher temperature of combustion reaching nearly 3000 K, with only water molecules as the exhaust particles. The smaller molecular mass in the exhaust results in a higher velocity of the exhaust jet, but in consequence the liquid hydrogen has a lower density and so a larger payload is required than when other denser aviation fuel propellants are used. This has resulted in liquid jet fuel propellant and liquid oxygen becoming the preferred mixture during the initial takeoff of the rocket, and a liquid hydrogen propellant then used during the following required rocket burst. The liquid rocket fuels make it easier to control the amount of required velocity provided from the exhaust jet, and that equates to the force causing the rocket to lift off from the ground and then counter the gravitational forces of the earth. The amount of energy is required for propelling the rocket, as the specific impulse or forces that move a rocket, with a given mass of solid or liquid rocket fuel, into the atmosphere and beyond.

The specific impulse is different for solid as opposed to liquid fuels. For solid rocket fuels, the energy density (Joules per kilogram) is greater than that for liquid propellants, resulting in very large specific impulses: (impulse (Newton second) per kilogram propellant)
for the liquid versus solid fuel rockets. The liquid fuel rocket (4500 N s kg^{-1}) has almost twice the specific impulse as solid fuel rockets (2500 N s kg^{-1}) and both can be used to get rockets out of earth's gravitational fields, and even propel them far enough to begin touching the edge of our solar system, like the recent launching of the New Horizon spacecraft. The New Horizon is presently within the orbit of Pluto, which is considered a dwarf planet and the 9th farthest planet from the sun [11]. It was launched into space in 2016, and it is continuing to use its rockets for trajectory adjustments. This is a non-renewable energy source for the spacecraft, and in the outer reaches of space, the inability to generate more fuel is the greatest limiting factor, whether using either solid or liquid rocket fuel, for the propulsion of the spacecraft while it is remaining in space. We need to go to a radioactive energy source, driving a more efficient engine utilizing magnetic and electrical impulse thrusters, rather than the mechanical thrust of combustion engines.

We are already using radioactive heavy metals to drive our energy plants to produce electricity. Most nuclear power plants use uranium as their energy source to heat a coolant, which is subsequently pumped through massive heat exchangers, causing the production of steam in the adjacent chamber. This steam then propels turbines, driving electromagnetic generators producing electrical power [12]. This electrical power is not without consequences, though, and the main problem is that the waste generated in today's uranium power plants is highly toxic, and will remain so for many thousands of years after its initial disposal. The alternative has been to use the Thorium-232 that is highly plentiful on the earth and produces a substance that is dramatically less toxic than the conventional nuclear power plant waste material [13]. This nuclear method for generating electrical power requires massive bodies of coolant and water to make it a viable propellant for vessels like submarines and aircraft carriers. But for spacecraft that are required to elevate above the ground, having a huge mass of water is prohibitive and impractical for interstellar space travel. The use of radioactive heavy metals, however, can be used directly and in electrolytic fluids to capture and harness the energy being released through their specific breakdown kinetics. The energy losses from radioactive heavy metals are in the form of alpha particles, beta electrons, and gamma rays.

The alpha particles (α) are the release of 2 protons and 2 neutrons (2p2n), which is a helium nucleus with a charge property of +2 [14], but an α particle has a mass of approximately 6.64424 $\times 10^{-27}$ kg and a kinetic potential of 3 to 7 $\times 10^9$ eV (3-7 MeV). This makes the α particles highly attractive to materials, but retain a limited travel distance in air of approximately only a few inches, and the inability to travel through barriers thicker than the thickness of a piece of paper. The speed of an α particle is about 5% the speed of light due to its large mass [15]. Thorium-232 is one radioactive metal that has been studied and is becoming a theoretical substitute for the extremely dangerous Uranium 238, presently used in nuclear reactors and electrical power plants [16]. The higher energetic and lighter released atomic particles are the beta particles.

The Beta particles (β) can be either an electron (e-) or a positron (p+) emission, depending upon the original radioactive atom, or upon a heavy metal [17]. Either particle is extremely small, so it can have a max energy of 1.7 MeV from the β- decay of P-32. The use of electron and positron release, from radioactive uranium, was first discussed by E. Rutherford in 1898 where he showed that a photographic plate was impacted by both positive and negative electrification, to an equal degree [18]. The distances that a β- particle can travel in the air will depend on its initial β- release energy and, in the case of P-32, the travel distance could be as much as 6 m in air [19]. This is many times more distance than an α particle, but a β- does not have the energy to penetrate through a piece of paper. This is not the case for gamma rays.

Gamma (γ) rays have no mass and considered photons of light. The γ rays are electromagnetic radiation and are the most energetic type of nuclear radiation. The release of γ rays by unstable nuclei can produce kinetic potential energy between 0.01 to 10 MeV depending on the parent atom (20). The traveling distance for γ rays can be as much as 12 ft. in air, but it can penetrate many types of materials (21). These three types of radiation are only one aspect we are considering when it comes to the design and development of a spacecraft that has antigravity properties.

The use of radioactive heavy metals is a great source of energy. If that is, the energy can be trapped and then is transferred, through a medium having an extreme flash point temperature. The substance mercury is fluid at both high and low temperatures and it can be electrically influenced from outside materials. By using radioactive heavy metals with a liquid heavy metal, you can generate an extreme amount of energy, by the release of any of the nuclear radiation listed above – the real question is:

How does this energy transmit to the disequilibrium of the molecules to be able to cause the force required to lift the spacecraft past the force of gravity?

The description of an impulse electromagnetic propulsion engine is beyond the depth of this chapter, but its description in detail will be in the following publication. The next level of understanding of an impulse system is a physical rotation of the radioactive heavy metal liquid slurry, which is fundamental, in the production and harvesting of the electromagnetic energy. The rotation is considered the 'frequency' of the system and by adjusting that frequency to generate the proper stimulation frequency will theoretically produce the compounding factors causing the required antigravity properties. When we take the dimensions of the system's rotating chamber (wavelength; λ) and multiply it by the frequency (f) you get a velocity (v) that has been shown to be 1/137 the speed of light (c) (22):

Equation 9.

$$\lambda = \frac{c}{f}$$

Equation 10.

$$c \frac{1}{137} = \lambda f$$

By using the adjustment defined above, Frank Znidarsic, who worked on antigravity concepts, showed a new physical interpretation of the fine-structure constant, as a ratio between the speed of light in a vacuum and the speed of light upon entering an atom (23). The fine-structure constant (α), or Sommerfeld's constant, is the fundamental physical constant of the elementary charged particles, relating to the strength of the electromagnetic interaction. The equation is shown below:

Equation 11.

$$\alpha = \frac{e^2}{(4\pi\varepsilon_0)} \hbar c = \frac{V_\perp}{c}$$

Where the elementary charge (e) squared is divided by 4*π that is multiplied by the electrical constant (ε0), Planck constant (ℏ) and the speed of light in a vacuum (c). The velocity (Vt) divided by the speed of light (c) produces a fine-structure constant of α-1 = 137.035. This velocity (Vt) is the velocity of a photon once it has slowed down after entering an atom. The velocity is proportional to the frequency multiplied by wavelength. The product of the frequency and wavelength reveals the transitional quantum velocity that is equal to the radii of the specifically associated atom. It may also be thought of as the velocity of sound in a medium when considering an antigravity system and mapping a heavy proton's movement. The transitional quantum velocity equation;

Equation 12.

$$V_t = f_t \, 2\pi \, (n \, r_p)$$

The frequency (ft) is multiplied by 2*π, and the photon wavelength (rp) multiplied by some integer (n), you produce the velocity at that specific transition. Through the rearrangement of Coulomb's constant in the form of a spring constant as described below:

Equation 13.

$$F = K \, x$$

The force (F) of the spring is equal to the kinetic energy (K) over a distance (x). The kinetic energy of an electron (K-e) in an atom expressed as a spring below:

Equation 14.

$$K_{-e} = \frac{F_{max}}{r_x}$$

The maximum force (Fmax) is divided by the radii (rx) is equal to the kinetic energy of an electron. To further determine the elastic energy of an electron:

Equation 15.

$$E = \frac{1}{2} K_{-e} \, (2 \, r_p)^2 = \frac{Q^2}{2c}$$

The frequency (rp) of an emitted photon matches and the wavelength, when put into the configuration as an electric charge (Q), gives the energy of the photon (E). Since the electron energy state has binding properties, these fields have to be reconfigured, when transitioning from the parent state into the daughter state. During this transitional phase, the electron travels at the above specific velocity (Vt) and has been shown to be 1.094 x 106 Hertz-meters. This velocity is the same for all electrons the frequency and wavelength will vary depending on the density of the atom as shown in velocity of transition equation:

Equation 16.

$$V_t = \frac{1}{2}\pi \sqrt{Z \left[\frac{F_{max}}{2nr_x}\right] / n \, M_n} \, * \, 2 \, nr_n$$

The number of protons (Z) along with the coulomb density (Fmax/2nrx)/nMn) will change the velocity of the transition electron, while the ground state radius (rn) will prove to produce the same velocity for all atoms. OK – I have given a few mathematical equations that relate to how a photon of light entering an atom whose velocity will be changed to match the atom, and the resulting frequency and wavelength can thus be calculated. Since the frequency and wavelength are relative to the velocity the specific energy of an electron can be calculated. By giving described values to these parameters we can start to evaluate the energy requirement and material specific atom density required by the radioactive electromagnetic slurry in the antigravity spacecraft to achieve enough energy to escape earth's gravitational field.

The energy for antigravity will require atoms to be altered by either a vibrational element at extreme frequencies or the bombardment of the slurry with nuclear particles (Diagram 1), in order to cause the artificial quantum transitional state of a proportion of the liquid slurry (24). Additionally, electrical energy is pumped into angular windings of the reservoir that result in a vortex within the slurry, or more simply a stirring at controllable and specific frequencies can be used to control the kinetic electromagnetic energy being generated and released.

Diagram 1.

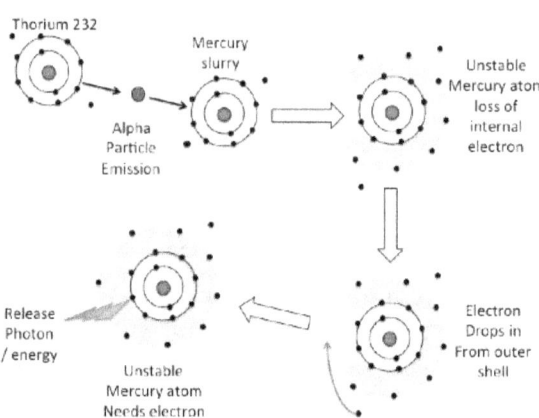

The greatest difficulty is how you control the release of energy. When we consider we are putting high-frequency electromagnetic energy in and getting gravitational propulsion out that is directly dependent on the rotational frequency causing the alignment of the electron Compton wave, and its reflective wave Doppler-shifted component. The electron Compton wave frequency is the natural vibration that all matter exhibits, which was proposed by Prince Louis Victor Pierre Raymond, 7e duc de Broglie in 1924 (25). The Doppler-shifted wave is an opposite of the reflected wave with a frequency shift. When these two waves combine for a doubling effect in the disturbance and thus form the De Broglie wavelength of matter. The De Broglie wavelength is double the energy output from the radioactive electromagnetic slurry, specifically increasing the antigravity velocity. The above theory is able to generate an antigravity velocity and further writings will determine the size of slurry container which is required to elevate a spacecraft with a specific mass, develop the directional electromagnetic propulsion components used to steer the spacecraft, and the development of adequate shielding to ensure the pilot and navigator survive the radioactivity and electromagnetic fields.

VIII. Vimana Prototypes

MRV UAV

A 15/15 cm prototype of Rukma vimana has been designed by our SWASTIK team.

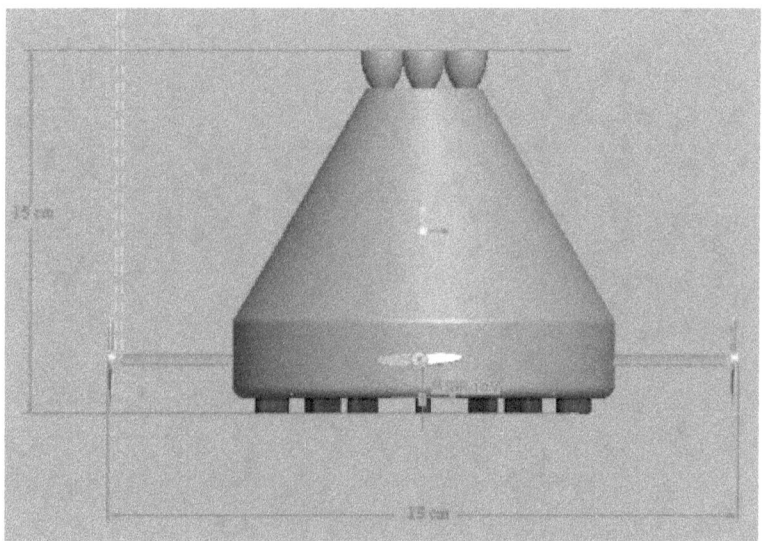

Rukma vimana Prototype model design

In order to do practical works on Rukma Vimana and to implement the advanced features of Rukma vimana, in modern day UAV, starting with small-scale projects, MRV UAV (Mini Rukma Vimana Unmanned Air Vehicle), using the scaled dimensions of Rukma vimana and propellers lift mechanisms with motors, propellers, and batteries. Our Works on Rukma Vimana Prototype, MRV UAV Project was selected to National Aerolympics 2014, Aeronautical Society of India and Won best team award.

Parts of MRV UAV:

1. Lift fans: Lift fans are the main part of the UAV. They can be used to go through mountainous regions. And fans are preferred other than wings for Vertical take-off. The lift fans configuration is similar to Rukma vimana Mentioned in Vimanika shasthra.

2. Camera: The cameras can be used to record the video of the path the UAV is going through or any disastrous incidents can be captured and help the situations.

3. Object collector: They can be used for defense purposes, delivering the medicine to the soldiers through object collectors.

4. Landing gears: The main purpose of landing gears is to make the UAV land in any desired location.

Advantages of MRV UAV in comparison with other UAVs are as follows:

- Based on Analysis for VTOL, UAVs are having lift fans embedded in Wings. But this MRV UAV has a simpler configuration, enabling the UAV to lift off with fans provided at the top of the UAV directly connected to the base of UAV with the help of Ducts.
- The Direction control can be achieved by operating the maneuvering fans acting as propellers.
- The UAV can move 360 degrees in at mid-air in a single position.
- It can not only land anywhere but also move on ground for various purposes.

Mechanisms: The Vimana prototype is having 8 propellers. the propellers at the lower part, side fans are having movable mechanisms to make the VTOL of the Vimana prototype. The upper propellers or the lift fans are also enabled to produce lift. The wings add more lift to the prototype. Once the prototype is in air, the Wings can move to and fro to reduce drag force and maneuver as mentioned in the table .

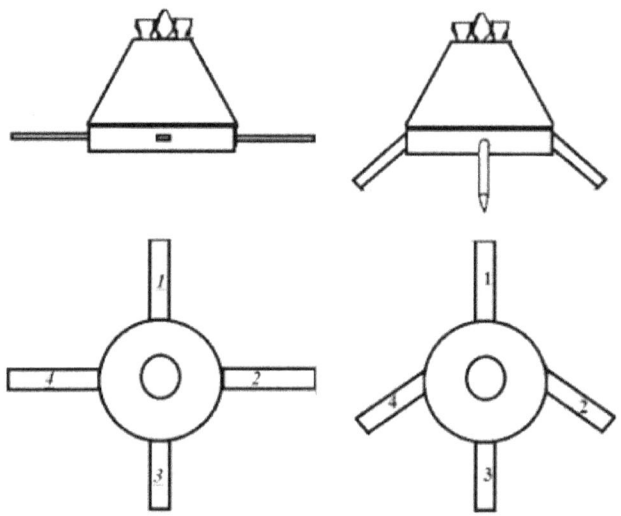

Rukma Vimana Prototype - maneuvers mechanisms

Rukma Vimana Lift calculations

Rukma vimana has Lift fans at the top, each 6 feet diameter. Calculations are as follows:

[EQ1]: PL [hp/ft^2] = power / A
[EQ2]: TL [lb/hp]= 8.6859 * PL^(-0.3107)
[EQ3]: Lift = TL * power >>>[lb] = [lb/hp] * [hp]

Using equation **one**, we calculate power loading (PL) of a 6 foot diameter (72") disk with 300hp absorbed.
PL = 300 / (pi*3'^2)
PL = 300 / 28.27
PL = 10.61 hp/ft^2

Using equation **two**, we calculate the thrust loading. Typical communication/interpretation error is in the negative exponent of the equation, X^(-Y) is the same as 1/(X^Y).
TL = 8.6859 * (10.61^-.3107)
TL = 8.6859 / (10.61^.3107)
TL = 8.6859 / 2.083
TL = 4.2 [lb/hp]

Using equation **three**, we calculate the lift/thrust.
Lift = TL * power
Lift = 4.2 * 300
Lift = 1,251 pounds

Lift = 5564.73 Newtons for Single Lift fan of Rukma vimana.

All lift fans; For 4 lift fans, the lift is; 4 x 5564.73 = 22258.92 Newtons

The side fans of Rukma vimana are 30 feet diameter each. 300 hp, 30' rotor develops 3,400 pounds of lift. 15123.95 Newtons lift for each propeller (side fan of Rukma vimana) All side fans; For 4 side fans, the lift is; = 4 x 15123.95 = 60495.8 Newtons

Complete Vimana lift is Lift fans + Side fans = 22258.92 Newtons + 60495.8 Newtons = 82754.72 Newtons In order to lift 100/100 feet vimana,

Now looking at the Modern day Heavy Transport Helicopter. Let us see the dimensions as mentioned in the following table.

Table: Dimensions of Heavy Transport Helicopter

Rotor Blades (main/tail)	8/5	
Tail Rotor Diameter	7,61 m	24 ft 12 in
Rotor Disc Area	804,2 m²	8656 ft²
Length	40,00 m	131 ft 3 in
Height	8,10 m	26 ft 7 in

Rukma vimana have wing structures, there are 4 wings which provide additional lift for the vimana. Moreover, the drag caused in helicopter type vehicle is not there in the case of Rukma vimana because the lift fans are able to send the air on to hit the ground. As well as the side fans tilt and the air hit the ground directly creating good lift force. Another important factor we need to consider is the Antigravity propulsion of Rukma vimana.

RESULTS AND DISCUSSION

When there are Reinvented devices and materials from vimana shasthra, then the Vimana technology cannot be neglected as it gives mankind to achieve more advanced technology. Some attempts of antigravity may fail due to improper procedures and materials, but when we follow the Vimana shasthra step by step and execute the process correctly, then we may decode the advanced technology. Modern scientists are able to understand the devices in vimana shasthra now and are reinvented. To Re-invent the vimana technology, modern humans need more research and development in order to reach the advanced space technology. In order to benefit the mankind by giving advanced technology and Free electricity from nature, and to make vimana initially in small-scale as prototypes Lab tests, 3D printing of vimana prototypes, our SWASTIK team is looking for encouragement and support for solving financial constraints to complete our dream and innovation by funding our research work. The vimana prototypes planned by our SWASTIK team are Hybrid models of advanced ancient vimana technology and the modern cutting edge technology. After Analysing the 3D model, CFD analysis and Thermal analysis of Rukma Vimana, we can understand that perfect Reinvention of Ancient vimana may be impossible due to the lost resources (raided libraries) and unavailable materials such as extinct species of the Flora. Thus replacing this absence with modern technology we can make the vimana 50% modern and 50% ancient, resulting in an advanced space vehicle.

MRV UAV works can be carried out in following ways:

1) 3D modeling design of MRV UAV
2) Prototype with modern technology: 3D printing
3) Wind tunnel tests and Improvements based on results
4) Levitation/ Antigravity tests using Tesla lift concept (Electromagnets, electricity & lift)

COMMENTS

Rukma vimana Prototype, MRV UAV is going to be a Successful VTOL. The three floors mentioned in the Rukma vimana were used for passengers, but the mini Rukma vimana UAV can make arrangements of using such place in a scaled dimension, to embed the weapons, or missiles used for Defence purposes. In scaled dimension of UAV, the lower floor can be used for location of retractable landing gears, the middle floor can be used for locating missiles and weapons systems inside the UAV and top floor can be used to set the cameras. The main advantage is the 360 degrees rotation of MRV UAV which becomes easy to attack the enemy planes from the back of the UAV. This can make the UAV to attack easily and also escape from enemy planes. Not only for defense purposes but also the same MRV concept can be used for other missions such as disaster rescue mission, the place inside the UAV can be utilized for Medical equipment (First aid) objects and also food for the victims. The bottom of the UAV can be designed such a way that it can be opened at the central part (other than landing gear belly) in order to unlock the container of medicine or food packets. Farmers have been facing many problems due to the unusual Weather conditions, there have been UAVs helping for this purpose also. But the main advantage using MRV UAV is, it has enough space inside it and simple configuration with VTOL can provide much better facilities without the damage of UAV in worst weather conditions.

TRIPURA VIMANA PROTOTYPE

For propulsion, we can replace the electric motors and generators with Solar panels at top of airship. The design of this vimana is such a way that the top floor has a slanting flat top surrounding the plate kind of structure having electric generator and motor. So we can keep solar panels on that. Airship using solar panels would be excellent. And structure we can make it a 3 in 1 usable airship. The wheels altogether linked to one single link, which can be kept in the bottom floor of the airship. The upper floor of the airship can be filled with hydrogen or helium for flight. The middle floor has storage of this hydrogen or helium gas, which can be interchanged between cylinders and top floor. If gas is in the top floor, the airship flies. If gas is in cylinders, the airship becomes a small car type thing with landing gear as wheels. If landing gears are retracted, it can go on the water. And the rudder and the propeller are used in all cases. Those are powered by solar panels. Mini Tripura Vimana can be a greener airship because it uses solar panels to power the airship. And in other cases, an arrangement can be made to charge for electricity. It can be a better airship than modern day airships because it is a 3 in 1 airship.

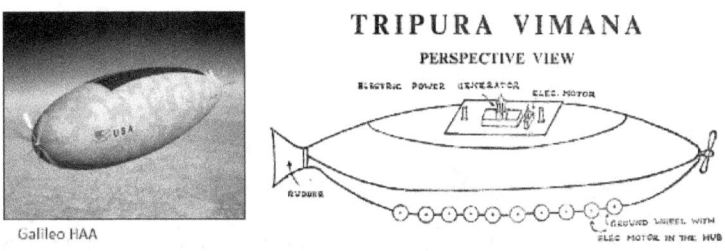

NASA's solarpowered airship and Tripura Vimana

MTV (MINI TRIPURA VIMANA) AIRSHIP

Mini Tripura vimana airship is hybrid green airship using the ancient as well as modern technology in order to make transport much easier and flexible. Instead of several vehicles for land, water, and air, we can use Mini Tripura vimana airship for all three medium of transport.

Mini Tripura vimana airship is similar to Tripura vimana which is a metal-clad airship which can also travel on land and on water. It is also similar to modern airships using solar panels, the airship stores energy and uses for traveling.

Table : Difference between Tripura vimana and Mini Tripura vimana airship

Tripura vimana	Mini Tripura vimana airship
Electric motors and generators at the top floor of vimana.	Uses solar panels at the upper layer of the airship.
Vimana is 100 feet wide and 100 feet height.	Designed Dimensions depend upon the purpose of airship such as military, passenger transport etc.
Can travel under water, it is used as a submarine.	Can travel only on the surface of water bodies but not like a submarine.
Can travel on land with the use of retractable ground wheels (landing gears) located at ground floor of vimana.	Can travel on land with the use of non-retractable landing gears located at the bottom of the airship.

Making of Mini Tripura vimana airship

Material: The material used to manufacture Tripura vimana was Trinetra loha. Indian scientists are working on this deciphered material partially produced in laboratories. Trinetra loha Shines like a peacock feather, fire proof, unbreakable, weightless. It is Impregnable by water, fire, air and heat, and indestructible.

Considering similarities of ancient with modern airship for MTV airship design configurations:

1. Propeller: Modern airships such as Galileo High altitude airship having similar propeller configuration, location design to that of Tripura vimana. Galileo HAA is a solar powered airship, MTV design is based upon the solar panels similar to it.
2. Landing gears: Similar to Tripura vimana, Modern airship such as solar ship having landing gears as well as passenger cabin at the bottom of the solar ship. MTV airship can be designed considering this design of landing gears and passenger cabin.
3. Water travel: A Chinese inventor has come up with an incredible new design for the car that will work on the tarmac, sand, ice, and even water. The all-terrain vehicle, which has a top speed of 62mph and works like a hovercraft, can move seamlessly between different surfaces. MTV airship can use similar techniques for traveling on water bodies. Volkswagen Aqua car would be capable of traveling on road, sand, water and ice. By considering above three modern inventions, we can make a hybrid airship that can travel on land, air, and water.

IX. Nikola Tesla's Vimana and Yantras

Tesla influenced by Swamy Vivekananda to work on Vedic texts

Tripura Vimana mentioned in Vimanika shasthra was Aerial Transportation vehicle which also had the ability to travel on land, in and on water. An attempt to create such Vimana, Nikola tesla designed Aerial Transportation vehicle which can travel on land and in air. The following content is a description of aerial transportation vehicle submitted by Nikola tesla for US patent office: Tesla took patent regarding apparatus for aerial transportation. Application filed October 4, 1927. Serial No. 223,915.

After observing the document of Nikola Tesla on an aerial transportation vehicle, we can find many similarities between Tripura Vimana, Rukma vimana, and Tesla's aerial vehicle.

Tripura vimana has ground wheels or wheels designed for transportation on land. Nikola Tesla also had designed the similar vehicle which can be used on the ground and also in the air.

Tripura vimana is the ancient vehicle which had the ability to travel on land in and on water and also for aerial transportation.

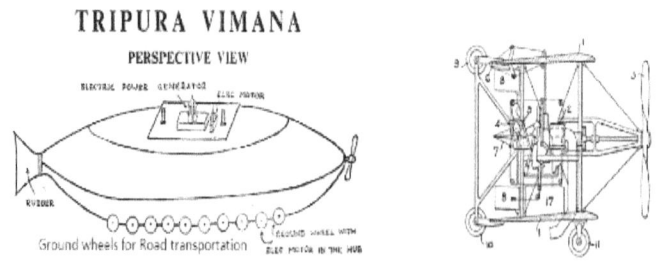

FIGURE: TRIPURA VIMANA AND TESLA AERIAL VEHICLE

Rukma vimana uses 4 lift fans at the top of vimana, to lift off the ground. Rukma is a VTOL vimana which used free energy from the sun to run the motors and electromagnet generator at the ground floor of vimana was used in space for antigravity propulsion. Tesla's vehicle also used the same VTOL methodology using lift fan at the top of the vehicle. The shape of both Rukma vimana and Tesla's vehicle are much similar. There were attempts to create free energy generators by Nikola tesla using gravitational force of earth. This propulsion has been described in Vimana shasthra, Yantra sarwaswa and other Sanskrit texts also describe its existence in ancient times.

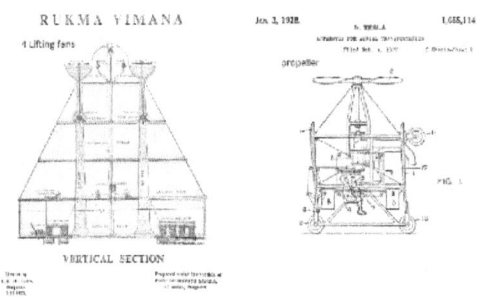

Figure: Rukma Vimana and Tesla Aerial Vehicle

Nikola Tesla Took Inspiration from Swamy Vivekananda and Indian Vedas for His World Acclaimed Work:
After his lab was burned down and his life's work had vanished. Nikola Tesla studied the concept of Prana and Akasha to work on FORCE and MATTER. He developed a new perspective on the world and started viewing the world in terms of frequencies and energy which resulted in him establishing his concepts on energy. This information is written not to take sides or argue against anyone's beliefs but only to give a small idea on the intensity of the knowledge and imagination of our ancestors. They even had the concept of sustainable energy, projectile science, and many others like Thrust, momentum, Thermodynamics, Astrophysics etc to name a few.

Similar to the descriptions mentioned in vimana shasthra, there are some works done by tesla, the following are some of those important discoveries.

Here is just a partial list of some of Tesla's contributions to the world based on his research and experiments from Vedas, that he had not been given credit for:
- Alternating Current -AC electricity (Thomas Edison literally stole his ideas from him and took the credit for it).

- Radio (Marconi just robbed the ideas and work of Tesla and got the credit for it)
- Hydro-electricity (Tesla Built the first Hydro-electric power plant at Niagara falls, As a result, we see what's there now)
- X-rays
- transistors
- Resonant frequency (everyone else figured it out 50 years later)
- Fluorescent and Neon lighting
- The induction motor
- The rotating magnetic field (precursor to gyroscope)
- Arc lighting
- Tesla coil
- Oscillators
- Encryption technology and scrambler
- Wireless communication and power transmission
- remote control
- Telegeodynamics (a way to search for metals and minerals)
- Tachometer and speedometer
- Refrigeration machines
- Bladeless turbines and pumps
- Cryogenic engineering
- reactive jet dirigible (precursor to Harrier jet)
- Hovercraft Flivver plane (precursor to Osprey helicopter/aircraft)
- Particle-beam weapons (precursor to Star Wars)

Likewise in ancient India, some selective greatest teachings were passed to the deserving Sages verbally, which remain stored in their memories, so that the power of knowledge is not misused by demonic beings referring to texts if it were written. Only the knowledge which was helpful for mankind were passed to common human beings. True to this belief of Vedas, all Tesla's engineering was done in his head, he never worked things out on paper or used scale models to come to a functioning final result. Things would appear in his head and he would simply record it exactly as it came to him, similar to Beethoven. Tesla died with the BrahmaAstra (Scalar Interferometer) knowledge in his head, as he did NOT want to give this destructive Vedic technology to the US military and Vatican, as promised to Vivekananda. It was to help mankind, NOT to destroy or enslave.

Death Ray

Nikola Tesla claimed to have invented a "death beam" which he called Teleforce in the 1930s. The device was capable of generating an intense targeted beam of energy "that could be used to dispose of enemy warplanes, foreign armies, or anything else you'd rather didn't exist".

The so-called "death ray" was never constructed because he believed it would become too easy for countries to destroy each other. Tesla proposed that a nation could "destroy anything approaching within 200 miles… [and] will provide a wall of power" in order to "make any country, large or small, impregnable against armies, airplanes, and other means for attack". He said that efforts had been made to steal the invention. His room had been entered and his papers had been scrutinized, but the thieves, or spies, left empty-handed.

Tesla's Oscillator

In 1898, Tesla claimed he had built and deployed a small oscillating device that, when attached to his office and operating, nearly shook down the building and everything around it. In other words, the device could allegedly simulate earthquakes. Realizing the potential terrors such a device could create, "Tesla said he took a hammer to the oscillator to disable it, instructing his employees to claim ignorance to the cause of the tremors if asked". Some theorists believe the government continues to use Tesla's research in places like the HAARP facility in Alaska.

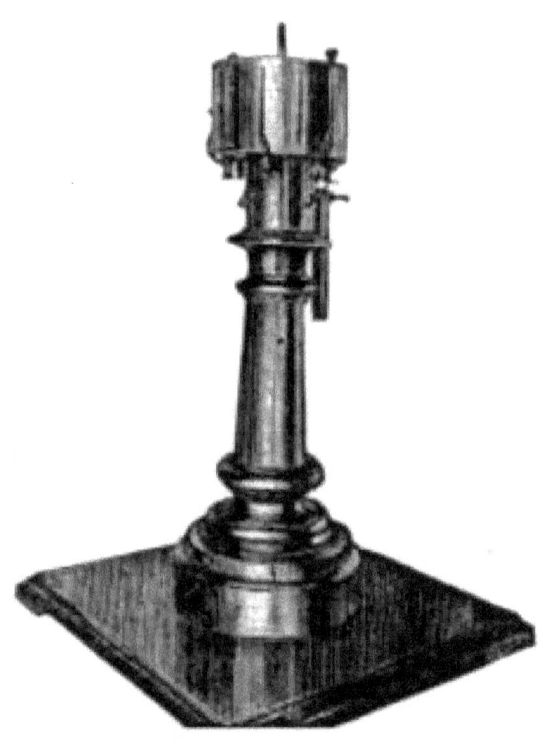

TESLA'S OSCILLATOR AS SHOWN AT THE
COLUMBIAN EXPOSITION.

Free Electricity System

With funding from JP Morgan, Tesla designed and built Wardenclyffe Tower, a gigantic wireless transmission station, in New York in 1901-1902. Morgan thought the Wardenclyffe Tower could provide wireless communication across the world. However, Tesla had other plans.

Tesla intended to transmit messages, telephony and even facsimile images across the Atlantic to England and to ships at sea based on his theories of using the Earth to conduct the signals. If the project worked, anyone could have electricity by simply sticking a rode into the ground. Unfortunately, free electricity is not profitable. And this system could be incredibly dangerous for the global elite because it could profoundly change the energy industry. Imagine how different the world would be if society didn't need oil and coal to function? Could the great world powers maintain control? Morgan refused to fund the changes. The project was abandoned in 1906 and never became operational.

The Flying Saucer

In 1911, Nikola Tesla told The New York Herald that he was working on an anti-gravity "flying machine".

"My flying machine will have neither wings nor propellers. You might see it on the ground and you would never guess that it was a flying machine. Yet it will be able to move at will through the air in any direction with perfect safety, at higher speeds than have yet been reached, regardless of weather and oblivious of "holes in the air" or downward currents. It will ascend in such currents if desired. It can remain absolutely stationary in the air, even in a wind, for a great length of time. Its lifting power will not depend upon any such delicate devices as the bird has to employ but upon positive mechanical action."

Tesla's flying saucer was powered by free energy system at a time when the fledgling aviation and motor car industry depended on oil and petroleum. His invention met the same fate as his free energy system.

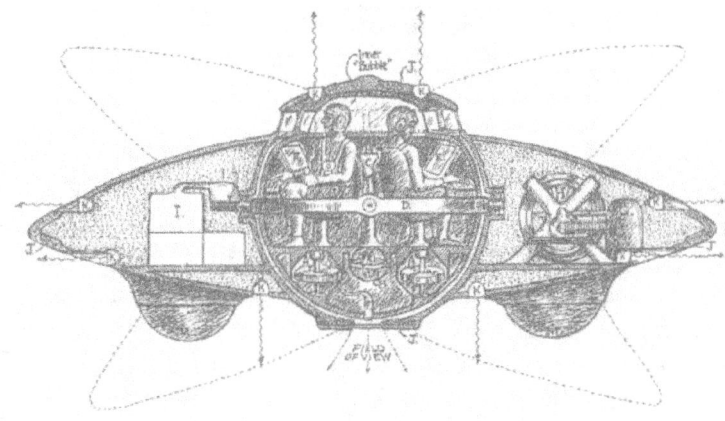

Improved Airships

Tesla proposed that electrically-powered airships would transport passengers from New York to London in three hours, traveling eight miles above the ground. He also imagined that airships might draw their power from the very atmosphere, never needing to stop for refueling. Unmanned airships might even be used to transport passengers to a preselected destination or for a remote aerial strike. He was never given credit for his invention. However, today, we have unmanned drones carrying out combat missions, supersonic airplanes that fly at amazing speeds and space shuttle technology that can circle the Earth in the upper atmosphere.

It was long suspected that the FBI literally stole all of his work, research, and inventions that he had in his possession when he died. This rumor has now been confirmed by recent, heavily redacted Freedom of Information Act requests released by the FBI.

X. Talpade's Vimana

Shivkar Bapuji Talpade is vedic scholar who did research on vimana shasthra, which he is aware due to his contact with Subbaraya Shasthry. He tried to implement in practical manner with whatever available with him. Due to the situations in his time, he could not get financial support and people near his living area claim that he has been cheated by evil minded humans. He has written many books on vimanas in Marathi, Bengali, gujrathi language which are soon going to be translated by modern Indian research scholars.

MARUTSAKHA VIMANA

Shivkar Bapuji Talpade tried to reconstruct one of ancient Indian Vimanas.

Fig. Marutsakha Vimana Photo

       This above picture is Marutsakha vimana by Talpade. It looks exactly like Tripura vimana. Detailed investigation of his works reveals that he tried to re-construct Tripura vimana.

Tripura vimana is three-way usable vehicle which travels in and on water, on land and in air. The name marutsakha given by Talpade to his vimana could lead us to the understanding that he tried to use one of the three uses of this ancient vehicle. The word Marutsakha or Marutsakthi meaning friend of air or power of the air, and thus we also see similar propulsion engines in Tripura vimana. Interesting fact of this vimana is that there is a device called "Ganapa Yantra" which is related to vimana antigravity generator and it is antigravity device mentioned in next chapter of this book.

XI. Hitler's De glocke

Vimana shasthra have descriptions of many weapons devices or Yantras, which might also be common in other civilizations during the time of vedic period. Hitler collected wide range of ancient texts and had tested experimented many of the descriptions.

As shown in Ancient Aliens TV show which air in History channel, the Hitler's antigravity device has been taken reference from secret documents and made a model. A wind Tunnel Test has been done on the model.

Secret behind Hitler's advanced weapons and antigravity machines:

At Poland former headquarters, Hitler's subterranean weapons project was carried out. Hitler and his team of scientists showed interest on ancient wisdom, advanced extra-terrestrial technology. Nazi engaged their top scientists to develop Vunda Wafu or wonder weapons. Weapons that were so diabolical and so scientifically incredible. According to testimony collected after war one of top secret project was names de Glocke or bell, which is a serious attempt to create time travel device which soar from sky and reach speed of light. Following war, no trace of bell was found but a handful of drawings. Device based on designs left behind by ancient astronauts. During 2nd World War, widely believed German army possessed most technologically advanced military on planet. Including guided missiles jet aircrafts, precision bombers. Hitler planned to use this weaponry to defeat the allied forces and create empire that would rule world. After defeated in 1945, And some believe that these diabolical minds brought secret knowledge that have extra-terrestrial origin. In 1997 a military journalist, was shown original documents by a police intelligence officer considering top secret Nazi project Chronos in mysterious weapon known as de Glocke or bell. Chronos ancient Greek titan god of time, project involved Time travel and machine is time machine.

The Henge is found at former Nazi science lab, head quarter for project Chronos. We find Medium size city underground.

Henge is made of concrete and very interesting structure, which appears 1st time on aerial 1944 built shortly before end of ww2. There are Power plants located all around the Henge, and consumed 1000 tons of coal a day. Henge was used to test de Glocke or bell shaped craft, so that it shouldn't fly away while testing. Prisoners of war were forced to participate in hi-tech experiments. Many suffered nerve damage and skin coming off of faces, exposed to high level radiations. Bell is, vortex propulsion type of antigravity propulsion.

According to the reports discovered after war, German scientists experimenting on antigravity propulsion, they had been developing a type of

vortex engine powered by mercury. Mercury will spin in gyroscopic manner within a closed system and create lift seemingly defying gravity.
Blue prints show, the bell shaped time machine device embed in a UFO shaped flying object.

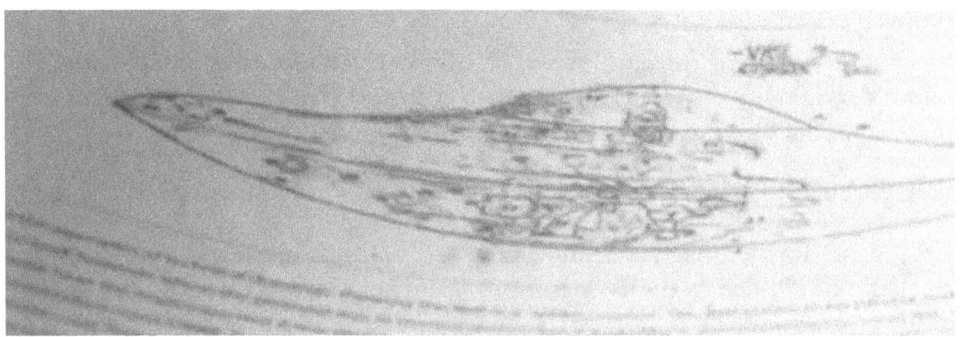
Figure: Blue Prints of Antigravity Flying machine

One may induce gravity from Electro Magnet forces. Bell disappeared at end of war with Hans Kammler, both escaped in window of time.
Nazi documentation describes combination of flying machine with advanced weapons of mass destruction, Bell is main part of propulsion system.
Nazi have fanatical desire to locate and study ancient texts and artefacts, possessed great power. Much of ancient knowledge was lost after war, much of these taken by allies as spoils of war. Served as inspiration of many advanced communication and weapons systems used today. Generating gravity, scientific discovery they made, they have document designed by general, and it was engaging all forces of assassin at end of war.
Vimanas, airships that could stop on a dyne and change direction at top speed, defying laws of physics. Mahabharata, earthly arrival of ancient spacecraft, which curiously resemble modern day UFO accounts. Today we identify it as history, 3 years before start of WW2 in 1936 UFO crash, UFO crash, Nazis rediscovered the ancient technology. Reverse engineered ancient technology.
Stupas are ancient bell shaped design, found at Buddhist temples. We humans are always copiers, we saw some Vimanas shape of stupas, and then we made copies of stupas. Stupa dates back 1000s of years and believed to have inspired by Vimanas.
Ancient extra-terrestrial technology, stupa is found Mahabharata. Vimanas is the shape of stupa. Building temples, at the top of temple, people always copied stupa, because it has something to do with the gods or extra-terrestrials. Vimanas were flying chariots of gods, made of gleaming material with amazing manoeuvrability in the air with powerful weapons that could annihilate the enemies from the far. They are able to cloak themselves.

Extreme scholars interpret this events in divine and ancient astronaut theory suggest that this is ruined and misunderstood the technology witnessing extra-terrestrial flying craft. In that ancient stupa buildings, every stupa is not empty, in every stupa we find Buddha. His hands are trying to manipulate some control. According to their belief, stupa is the small vehicle with which we reach the big vehicle. The big vehicle is surrounding the planet, small vehicle comes down to earth and goes again to the big vehicle, so Buddha have to manipulate some controls. On top of every Indian temple there is representation of Vimanas. It is not mythology, it is Reality. Vimanas came from Cities in the sky. Nazi inspired by Sanskrit texts, they studied and influenced. David Hatcher Childress, met a Nazi scientist who said that he worked on stupa and he never finished the work, if he has finished he would have won the war.

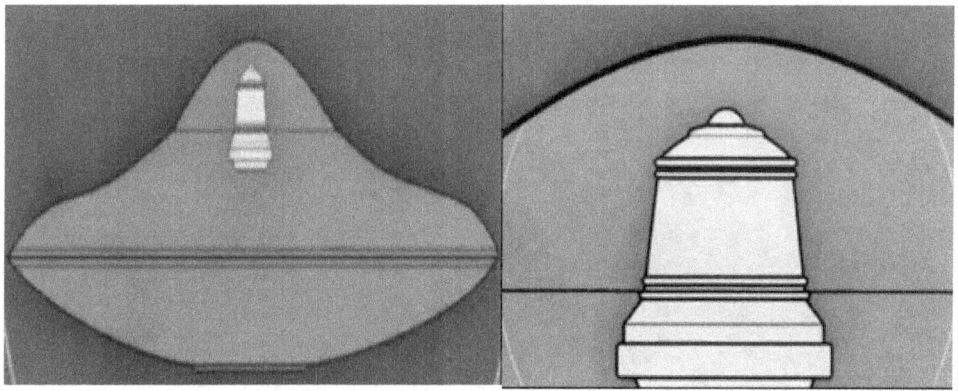

Fig: Bell shaped object embedded inside a UFO shaped flying object

Tesla worked on these vimanas, and his works also show the similar descriptions of flying airship with electricity which cause antigravity lift to the craft. Antigravity technology from Vedic texts was also followed by Hitler. His device Die Glocke is doubted to be similar to the Ganapa yantra which causes antigravity connecting to the Vimana mercury vortex generator.

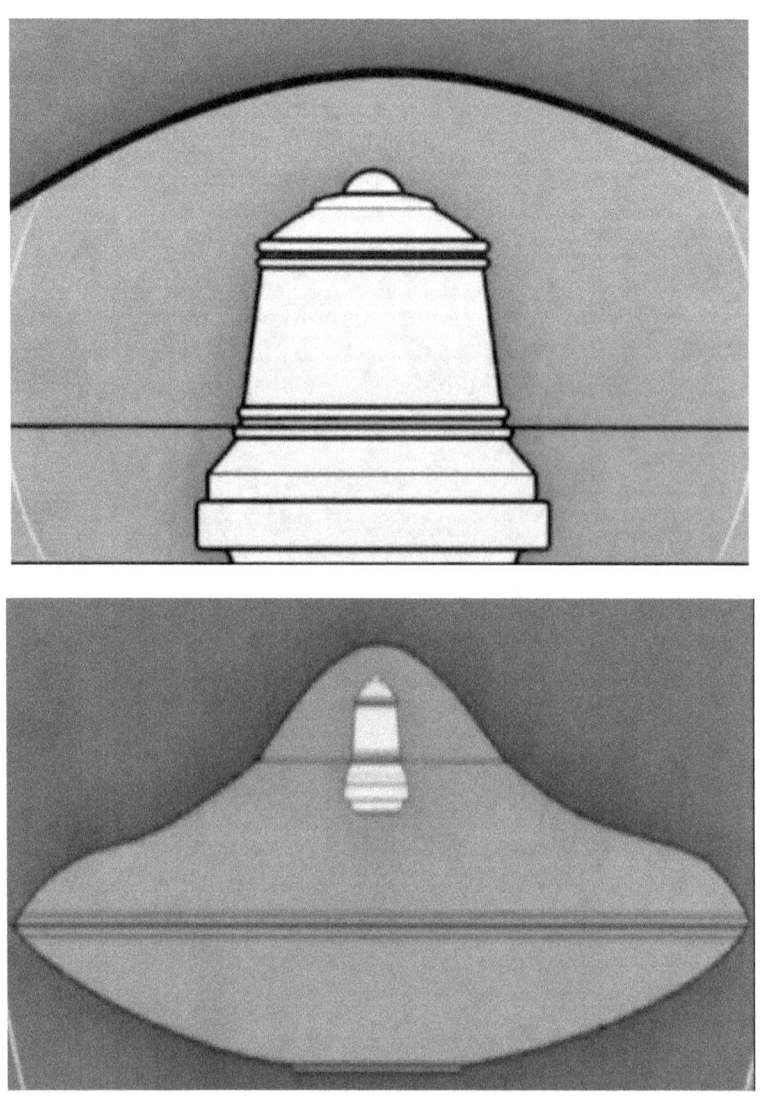

Bell shaped object embedded inside a UFO shaped flying object

XII. Marutsakha Vimana, 3D Printing and Wind Tunnel Testing

As shown in Ancient Aliens, History Channel

Marutsakha vimana was made by Talpade by referring to Tripura vimana. The Vimana dimensions are modified slightly to make it look like one of the pictures drawn by Ellappa in Vimana shasthra, which looks similar to UFO shape. This model is going to be 3D printed to test in Wind tunnel. Also, the dome shape is added on top of the motor and generator as per my interpretation of vimana descriptions mentioned in vimana shasthra and also as per the attempt make by Talpade through his Marutsakha vimana. Tripura vimana was attempted by Nikola Tesla. Tripura vimana uses Ganapa Yantra and vortex generator for antigravity, which works similarly with concept idea mentioned by Tesla. Ganapa Yantra and generator of vimana is similar to die Glocke, which was attempted by Hitler. The Ganapa Yantra, Vimana generator and Die Glocke are similar devices used to activate antigravity, located at similar locations inside the flying crafts.

Above picture is 3D model design by Kavya, Default view in 3D modeling software.

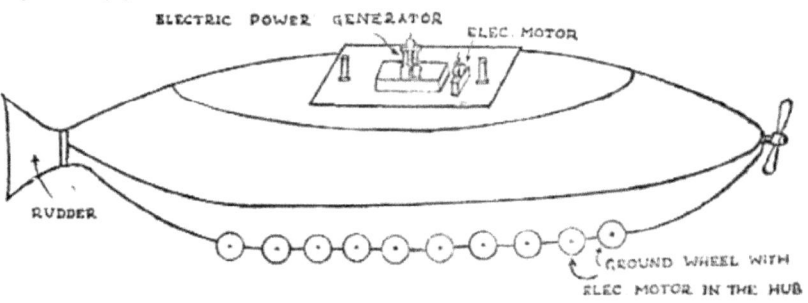

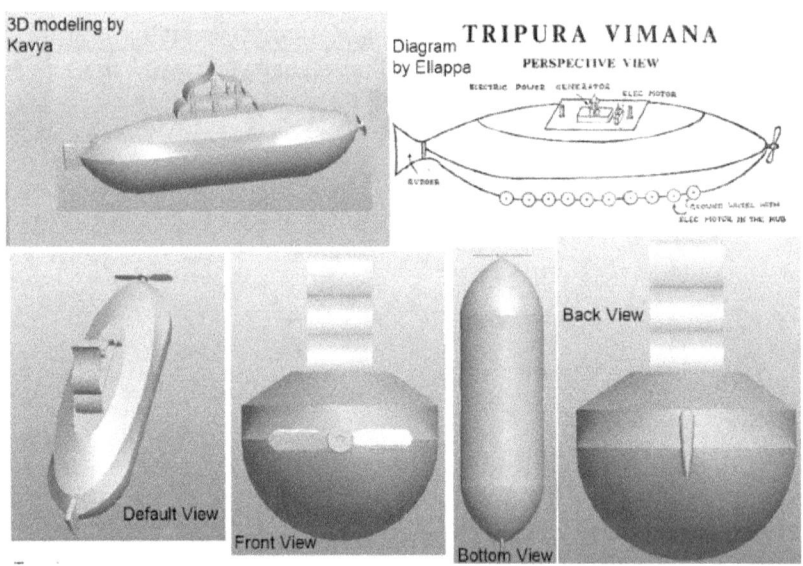

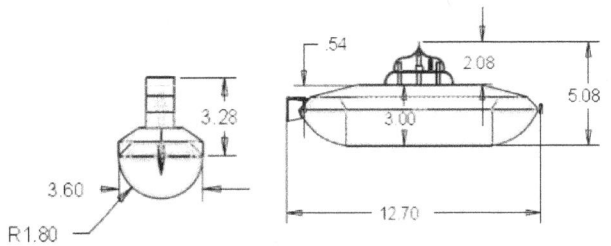

Right side, Default view

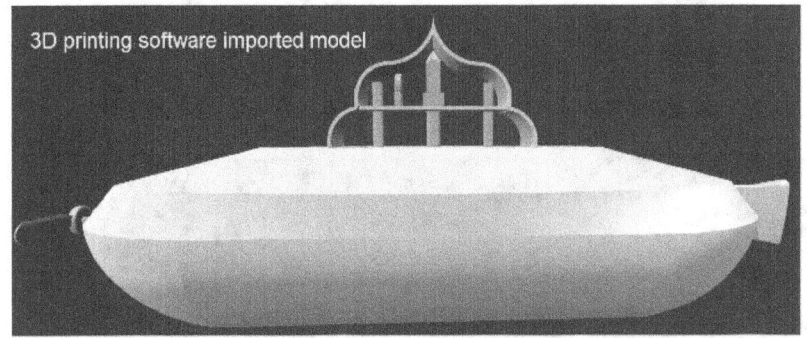

3D printing software imported model

Kavya's Design on Marutsakha Vimana in History channel

Wind Tunnel Testing in California University Irvine

3D model has been shared by Kavya to Aerospace systems engineer Dr. Travis Taylor and Giorgio Tsoukalos.

Vimana in the wind tunnel

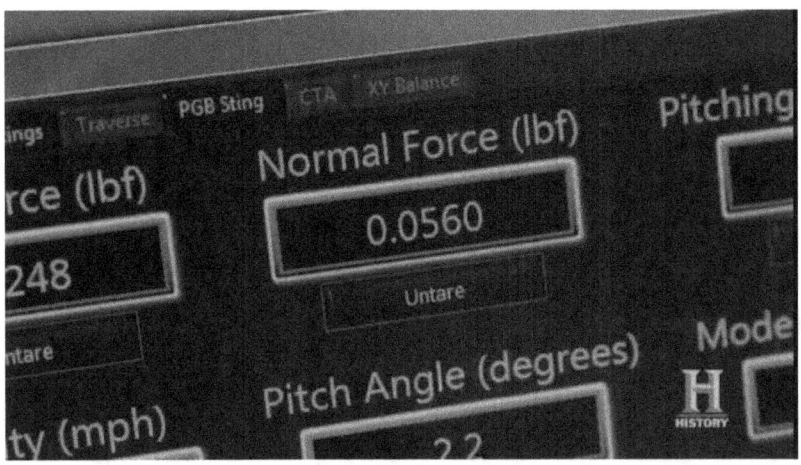

Results of the wind tunnel experiment

Ancient vimana showing positive results as an aircraft would.

This is a good scientific experiment performed on Vimana model and the results of this vimana made from the texts is a viable aerodynamic structure. It is flight worthy

XIII. 3D and 4D Printing of Ancient Indian Vimanas

By **Eshwar Reddy Cholleti**
Founder of 3D Srishti Pvt Ltd
and
Kavya Vaddadi
Researcher and Founder,
SWSTIK – Scientific Works on Advanced Space Technology Investigators for Knowledge,
Design Engineer and CFD analyst at VEDAS Company,
Research Service Provider at Vedic Scientific Research Foundation.

Overview of Chapter:

The intention of this chapter is to present possibilities of 3D printing of Ancient Vimanas and future scope of how 4D printing changes the idea of manufacturing of Ancient Vimanas in future generations. In this chapter, we discuss the attempt of manufacturing and prototyping of ancient Vimanas by 3D printing. The description of steps involved in the manufacturing of ancient Vimanas by present advanced manufacturing 3D printing techniques, comparison of traditional machining of CNC with additive manufacturing processes, AM Materials, and Applications, AM data formats and software, steps involved in exporting an STL file from CREO. AM build setup preparation, a brief description of the fabrication of ancient Vimanas by seven 3D printing processes and scope of 4D printing involved in the functioning of some ancient vimana objects by expected behavior in response to surroundings as a function of time along with 3 Dimensions.

-

1. Introduction:

Ancient Alien Ancestors had advanced space technology, their civilization is much older and lived much longer than ours, and definitely we can understand how advanced they would have been. An old aircraft artifact model *(as shown in Figure a)* was found and researchers made its larger scale model prototype to check if the artifact really had flying abilities. Results were amazing, the artifact model aircraft prototype flew much better than expected. India is one among the places where ancient ancestors lived, Ancient documents such as Vimana shasthra are available with us, enabling us to study and understand the advanced space technology. Inspired from the foreign country researchers, we decided to make a prototype of Vimanas, by referring to ancient vimana texts

– Vimana shasthra *(as shown in Figure b)*. The 3D modeling and analysis of ancient Vimanas as mentioned in the previous chapter enables us to plan further process for practical research on Vimanas. 3D printing of Ancient Vimanas is one of our attempts to understand these ancient amazing Vimanas.

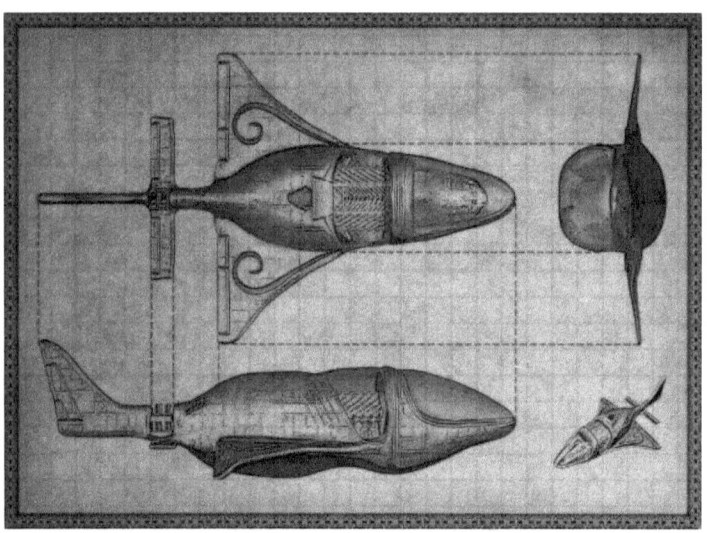

Figure a

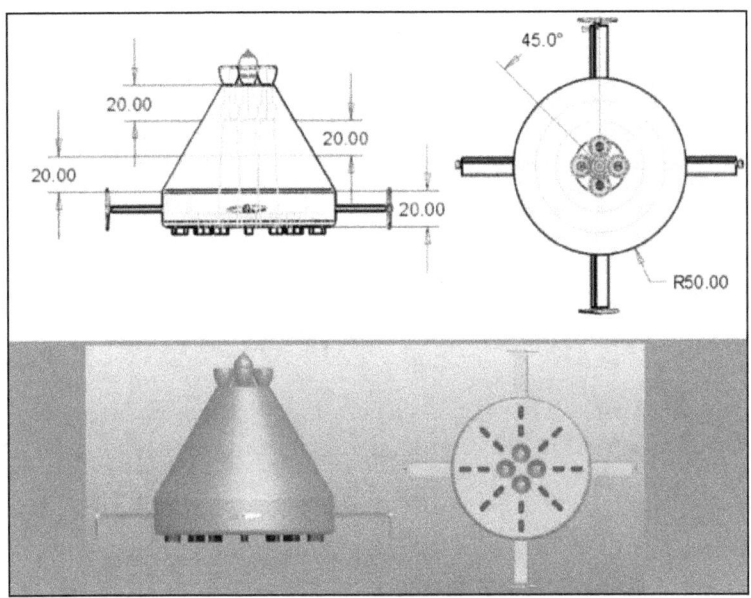

Figure b

1.2: Additive Manufacturing:

Additive Manufacturing (AM) is a novel technology, only three decades old, colloquially known as 3D printing and formerly known as Rapid Prototyping (RP), AM is defined by the American Society for Testing and Materials (ASTM) as "a process of joining materials to make objects from 3D model data, usually layer upon layer, as opposed to subtractive manufacturing methodologies".

AM enables us to manufacture complex structures which are not possible to fabricate by conventional methods of manufacturing that allow optimizing the component's weight, shape, and strength. As per ASTM standards, AM is classified into seven processes, Such as Vat photo polymerization, material jetting ,binder jetting , material extrusion, powder bed fusion , sheet lamination , direct energy deposition, however, each process has its own limitations as each AM process caters only limited materials.

Ancient vimanas can be manufactured by 3D printers, the present challenges to build ancient vimanas is availability of the ancient materials like Raja Loha, Trinetra Loha with different compositions, in order to overcome this materials barrier, we are using similar present existing materials compositions, like Lead, iron pyrites, iron, mercury, mica, silver.

The materials in various forms as per feasibility can be used in 3D printing processes such as photopolymer resins in VAT photopolymerization processes, metals, polymers and ceramics in binder jetting processes, the powder based materials such as common metals like titanium, aluminium, copper, mica, silver, steel, cobalt chrome in amorphous forms can be used in Powder bed fusion processes.

The description of basic terms like Additive Manufacturing (AM), 3D Printing, Rapid Prototyping(RP) is briefly described as follows.

Rapid Prototyping (RP): Is a technology that produces models and prototype parts from 3D CAD model data, CT and MRI scan data, and model data created from 3D object digitizing systems.

3D Printing: The fabrication of object through the deposition of a material using a print head, nozzle, or other printer technology.

Additive Manufacturing: Is new process of joining materials to make objects from 3D model data, usually layer upon layer, as opposed to subtractive manufacturing methodologies.

Synonyms of AM: additive fabrication, additive processes, additive techniques, additive layer manufacturing, layer manufacturing and freeform fabrication.

2. Steps involved in 3D Printing of Vimanas:

The procedure involved in 3D printing processes as mostly similar, so as a result, any type of ancient vimana should undergo similar steps in manufacturing processes, irrespective of the type of ancient vimana. The steps of fabrication are described as follows:

Step 1: CAD: The 3D printing of ancient vimanas starts with modeling the vimana parts with required dimensions and features describe the external geometry, This can be done by using almost any professional CAD solid modeling software, but the output must be a 3D solid or surface representation. Reverse engineering equipment (e.g., laser scanning) can also be used to create this representation.

The following are the designs of various ancient vimanas in small-scale modeled by using modeling software. Small scale Designed CAD models as shown below, All dimensions are in mm

a) Rukma Vimana

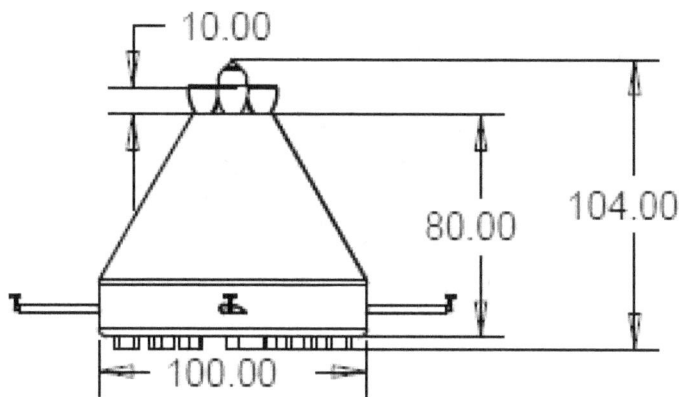

b) Sundara Vimana

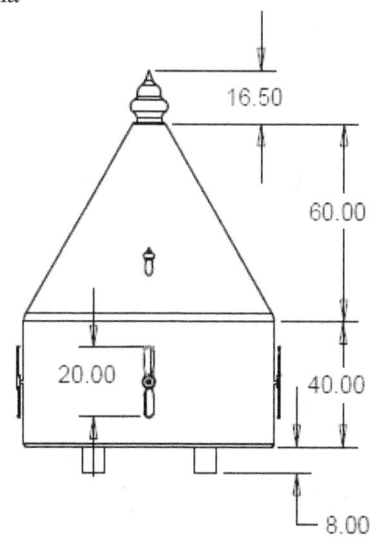

c) Shakuna Vimana

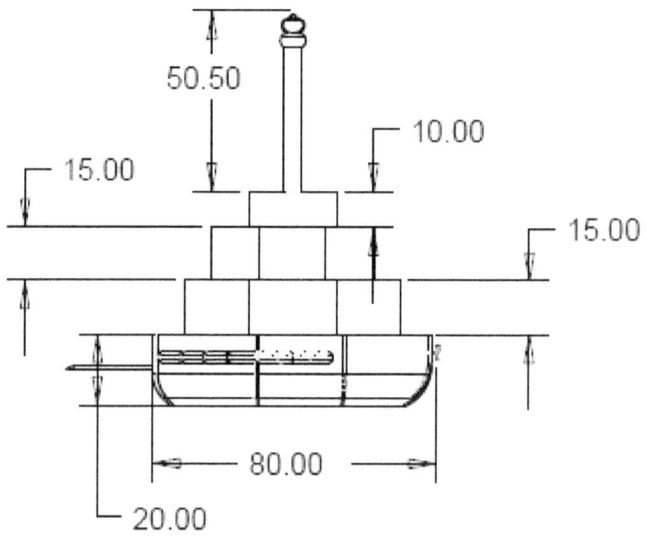

d) Tripura Vimana

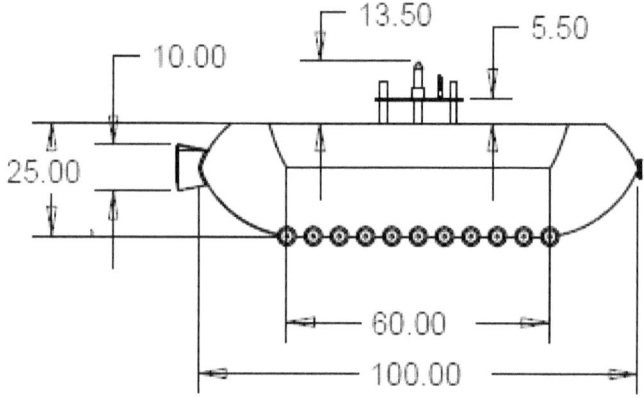

Step 2: Conversion to STL: The design file of ancient Vimanas has to be converted into STL file format, most of the all 3D printing machines accept STL file format, which has become a standard, and nearly every CAD system can output such a file format. This file describes the external closed surface of the original CAD model and forms the basis for calculation of the slices.

STL file format of Rukma Vimana shown below

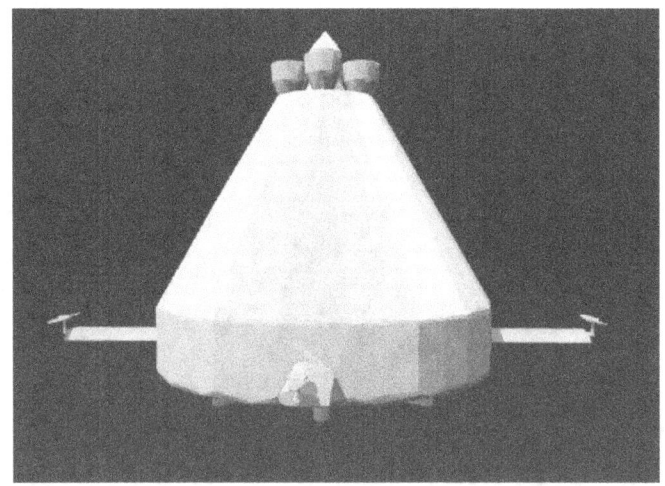

(All other three vimanas also similarly converted to STL format)

Step 3: Sliced layers: The STL file describing the vimana part must be transferred to the 3D printing machine. Here, there may be some general manipulation of the file so that it is the correct size, position, and orientation for building for a sliced layer of vimana part is developed.

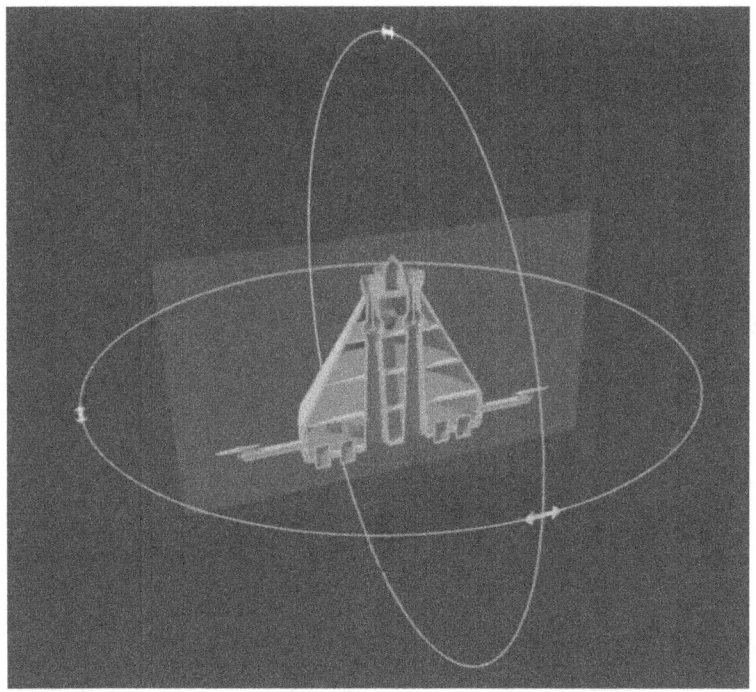

Step 4: 3D printing system: Building the Vimanas parts is mainly an automated process and the 3D printing machine can largely carry on without supervision. Only superficial monitoring of the machine needs to take place at this time to ensure no errors have taken place like running out of material, etc.

Step 5: End Part: Ancient vimana parts may now be ready to be used. However, they may also require additional treatment before they are acceptable for use. For example, they may require priming and printing to give an acceptable surface texture and finish of ancient vimana parts.

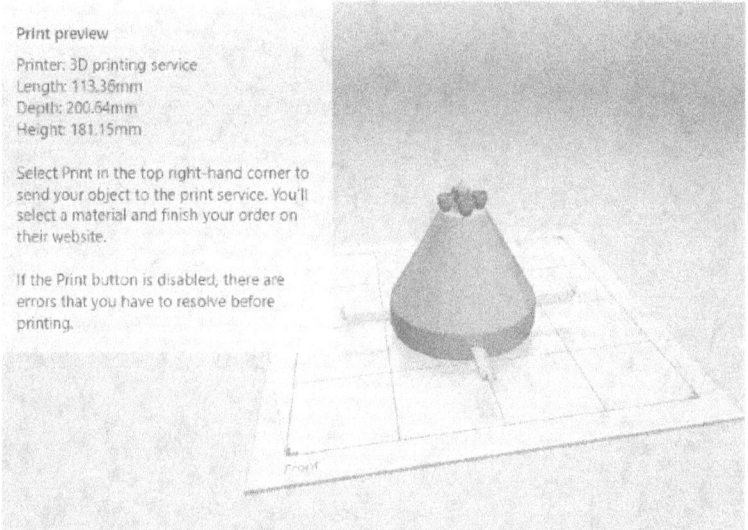

3. Tradational Machining of CNC Vs Additive Manufacturing Process

The process of building Ancient Vimana parts is performed in two possible ways, they are

By removing material layer by layer i.e., CNC machining and adding material layer by layer
i.e., Additive manufacturing process.

As we know AM is an Automated fabrication, AM shares some of its DNA with Computer Numerical Controlled machining technology. CNC is also computer-based technology that is used to manufacture products. CNC differs mainly in that it is primarily a subtractive rather than additive process, requiring a block of material that must be at least as big as the part that is to be made. This section discusses a range of topics where comparisons between CNC machining and AM can be made. The purpose is not really to influence the choice of one technology over another rather than to establish how they may be implemented for different stages in the product development process, or for different types of product.

3.1 Material Choice

AM technology was originally developed around polymeric materials, waxes and paper laminates. Subsequently, there has been the introduction of composites, metals, and ceramics. CNC machining can be used for soft materials, like medium-density fibreboard (MDF), machineable foams, machinable waxes, and even some polymers. However, use of CNC to shape softer materials is focused on preparing these parts for use in a multistage process like casting. When using CNC machining to make final products, it works particularly well for hard, relatively brittle materials like steels and other metal alloys to produce high accuracy parts with well-defined properties. Some AM parts, in contrast, may have voids or anisotropy that are a function of part orientation, process parameters or how the design was input to the machine, whereas CNC parts will normally be more homogeneous and predictable in quality.

3.2 Speed

High-speed CNC machining can generally remove material much faster than AM machines can add a similar volume of material. However, this is only part of the picture, as AM technology can be used to produce a part in a single stage. CNC machines require considerable setup and process planning, particularly as parts become more complex in their geometry. Speed must, therefore, be considered in terms of the whole process rather than just the physical interaction of the part material. CNC is like to be a multistage manufacturing process, requiring repositioning or relocation of parts within one machine or use of more than one machine. To make a part in an AM machine, it may only take a few hours: and in fact, multiple parts are often batched together inside single AM build. Finishing may take a few days if the requirement is for high quality. Using CNC machining, this same process may take weeks.

3.2 Complexity

As mentioned above, the higher the geometric complexity, the greater the advantage AM has over CNC. If CNC is being used to create a part directly in a single piece, then there are some geometric features that cannot be fabricated. Since a machining tool must be carried in a spindle, there may be certain accessibility constraints or clashes preventing the tool from being located on the machining surface of a part. AM processes are not constrained in the same way and undercut and internal features can be easily built without specific process planning. Certain parts cannot be fabricated by CNC unless they are broken up into components and reassembled at a later stage.

3.3 Accuracy

AM machine generally operates with a resolution of a few tens of microns. It is common for AM machines to also have variable resolution along different orthogonal axes. Typically, the vertical build axes correspond to layer thickness and this would be of a lower resolution compared with the two axes in the build plane. Accuracy in the build plane is determined by the positioning of the build mechanism, which will normally involve gearboxes and motors of some kind. This mechanism may also determine the minimum feature size as well. For example, SL uses a laser as a part of the build mechanism that will normally be positioned using galvanometric mirror drives. The resolution of the galvanometers would determine the overall dimensions of parts built, while the diameter of the laser beam would determine the minimum wall thickness. The accuracy of CNC machines, on the other hand, is mainly determined by a similar positioning resolution along all three orthogonal axes and by the diameter of the rotary cutting tools. There are factors that are defined by the tool geometry, like the radius of the internal corners, but wall thickness can be thinner than the tool diameter since it is a subtractive process. In both cases, very fine detail will also be a function of the properties of the build material.

3.4 Geometry

AM machines essentially break up a complex, 3D problem into a series of simple 2D cross-sections with a nominal thickness. In this way, the connection of surface in 3D is removed and continuity is determined by how close the proximity of one cross-section is with an adjacent one. Since this cannot be easily done in CNC, machining of surfaces must normally be generated in 3D space. With simple geometries, like cylinders, cuboids, cones, etc., this is a relatively easy process defined by joining points along a path: these points being quite far apart and the tool orientation being fixed. In cases of free-form surfaces, these points can become very close together with many changes in orientation. Such geometry can become extremely difficult to produce with CNC as shown in figure 2, even with 5-axis control or greater. Undercuts, enclosures, sharp internal corners and other features can all fail if these features are beyond a certain limit.

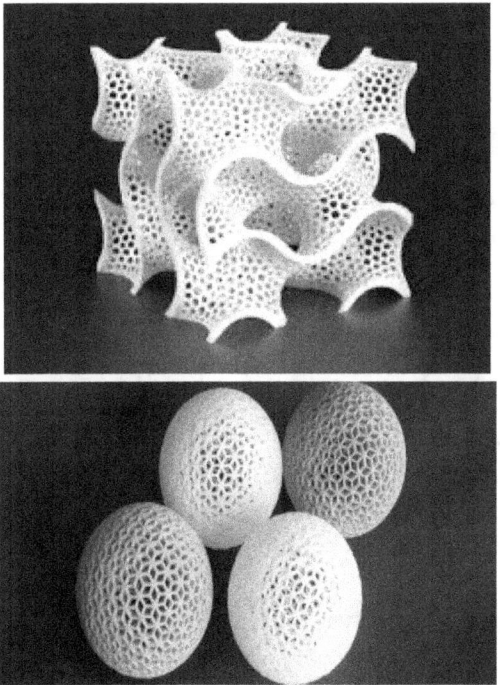

Figure 2: AM fabricated complex geometries [2]

3.5 Additive Manufacturing Materials

The industrial grade materials usage in each of the additive manufacturing process is rapidly growing by using nanomaterials to obtain better properties. The two major categories of AM materials are plastics and metals. However, a variety of filled and composite materials are also available, in addition to ceramics and ceramic-metal hybrids.

The plastic materials are majorly classified into two groups a)Thermo Plastics and b) Thermo Set Plastics, the various materials comes under this major groups of plastic materials is presented as follows:

a) Thermo Plastics:

Thermoplastics can soften upon heating and return to their original form. They are easily moulded and extruded into films, fibers and packaging, less rigid than thermosets, some of the major thermoplastic materials used in AM are as mentioned below:

- ABS
- PA
- PC
- PLA
- ULTEM
- Polystyrene
- Polypropylene
- Glass
- Carbon
- Aluminum filled PA or DuraForm Flex

b) Thermo Set Plastics:

Thermoset or thermosetting plastics are the plastic materials by which once cooled and hardened, these plastics retain their shapes and cannot return to their original form. They are hard and durable. The following materials fall into the thermo set plastics.

- Acrylate
- Acrylic
- Phenolic resins
- Epoxy based photopolymers.

c) Metals:

The growing number of metal materials available for additive-manufacturing systems is impressive. A designer can choose from a wide range, including:
- Titanium alloys
- Stainless steel
- Tool steel
- Commercially pure titanium
- Aluminium alloys
- Nickel-based alloys
- Cobalt-chromium alloys
- Copper-chromium alloys
- Gold
- Silver

As per ASTM F42 committee classification in January 2012, 7 processes in Additive manufacturing were identified and for all the 7 AM process Vs. The material is shown in the below Table 1.

Table 1: AM Process Vs Material

AM materials	Material extrusion	Material Jetting	Binder Jetting	Vat Photopoly-merization	Sheet Lamination	Powder Bed fusion	Directed energy deposition
Polymers, polymer blends	x	x	x	x	x	x	
Composites		x	x	x		x	
Metals		x	x		x	x	x
Graded/hybrid metals					x		x
Ceramics			x	x		x	
Investment casting patterns		x	x	x		x	
Sand molds and cores	x		x			x	
Paper					x		

Source: Wohlers Associates, Inc.

The above-mentioned materials are commercially available conventional AM materials, but in the case of manufacturing ancient vimanas the composition varies, the ancient vimanas material composition and their possibility of modifying them to use in AM are described as follows.

a) Tripura Vimana is made of Trinetra Loha.

Trinetra loha is explained by Shaakataayana, its composition is as follows:

Jyotishmatee loha 10 parts, kaanta-Mitra 8 parts, vajramukha loha 16 parts, these 3 to be filled in crucible, then adding tankana or borax 5 parts, trynika 7 parts, shrapanikaa 11 parts, maandalika 5 parts, ruchaka or natron 3 parts, mercury 3 parts, then filled in crucible in padmamukha furnace and heated to 631 degrees with trimukhee bellows, the resulting liquid, if poured into cooler, will yield a metal, shine like peacock feather, unburnable, unbreakable, weightless, impregnable by water, fire, air and heat, and indestructible.

By utilizing this above composition of various elements, we have to make the output material into fine grain amorphous form, this mixed grain powder in the powder feed supply is fed by power feed piston to power bed in which sintering of vimana parts will be done in order to obtain the required vimana part with more structural properties.

b) Shakuna Vimana, Sundara Vimana, and Rukma Vimana are made of RajaLoha

The composition of RajaLoha is as follows:

Praana-kshaara or ammonium chloride 4 parts, wild Bengal gram 32 parts, shashakanda (or lodhra) benzoin 18 parts, naaga or lead 20 parts, sea-foam 16 parts, maakshika or iron pyrites 6 parts, panchaanana or iron 20 parts, paara or mercury 15 parts, kshaara-traya or 3 kinds of salt: natron, salt-petre, borax, 28 parts, panchaanana or mica 20 parts, hamsa or silver 17 parts, garada or aconite 8 parts, and panchaamrita or 5 sweets--curds, milk, ghee, sugar, honey, these should be filled in the melter, and after boiling, and drawing the liquid through two outlets, fill in the crucible and place in furnace, and blow to 800 degrees' heat, and then transfer it to the cooler. That will be Raajaloha, pure, golden-coloured, tensile, and mild. The vimana, made out of this loha or alloy, will be very beautiful and delightful.

The all above compositions of various materials results in a material called Raja loha, this final raja loha material is sliced down into small pellets and fed through embedded pellet hoper on to the guided screw driven by a motor and melted by the heater as a result the rajaloha filament is obtained.

This rajaloha filament is fed into material extrusion 3D printing machine in order to build vimana parts.

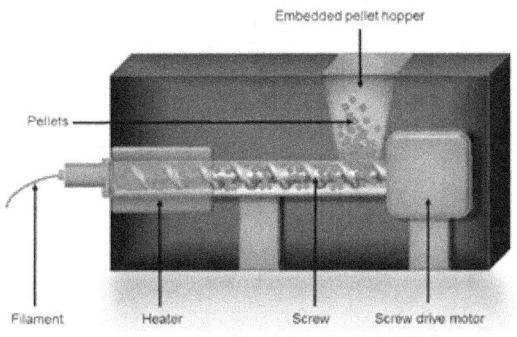

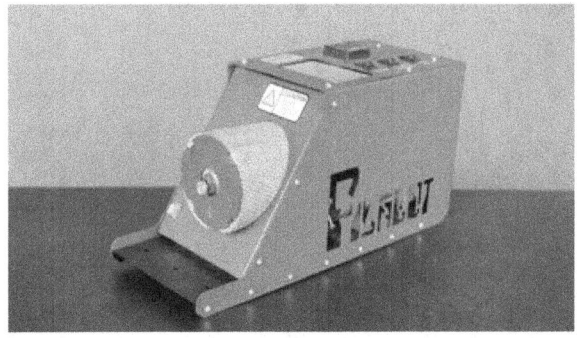

Diagram of a filament maker and a picture of the 'Filabot – Original' used for producing raja loha Filament.

Our SWASTIK research team and 3D Srishti Company in collaboration trying to synthesise the raja loha Filament and Trinetra Loha grain powder. This is in under research stage in order to overcome the various challenges, such as material homogeneity and amicability of composition.

3.6 Additive Manufacturing Applications

Prototyping was among the earliest applications of AM technologies and remains one of the most powerful tools for product development. As material properties, surface finish, and dimensional accuracy have improved, AM models have been increasingly used for functional prototyping and accuracy have improved, AM models have been increasingly used for functional prototyping and for fit and assembly testing. AM parts are also being used as patterns for tooling and metal casting processes. In addition to manufacturing ancient vimanas by Additive manufacturing process, there are a vast group of applications of AM in various fields mentioned below.

➢ **Prototyping**
- Visual Aids & Presentation Models

- Fit, Function, and Assembly Models
➢ **Tooling**
➢ **Metal Casting**
➢ **Architectural**
➢ **Medical**
  - Anatomical & Surgical Models
  - Custom Prosthetic Design
  - Virtual Surgical Planning & Personalized Surgical Instruments
  - Custom fabricated Implants
➢ **Dental**
➢ **Bio-Printing**
➢ **Electronics**
➢ **Direct Part Production**
  - Aerospace
  - Automotive & Motorsports
  - Medical
  - Avatars & Figurines
  - Furniture, Home, and Office Accessories
  - Applications in Music
  - Customer-created Products
  - Art & Jewellery
  - Gifts, Trophies & Memorials
  - Marketing & Advertising
  - Museum Displays
  - Fashion & High-performance Products

3.7 Additive Manufacturing Capabilities

The following are the AM capabilities.

➢ Easy to fabricate complex 3D models
➢ Customized medical implants
➢ Low cost & time
➢ High Accuracy

- Wide range of materials
- Functional testing

3.8 Additive Manufacturing Limitations

The following are the limitations of AM.

- Strength
- Tough for batch processes
- Post processing
- Initial equipment cost
- Skilled labour

4 AM Data Formats & Software's

4.1 AM Data Formats

AM consists of several file formats in that some of them were used most commonly and some are used for specific applications. Out of all the AM data formats, some of the commonly used data formats are listed below.

- STL (Stereolithographic) format
- STEP (Standard for the Exchange of Product Model Data)
- IGES (Initial Graphics Exchange Specification)
- HP/GL (Hewlett-Packed Graphics Language)
- CT (Computerized Tomography)/MRI (Magnetic Resonance Imaging)

Apart from these some of the newly proposed data formats are

- RPI (Rapid Prototyping Interface)
- SLC (Stereo lithography Contour)
- SLI (Stereo lithography Interface)
- LEAF (Layer Exchange ASCII Format)
- CLI (Common Layer Interface)

Among all the above formats STL is globally accepted by all the RP machine manufacturers. Hence this is the more popular format, even though STL has a number of problems in it and which are discussed as follows.

Representation methods used to describe CAD geometry vary from one system to another. A standard interface is needed to convey geometric descriptions from various CAD packages to rapid prototyping systems. The STL (Stereo Lithography) file, as the *de facto* standard, has been used in many, if not all, rapid prototyping systems. The STL file, conceived by the 3D Systems, USA, is created from the CAD database via an interface on the CAD system. This file consists of an unordered list of triangular facets representing the outside skin of an object. There are two formats to the STL file. One is the ASCII format and the other is the binary format. The size of the ASCII STL file is larger than that of the binary format but is human readable.

In an STL file, triangular facets are described by a set of X, Y and Z coordinates for each of the vertices and a unit normal vector with X, Y and Z to indicate which side of the facet is an object. An example is shown in figure 2.

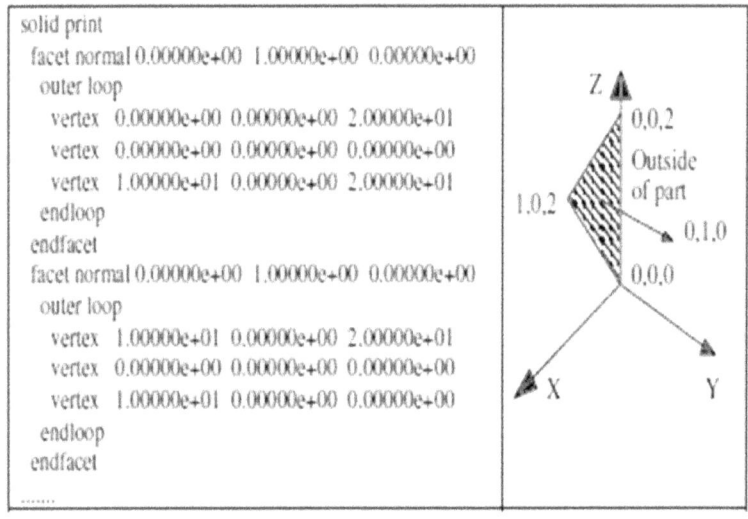

Figure 2: Sample STL file.

Because the STL file is a model derived from precise CAD models, it is, therefore, an approximate model of a part. Besides, many commercial CAD models are not robust enough to generate the facet model (STL file) and frequently have problems. Nevertheless, there are several advantages of the STL file. First, it provides a sample method of representing 3D CAD data. Second, it is already a *de facto* standard and has used by most CAD systems and rapid prototyping systems. Finally, it can provide small and accurate files for data transfer for certain shapes. On the other hand, several disadvantages of the STL file exist. First, the STL file is many times larger than the original CAD data file for a given accuracy parameter. The STL file carries much redundancy information such as duplicate vertices and edges are shown in Figure 3. Second, the geometry flaws exist in the STL file because many commercial tessellation algorithms used by CAD vendor today are not robust. This gives rise to the need for a "repair software" which slows the production cycle time. Finally, the subsequent slicing of large STL files can take many hours. However, some RP processes can slice while they are building the previous layer and this will alleviate this disadvantage.

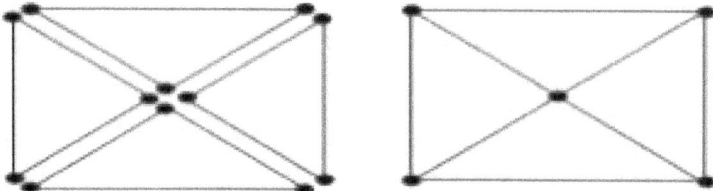

Figure 3: Edge and vertex redundancy in STL format.

When exporting to STL in your CAD package, you may see the parameters for chord height, deviation, angle tolerance, and poly count are the some of the parameters that affect the faceting of the STL.

Some of the problems plague STL files and due to their very nature of STL files as they contain no topological data. Many commercial tessellation algorithms used by CAD vendors today are also not robust, and as a result, they tend to create polygonal approximation models which exhibit the following errors:

➢ Gaps (Cracks, holes, punctures) that is, missing facets.
➢ Degenerate facets (where all its edges are collinear).
➢ Overlapping facets.
➢ Non-manifold topology conditions.

4.2 AM Software's

It is necessary to translate CAD models into a format that AM machine can process. The most popular format, used by all commercial AM systems, is the STL file developed by 3D Systems, The STL format is popular because it is a simple definition of a 3D model user defined parameters in the CD Software determine the size of the facets and accuracy of the approximation. The production of a good STL file requires good source CAD data. The terms closed volume or water-tight are often used to describe good STL data, indicating that the skin of such data would not leak if the volume of the model were filled with fluid.

If the STL file is generated from flowed CAD data that contains gaps, overlaps, inconsistent surface normal, and so forth, these problems include missing features, voids, and unattached areas, even build failures. Many CAD systems do not offer tools for shaded visualization of the STL files they generate consequently, CAD users often unknowingly create and distribute flawed STL files. These issues have spawned the development of software products that permit one to view and repair STL data as an alternative to going back to the source CAD to do so. The list of AM software's is listed below.

- Magics
- Mimics
- 3-matic
- Velocity 2
- 3D-Doctor
- VoXim
- Simplant
- SurgiGuide
- Solid View
- View Expert
- 3Data Expert
- Rhino
- Stlview
- Rapid Prototyping Module (SDRC)

5. Exporting an STL file from Creo

Here you will be guided by the parameter and steps for exporting an STL file from Creo. Important things to be noted before converting to STL are as follows

5.1 Deviation control

The deviation Control settings in the Export STL dialog box affect the accuracy of the model and the size of its file.

Open Chord Height (chordal tolerance)

a. This setting specifies the maximum distance between the surface of the original design and the tessellated surface of the STL triangle (the chord).

b. Chord height controls the degree of tessellation of the model surface.

c. The smaller the chord height, the less deviation from the actual part surface (but the bigger the file).

5.2 Angle control

This setting regulates how much additional tessellation occurs along surfaces with small radii. The smaller the radii, the more triangles are used. The setting can be between 0 and 1. Unless a higher setting is necessary, to achieve smoother surfaces, 0 is recommended.

Once you have reviewed the above the controls and adjusted your settings, click Apply > OK to create STL file.

5.3 Steps to convert to STL file.

Step 1: Open the File pull-down menu and Save as then Save a Copy.

Step 2: Change the Type option to Stereolithography (*.stl).

Step 3: Accept the default filename or type in a new one.

Step 4: Click on Ok to close the dialog a new Export STL dialog will open.

Step 5: Change the **Chord Height** Value to 0 (Zero).

Note: In Creo, we will enter the smallest practical value for Chord Height. The resulting number is based on the size and complexity of the part.

Step 6: Change the Angle Control values to 0 (Zero).

Step 7: On OK to close the dialog and the STL file be created in the working directory.

The display will change to show the faceted surface of the STL file, zoom into curved corner and the triangular surfaces created by the STL conversion.

6. AM Build Setup Preparation

In current days the AM systems are accessible to almost each and every person. This all is possible due to ease of access with open source software's which are listed in the below and for an example out of all open source software's Cura is explained in detail.

- Cura
- MeshLab
- Skeinforge
- Slic3r
- Repsnapper
- ConvertSTL
- IVCON
- Creation Workshop

6.1 Cura

Cura is an open source program which prepares your object into a file (g-code) that can be read by the 3D Printer.

6.2 Loading of file in Cura

Default home screen of Cura is as shown in figure 1 below.

Figure 1: Default Cura screen

Click on the "Load" icon to input the STL file as shown in figure 2 below.

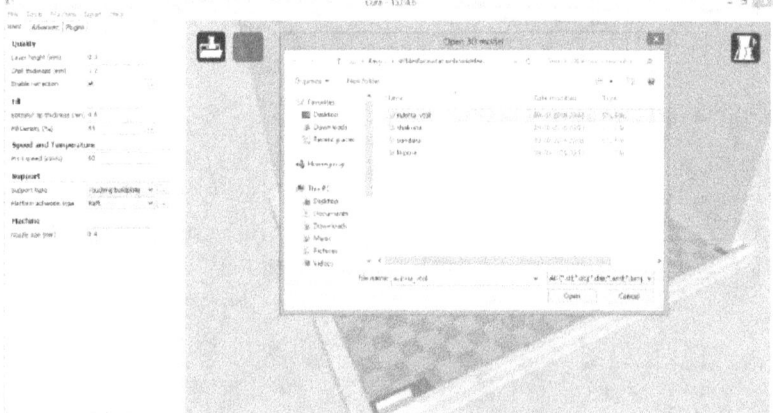

Figure 2: Loading of STL file in Cura

After selecting the STL part will be loaded into the build platform as shown in Figure 3.

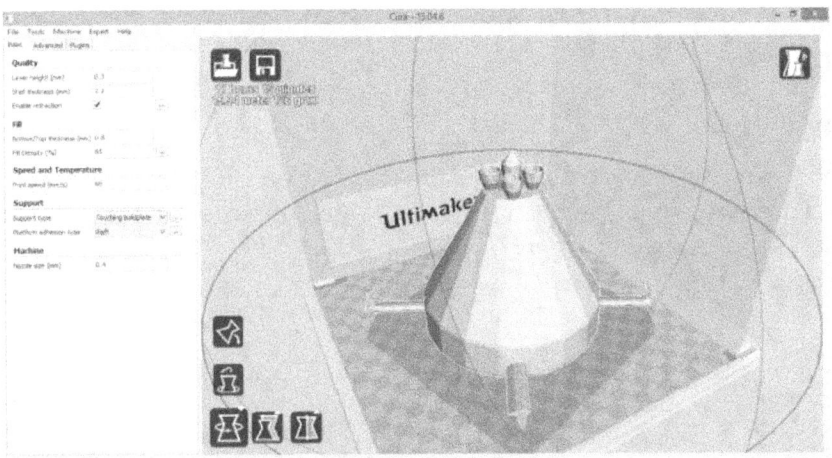

Figure 3: STL model in build platform

6.2.1 Slicing/ Layer Thickness / Slice Thickness

After loading the model Cura will provide the option of layer thickness from 0.06mm to 0.25mm (This also depends on machine capability). 0.06mm will give the finest quality of the part and 0.25mm will finish the printing of the part in less time. For optimum time and quality select layer thickness as 0.1mm as shown in Figure 4 below.

Figure 4: Layer thickness 0.1mm and time taken 28 hours 0 minutes.

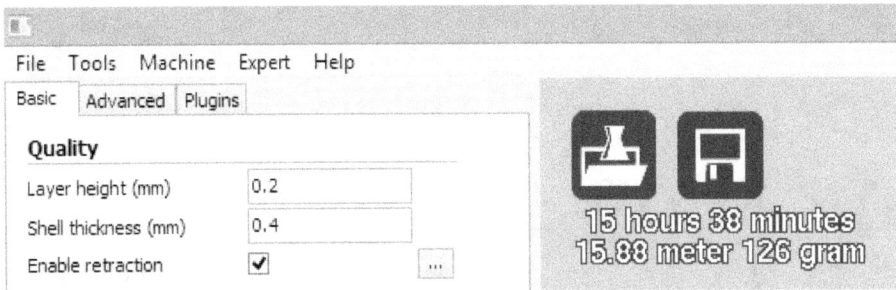

Figure 5: Layer thickness 0.2mm and time took 15 hours 38 minutes.

Note: Figure 4 and 5 shows Build time inversely proportional to Layer Thickness.

6.2.2 Support Generation

Supports are automatically generated depending on the option selected in the Cura. The function of supports is to provide supports for the features like overhanging as shown in Figure 6.

Figure 6: Marron color is a model and the dark blue color is support.

In Cura, support generation has three options as shown in figure 7 below.

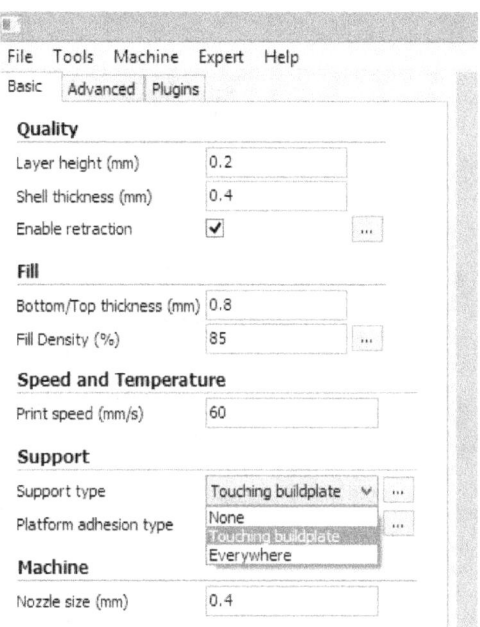

Figure 7: Options for support type in Cura

6.2.3 Layer Views

Cura provides flexibility to view the individual layers as shown in Figure 8 and 9.

Figure 8: View mode   Figure 9: 378th layer out of 621 layers.

6.2.4 Printing Speed and Temperature

Generally, ABS and PLA filaments will have printing temperature between 200°C to 230°C.

The speed of the printing is inversely proportional quality. The setup for printing speed and temperature are shown in Figure 10 below.

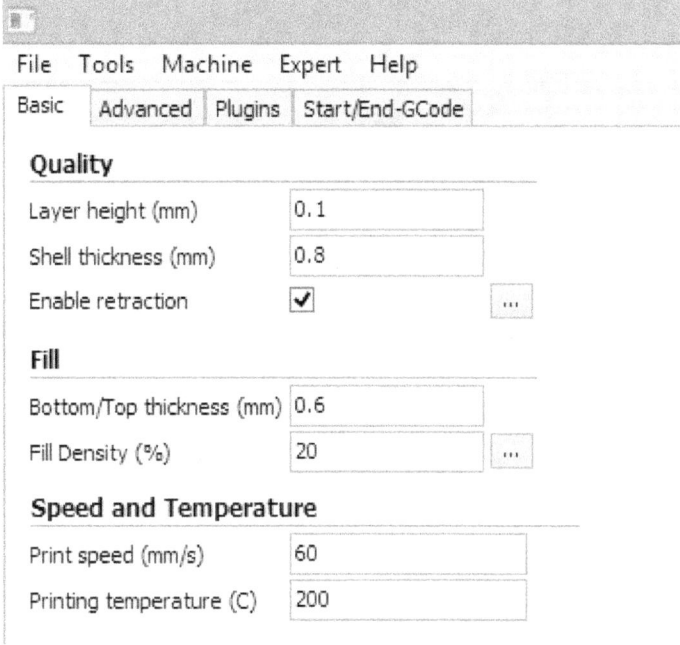

Figure 10: Default print speed 60 and temperature 200 ºC

6.2.5 Filament

Generally for FDM machines filament diameter is of 1.75mm and 3mm diameter in standard sizes. In Cura, the filament options are shown in Figure 11 below.

Figure 11: Filament settings in Cura.

6.2.6 Fill Density

Fill density of the model option is shown in figure 12 below.

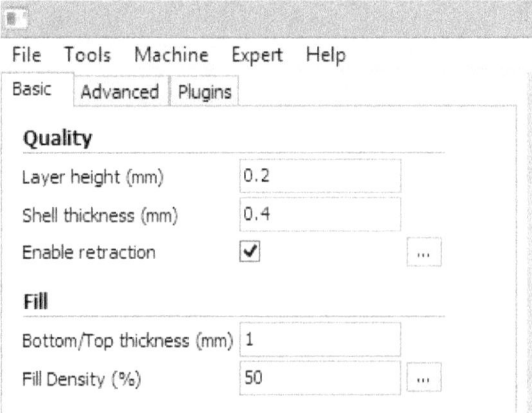

Figure 12: Fill density in Cura

6.2.7 Machine Setup

Cura being the open source software allows configuring the assembled and available low-cost 3D Printers.

The setup of the machine can be done using the "machine" tab as shown in figure 13 below.

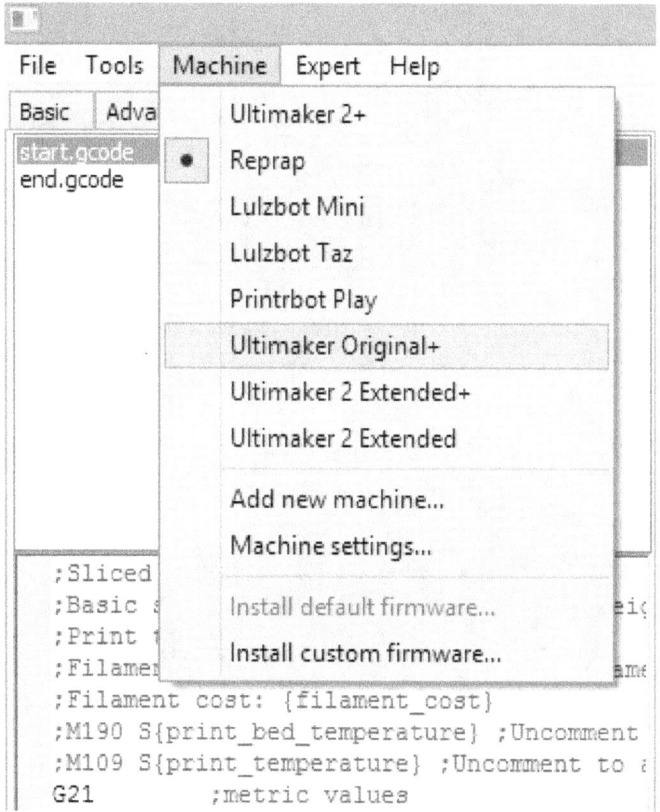

Figure 13: Machine setup in Cura

6.2.8 G code

The G-code will be generated for the vimana model after providing the basic parameters like Layer thickness, Fill density, speed, Temperature, Flow rate and etc. The G-code generated for Vimanas is more 13000 pages or G code is generated, so, it is not possible to present whole G-Code, so a small beginning of the G code is presented in the below Figure14.

```
File Edit Format View Help
;FLAVOR:UltiGCode
;TIME:51020
;MATERIAL:70198
;MATERIAL2:0
;NOZZLE_DIAMETER:0.400000
;NOZZLE_DIAMETER2:0.400000

;Layer count: 1241
;LAYER:0
M107
G0 F9000 X124.029 Y63.556 Z0.300
G0 X125.345 Y63.140
;TYPE:SKIRT
G1 F1200 X131.559 Y61.162 E0.78255
G1 X132.540 Y64.271 E1.17376
G1 X133.949 Y68.697 E1.73114
G1 X129.983 Y76.352 E2.76571
G1 X128.667 Y76.768 E2.93133
G1 X125.335 Y77.829 E3.35095
G1 X125.404 Y78.046 E3.37828
G1 X126.813 Y82.471 E3.93555
G1 X122.847 Y90.126 E4.97011
G1 X121.524 Y90.544 E5.13661
G1 X115.316 Y92.520 E5.91839
G1 X114.333 Y89.405 E6.31036
G1 X112.926 Y84.986 E6.86687
G1 X116.893 Y77.331 E7.90150
G1 X118.209 Y76.915 E8.06712
G1 X121.541 Y75.855 E8.48670
G1 X121.470 Y75.631 E8.51490
G1 X120.062 Y71.211 E9.07156
G1 X124.029 Y63.556 E10.10618
G1 X125.345 Y63.140 E10.27180
G0 F9000 X99.051 Y73.583
G1 F1200 X98.396 Y74.822 E10.43998
G1 X97.150 Y77.232 E10.76555
G1 X98.433 Y78.237 E10.96112
```

Figure 14: Sample G-code generated in Cura.

7. AM Build Processes

Most AM processes have a lot in common. Especially, input to the systems consists of 3D model data, and fabrication occurs by joining materials in successive layers. However, a wide range of discretely different processes is available under the AM umbrella.

Before we go into the manufacturing of vimanas by 3D printing, we have to know the various manufacturing processes available in 3D printing technology. As per ASTM International committee, F42 in January 2012 on Additive Manufacturing Technologies approved a list of AM processes category names and definitions from the specification "standard Terminology for Additive Manufacturing Technologies". The ASTM approved a system of 7 processes described briefly as follows:
1) Material extrusion- an additive manufacturing process in which material is selectively dispensed through a nozzle or orifice.

2) Material jetting- an additive manufacturing process in which droplets of build material are selectively deposited.

3) Binder jetting- an additive manufacturing process in which a liquid bonding agent is selectively deposited to join powder materials.

4) Sheet lamination- an additive manufacturing process in which sheets of material are bonded to form an object.

5) Vat photopolymerization- an additive manufacturing process in which liquid photopolymer in a vat is selectively cured by light-activated polymerization

6) Power bed fusion- an additive manufacturing process in which thermal energy selectively fuses regions of a power bed.

7) Direct energy deposition- an additive manufacturing process in which focused thermal energy is used to fuse materials by melting as the material is being deposited.

By using each of the above processes we can build ancient vimanas as per our requirement and use such as for fabricating static prototype or the dynamically scaled model of ancient vimanas, the possibility of building ancient vimanas by each of the processes is described in detail as follows.

7.1 Material extrusion

Material extrusion is an AM process in which material is selectively dispensed through a nozzle or orifice. Material extrusion machines force material through a nozzle as the extrusion head or the build platform moves in the x-y plane. After a layer is completed, the build platform moves down, or the extrusion head moves up, and the next layer is extruded and adhered to the previous layer. The schematic for material extrusion is shown in figure 1 below.

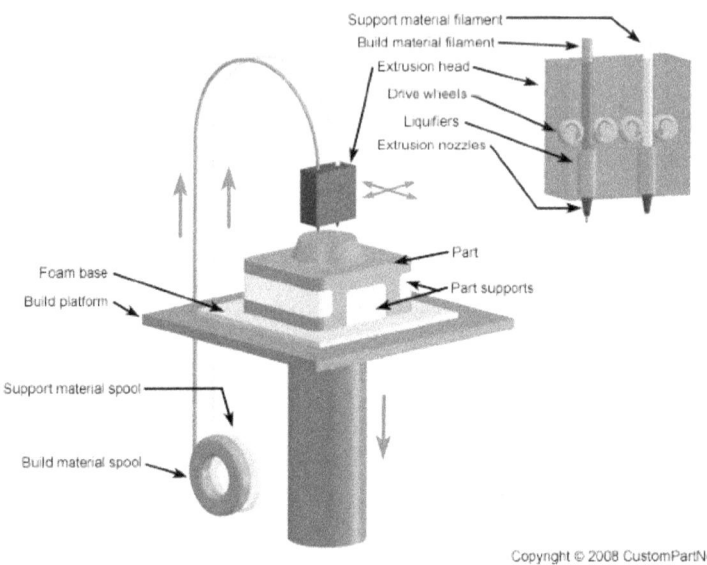

Figure 1: Schematic diagram for Material Extrusion [3].

Material extrusion systems represent the largest installed base of AM machines. The process was pioneered by Stratasys Inc., which introduced the first material extrusion-based fused deposition system (FDM) in 1991. Now called Stratasys Ltd., the company has manufactured and sold more than 21,000 FDM systems through the end of 2012. FDM systems use two spools of material, one for the build material and the second for the support material.

The other large group of material extrusion systems is comprised of low-cost personal 3D printers derived mostly from RepRap open-source project. A few of the companies that offer these systems include Aleph Objects, Fabbster, MakerBot Industries, Delta Micro Factory Corp., 3D Systems and 3D Srishti Pvt Ltd. Compared to other AM processes, material extrusion systems are relatively inexpensive. Most of the personal 3D printers that use material extrusion sell from about $500 to $4,000. Some versions sell as a kit that customers must assemble. FDM systems from Stratasys sell from about $9,500 to about $500,000. The most expensive FDM systems are capable of producing large parts in a wide range of materials.

7.1.1 Some of the Material extrusion Parts

Below parts (as shown in Figure 2) are printed using Material Extrusion process.

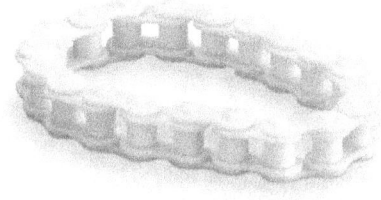

Figure 2: Parts fabricated using material extrusion process [4].

7.1.2 Materials used in of Material Extrusion

- Polymers: ABS, Nylon, PC, AB.

7.1.3 Advantages of Material Extrusion

- Robotically guided plastic extrusion.
- Thermoplastic materials.
- Low temperature/high-temperature materials.
- Water soluble or wax supports.

7.1.4 Limitations of Material Extrusion

- The nozzle radius limits and reduces the final quality.
- Accuracy and speed are low when compared to other processes and accuracy of the final model are limited to material nozzle thickness.
- The constant pressure of material is required in order to increase the quality of finish.

7.2 Material Jetting

The material-jetting process uses inkjet printing heads to deposit droplets of build material. The droplets are dispensed selectively as one or more print heads move across the build area. Substances used in material jetting are typically photopolymers or wax-like materials that can be used as invest en casting patterns. The schematic for material jetting is shown in Figure 3.

3D Systems offers a material-jetting process called multi-jet modeling. The company's Project brand of systems includes material jetting, but also three other types of process technology. Machines from Solids cape produce wax patterns for casting small metal parts using a proprietary thermoplastic ink-jetting process combined with high-precision milling of each layer.

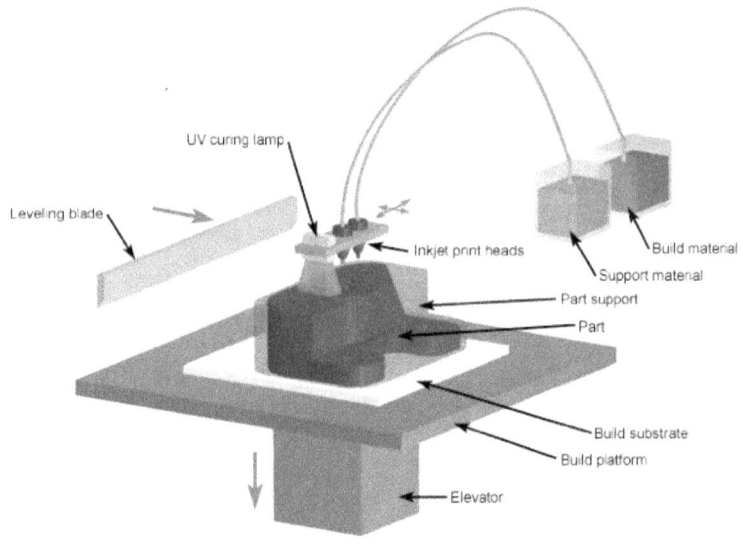

Figure 3: Schematic diagram for material jetting [5]

3D Systems offers a material-jetting process called multi-jet modeling. The company's ProJet brand of systems includes material jetting, but also three other types of process technology. Machines from Solidscape produce wax patterns for casting small metal parts using a proprietary thermoplastic ink jetting process combined with high-precision milling of each layer.

The material jetting systems from Soildscape range in price from $26,000 to $46,000. The Stratasys Ltd. Material-jetting systems that use a single build material are $20,000 to $173,000.

The Connex multi-material systems cost $160,000. The ProJet material-jetting systems from 3D Systems are priced from $60,000 to $160,000.

7.2.1 Some of the Material Jetting parts

The below parts (shown in figure 4) are fabricated using material jetting principle.

Figure 4: Parts fabricated using material jetting process[6]

7.2.2 Materials used in Material Jetting

Polymers: Polypropylene, HDPE, PC, PMMA, PC, ABS, HIPS, EDP.

7.2.3 Advantage of Material Jetting

- Combines high precision milling.
- No ash content/thermal expansion.
- Wax patterns for metal castings.
- High level of precision, accuracy and surface finish ($\pm$ 0.001 mm).

7.2.4 Limitations of Material Jetting

- The support material is often required.
- A high accuracy can be achieved but materials are limited and only polymers and waxes can be used.

7.3 Binder Jetting

Binder jetting is a process by which a liquid bonding agent is selectively deposited through inkjet print head nozzles to join powder materials in a powder bed. Binder jetting is similar to material jetting in its use of inkjet printing to dispense material. The differences are that with binder jetting, the dispensed material is not built material, but rather a liquid that is deposited onto a bed of powder to hold the powder in the desired shape. The Schematic of binder jetting is shown in Figure 5 below.

The binder jetting process was originally developed by MIT and was called 3D printing (3DP). Licensees of the 3DP technology include the ExOne Company and 3D Systems (Z Crop. Was the original licensee, but Z Corp. was acquired by 3D Systems in 2012). Voxeljet Technology GmbH is a sub-licensee of 3D Systems. MIT has licensed the technology to other companies, as well. The ZPrinter systems, originally commercialized by Z Crop and now sold by 3D Systems as a ProJet x60 series, use plaster-based powders and a water-based binder. Several models can print in full color. Microjet Technology, a Taiwanese company, offers similar binder jetting systems with full-color options.

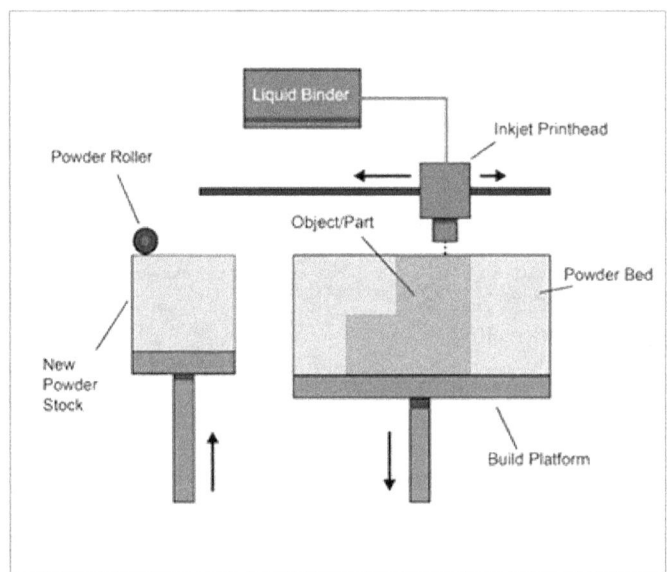

Figure 5: Schematic diagram for binder jetting [5].

ExOne has developed the technology for jetting a liquid binder onto the surface of metal or sand powder beds. Metal parts produced by binder jetting must be sintered and infiltrated with a second metal after the AM build process. The ExOne machines offer large build envelope for both sand and metal materials. These machines are capable of building parts at relatively high speeds, although for metal parts, the additional time for sintering and infiltration must be taken into account. Voxeljet offers large systems with wide print heads. The powder materials used by Voxeljet include an acrylic and foundry sand. The binder reacts at room temperature but must cure in the powder bed for a few hours before the parts can be removed.

The Swedish company Hoganas has developed a binder-jetting system for metals and ceramics, with a focus on stainless steel. Parts must be heat treated after they are built. Binder jetting systems vary widely in cost. The ProJet x60 series are office compatible 3D printers offering a range of build envelopes, build speeds and color options. They range in price from about $16,500 to $113,900. ExOne's systems are more expensive, given the large build boxes of its systems. ExOne's system range from $125,000 to $1.4 million. The build envelopes of the systems from Voxeljet are also large, so prices are relatively high in comparison to other polymer material AM systems. Base prices range from €120,000 to about € 1.4 million.

7.3.1 Some of the Binder Jetting parts

Below parts (Shown in Figure 6) are built by using binder jetting principle.

Figure 6: Parts fabricated using binder jetting process [7].

7.3.2 Materials used in Binder Jetting

- ➢ Metals: Stainless steel.
- ➢ Polymers: ABS, PA, PC.
- ➢ Ceramics: Glass.

7.3.3 Advantages of Binder Jetting

- ➢ Colored parts.

- The accuracy of ± 0.125 mm.
- No supports required.

7.3.4 Limitations of Binder Jetting

- Costly material.
- Not always suitable for structural parts.
- Additional post processing can add significant time to the overall process.

7.4 Sheet Lamination

Sheet lamination is defined as a process in which sheets of materials are bonded to form an object. Sheet materials can be adhesive-coated papers that form a plywood-like solid when laminated into a 3D object or metal tapes and foils that form metal parts. The first commercialized sheet lamination technology was laminated object manufacturing (LOM) from Helisys, which used a roll of craft paper coated with adhesive on one side and a heated roller to laminate successive layers. The schematic for sheet lamination is shown in Figure 7 below.

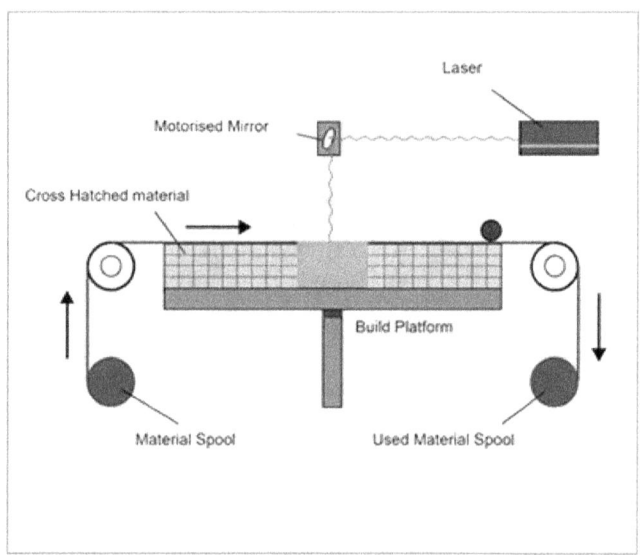

Figure 7: Schematic diagram for sheet lamination [5].

Mcor Technologies manufactures 3D printers that use standard sheets of paper as the build material. The machine holds three reams of A4 or letter-sized office paper and selectively dispenses a water-soluble adhesive that bonds the layers. A tungsten carbide blade cuts a profile in each layer. The cost of material is among the lowest in the industry. Prices range from $36,400 for the

Matrix 300+ to $47,600 for the full-color Iris model. Another sheet lamination technology is ultrasonic additive manufacturing (UAM) from Fabrisonic. Solidica originally marketed this technology as ultrasonic consolidation. UAM uses ultrasonic welding to bond layers of the thin metal tapes and foils. Layers are welded together by a combination of ultrasonic energy supplied by twin, high-frequency transducers and the compressive force created by the system's rolling "sonotrobe."

Fabrisonic offers systems that combine UAM with full CNC-machining capabilities. The company has announced three different UAM systems, which are in the late stages of development.

7.4.1 Some of the Sheet Lamination Parts:

Below parts (shown in Figure 8) are fabricated using sheet lamination process.

Figure 8: Parts fabricated using sheet lamination [8].

7.4.2 Materials used in Sheet Lamination

- Effectively any sheet material capable of being rolled. Paper, plastic and some sheet metals.
- The most commonly used materials are A4 paper.

7.4.3 Advantages of Sheet Lamination

- Benefits include speed, low cost, ease of material handling, but the strength and integrity of models is reliant on the adhesive used.
- Cutting can be very fast due to the cutting route only being that of the shape outline, not the entire cross-sectional area.

7.4.4 Limitations of Sheet Lamination

➢ Finishes can vary depending on paper or plastic material but may require post processing to achieve the desired effect.

➢ Limited material use.

➢ Fusion processes require more research to further advance the process into a more mainstream positioning.

7.5 Vat Photopolymerization

Vat photopolymerization is a process by which liquid photopolymer in a vat is selectively cured by light-activated polymerization. Stereolithography (SL), the first patented and commercialized AM process, uses an ultraviolet laser and x-y scanning mirrors on computer-controlled galvanometers to scan the top surface of a liquid photopolymer in a vat. Other vat photopolymerization technologies use a lamp or light-emitting diodes (LEDs) as the source of UV energy and digital light processing (DLP) technology, which employs an array of micromirrors to project an image onto the surface of a vat, curing the entire layer at once.

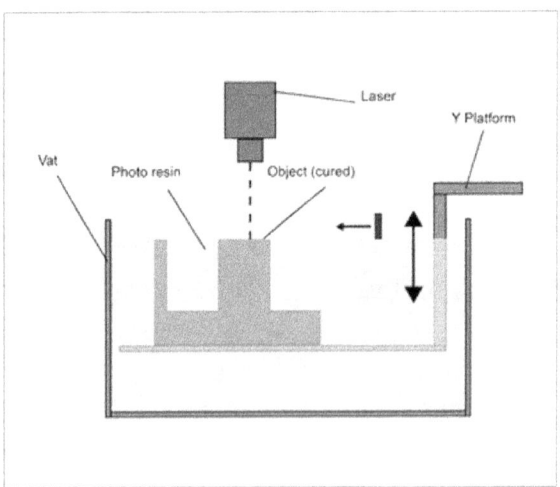

Figure 9: Schematic diagram for vat photopolymerization [5].

3D Systems commercialized stereolithography in 1988, and SL was the undisputed leading process in the early years of additive manufacturing. 3D Systems continues to manufacture SL machines, although unit sales are believed to be in steady decline as newer processes develop. Stereolithography systems were developed and commercialized by many companies in Japan in the early years, but unit sales of this type of systems have fallen precipitously in the past decade.

Most SL system manufacturers in Japan have abandoned the market. Vat photopolymerization systems are also manufactured in Europe, as well as in China and Korea.

The AM systems form Envisiontec GmbH use either a lamp or LED source and DLP for curing the photopolymer. Advances in DLP technology have enabled high resolution and finely featured parts. The company's Perfactory systems build parts from the top down (With the build platform above the part) by curing the photosensitive resin through a transparent build area. DWS is an Italian manufacturer of vat photopolymerization systems. Parts are built from the top down, similar to the Envisiontec Perfactory. The photosensitive resin is cured through a transparent tank using a solid-state laser.

7.5.1 Some of the Vat photopolymerization parts

Figure 10: Parts fabricated using vat photopolymerization process [9].

7.5.2 Materials used in Vat photopolymerization

- ➢ The Vat polymerization process uses plastics and polymers.
- ➢ Polymers: UV-curable Photopolymer resin.

7.5.3 Advantages of Vat photopolymerization

- ➢ High level of accuracy and good finish.
- ➢ Relatively quick process.
- ➢ Typically large build areas: object size: 1000mm x 800mm x 500mm and max model weight of 200 Kg.

7.5.4 Limitations of Vat photopolymerization

- ➢ Relatively expensive.
- ➢ Lengthily post processing time and removal from resin.
- ➢ Limited material use of photo-resins.
- ➢ Often requires support structures and post curing for parts to be strong enough for structural use.

7.6 Powder Bed Fusion

Powder Bed fusion is a process by which thermal energy fuses selective regions of a powder bed. The source of the thermal energy is a laser or an electron beam. The thermal energy melts the powder material, which then changes to a solid phase as it cools. Terms that are also used in the AM industry for powder bed fusion processes and systems include laser sintering, selective laser sintering, selective maser melting, direct laser sintering, and electron beam melting.

Both polymer and metal materials are available in powder bed fusion processes. For polymers, the unfused powder surrounding a part serves as a fixturing system, so no additional supports are usually needed. For metal parts, anchors are typically required to attach parts to a base plate and support down-facing surfaces. This is necessary because of the higher melting point of metal powders. Thermal gradients in the build chamber are high, which can lead to thermal stresses and warping if anchors are not used. Because powder bed fusion is a thermal process, warping, stresses, and heat-induced distortion are potential problems for all materials. Laser-based powder bed fusion systems generally produce a better surface finish and finer feature detail than electron beam systems. Electron beam systems are somewhat more expensive but are faster. Also, electron beam systems produce less residual stress in parts, resulting in less distortion and less need for anchors and support structures.

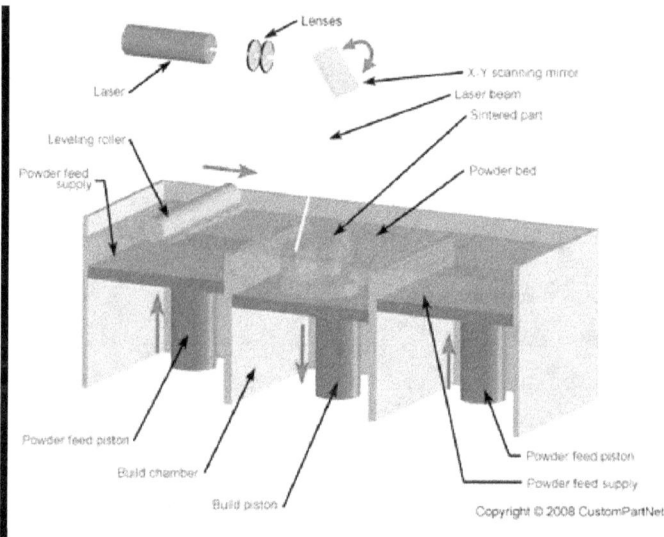

Figure 11: Schematic diagram for powder bed fusion [5].

Powder bed fusion systems are relatively expensive compared with most other AM processes, especially the machines that process metals. Operating costs are comparatively high due to the cost of materials, the recycling issues with polymer powders, and facility requirements for inert gas and safety. Parts made on these machines are being used increasingly for final products, so manufacturers have begun to include process control capabilities in the machines to ensure process quality and repeatability.

7.6.1 Some of the Powder Bed Fusion Parts

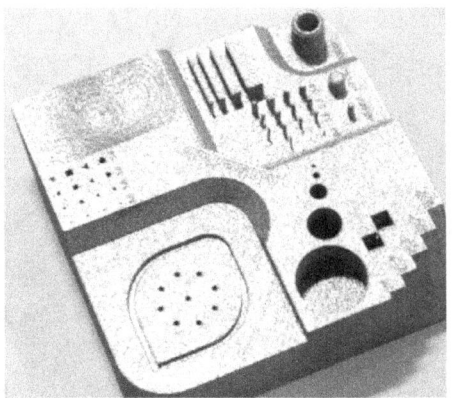

Figure 12: Parts fabricated using powder bed fusion [10].

7.6.2 Materials used in Powder Bed Fusion

➢ The powder bed fusion process uses any powder based materials, but common metals and polymers used are Nylon, Stainless Steel, Titanium, Aluminium, Cobalt Chrome, Steel, titanium, Cobalt Chrome, and Copper.

7.6.3 Advantages of Powder Bed Fusion

➢ Relatively inexpensive.
➢ Suitable for visual models and prototypes.

- Powder acts as an integrated support structure.
- Large range of material options.

7.6.4 Limitations of Powder Bed Fusion

- Relatively slow speed.
- Lack of structural properties in materials.
- Size limitations.
- High power usage.
- Finish is dependent on powder grain size.

7.7 Directed Energy Deposition

In the Directed Energy Deposition process, focused thermal energy is used to fuse materials by melting as the material is being deposited. In most cases, a laser is the source of the energy, and the material is a metal powder.

Directed energy deposition systems have had somewhat limited success in the AM market, although the process offers unique capabilities. More than one material can be deposited simultaneously, making functionally graded parts possible. Also, most directed energy deposition systems use a 4 or 5 axis motion system or a robotic arm to position the deposition head, so the build process is not limited to successive horizontal layers on parallel planes. This capability makes the process suitable for adding material to an existing part, such as representing a worn part or tool.

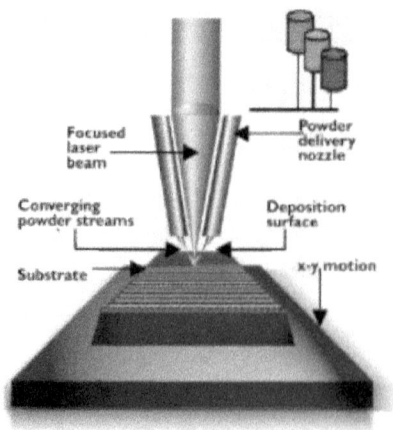

Figure 13: Schematic diagram for direct energy deposition [11].

Laser engineered net shaping (LENS) from Optomec is a directed energy deposition process that injects a metal powder into a pool of molten metal created by a focused laser beam. The POM Group manufactured direct metal deposition (DMD) systems, but the company was purchased by AM3D in 2012 and is currently selling parts only. The DMD process uses a CO_2 laser and a concentric powder nozzle to melt powder introduced into a melt pool. Be AM is a spinoff of Irepa Laser, and sells Easy CLAD machines, although few have been placed at customer sites. Sciaky Inc offers a directed energy deposition process called direct manufacturing (DM).

The energy source in DM is an electron beam, and material is introduced by wire feed instead of a powder deposition head. The DM electron beam welding system is fast, but parts produced by the process have a very coarse surface that requires extensive post-build finish machining. Direct energy deposition system is quite expensive, ranging from $350,000 to more than $1.5 million. Metal material processing, multiple material options, multi-axis motion, and process controls all contribute to the higher system cost.

7.7.1 Some of the Directed Energy Deposited Parts

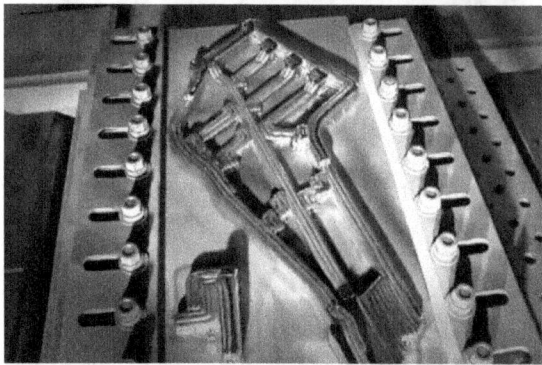

Figure 14: Parts fabricated using direct energy deposition process [12].

7.7.2 Materials used in Directed Energy Deposition

- ➤ The LENS uses metals and not polymers or ceramics. For example Metals like Cobalt, Chrome or Titanium.

7.7.3 Advantages of Directed Energy Deposition

- ➤ Ability to control the grain structure to a high degree, which lends the process to repair work of high quality, functional parts.
- ➤ A balance is needed between surface quality and speed, although with repair applications, speed can often be sacrificed for a high accuracy and a pre- determined microstructure.

7.7. 4 Limitations of Directed Energy Deposition

- ➤ Finishes can vary depending on paper or plastic material but may require post processing to achieve the desired effect.
- ➤ Limited material use.
- ➤ Fusion processes require more research to further advance the process into a more mainstream positioning.

8. Future scope in the advancement of manufacturing processes of Ancient vimanas:

The next generation manufacturing of ancient vimanas by 3D printing is 4D printing of ancient vimanas, this 4D printing can be described as an additive manufacturing process that integrates smart materials into the starting form of the printing material for 3D printed structures/components. After fabrication, the 3D object would respond in an intended manner to external stimuli from the environment or through human interference, resulting in a change in shape or physical properties over time.

4D printing enabled by usage of smart materials, smart materials defined as materials that would either change their shape or properties between different physical domains in a useful manner under the influence of certain stimuli from the environment, introduction of smart materials in the AM-fabricated components are able to alter their shape or properties over time (the 4th dimension) as a response to the applied external stimuli.

One important type of smart material that has been used widely is the piezoelectric material that is able to produce electrical charge or voltage when experiencing an externally applied stress and vice versa. The applications of piezoelectric material can be found in loud speakers, acoustic imaging, and energy harvesting, actuators, transducers and tissue regeneration. Piezoelectric polymeric materials have some unique characteristics as compared to other piezoelectric materials. These materials are suitable for systems that require mechanical flexibility, small active elements, biocompatibility and solution-based processability. Under the influence of light, the polymer would form cross-linking and direct covalent bonding with the chemical groups of the smart nanoparticles and this grafts the nanoparticles to the backbones of the polymer.

4D printing is achieved by the 3D printing of shape memory polymers (SMPs), actuators for soft robotics, self-evolving structures, anti-counterfeiting system, and active origami and controlled sequential folding.

The programming of physical and biological materials to change shape, change properties, to build themselves by using software called **cadnano**, that allows us to design three-dimensional shapes like nanorobots or drug delivery systems and use DNA to self-assemble those functional structures. The programmable materials that build themselves known as, self-assembly, is a process by which disordered parts build an ordered structure through only local interaction.

In 4D printing, materials and geometry have to be tightly coupled with the passive energy source like heat, shaking, pneumatics, gravity, magnetics, with smartly designed interactions and this interaction allow for error correction, and they allow the shapes to go from one state to another state.

The idea behind 4D printing is adding a new capability of transforming from one shape to another shape directly on their own by using 3D printed multi-materials, similar to robotics without wires or motors. So you completely print this part, and it can transform into something else. The Autodesk developing a software called Cyborg, this helps us to simulate self-assembly behavior to optimize which parts are folding when. Cyborg software helps us to design of nanoscale self-assembly systems and human scale self-assembly systems. So no human interaction. Program and transformation have been embedded directly into the materials themselves and it might just be the manufacturing technique that allows us to produce more adaptive infrastructure in the future. In future, the composition of the ancient vimana material will be modified in order to gain more compatible material as per existence in the earth crust with modified composition, this are termed as smart materials and this material can be used in 4D printing of Ancient vimanas by which a predicted behaviour of vimanas can be obtained with its surrounding environment.

RESULTS AND DISCUSSIONS

Vimana Research works: We approached the research with a creative view, merging the modern technology and taking advantage of ancient aeronautics which will be a benefit to humanity.

The vimana prototypes we planned are Hybrid models of advanced ancient vimana technology and the modern cutting edge technology. We can understand that perfect Reinvention of Ancient vimana may be impossible due to the lost resources (raided libraries) and unavailable materials such as extinct species of the Flora. Thus replacing this absence with modern technology we can make the vimana 50% modern and 50% ancient, resulting in an advanced space vehicle. When there are Reinvented devices and materials from vimana shasthra, then the Vimana technology cannot be neglected as it gives mankind to achieve more advanced technology. In order to benefit the mankind by giving advanced technology and Free electricity from nature, and to make vimana initially in small-scale as prototypes Lab tests, 3D printing of vimana prototypes, we are on a mission to complete our dream and innovation. For the first time in history a 3D model rendering of all four vedic Vimanas as Mentioned in the ancient Indian Vimanika Shastra. A detailed work of 3D modeling has been done and reverse engineered to know dimensions design. Our aim is to create new bridges in the discovery of ancient Indian technology and to bring that knowledge in present time for a better evolution of our society and world.

Future projects are being planned for Vimanas Prototypes, materials and propulsion. It could lead us to examine understand and study in-depth of vimana technology and to Reverse engineer ancient Indian Vimanas with the help of scientists.

Mercury Vortex Propulsion

By Kavya Vaddadi

Talpade was Vedic Sanskrit scholar who was in contact with Subbaraya shasthry. As per Vedic traditions, Oral transmission of Sanskrit shlokas was commonly done, where the knowledge is passed on to the students by the Guru, thus Talpade was able to get the knowledge from Subbaraya shasthry even before the Vimana shasthra book has released. With his interest in Vedic technology, He did research on vimana shasthra. Vimanas are described as huge 100 feet in dimensions, Talpade was the first man to fly, and implementing the vimana shasthra in small scale dimensions he made a prototype called "Marutsakha vimana". Talpade had a great interest in testing the technology by implementing into the practical manner, he worked on propulsion materials and all available materials with him and he made the Marutsakha vimana, which is said to have used mercury propulsion.

As per my investigation, Marutsakha vimana was similar to Tripura vimana. Tripura vimana could travel three ways, which is on and in water, on land, and in air. I have examined the vimana diagrams with scientist and we realized that this craft could fly with antigravity. And when I observed the texts, Tripura vimana have generator and it seems to be a mercury vortex generator connected to the ganapa Yantra which also could be responsible for antigravity of the vimana. Talpade could have used a similar device to fly the Marutsakha with the available equipment and materials he has.

Talpade wrote books on vimana shasthra in the Marathi language. It is clear that Talpade worked on all Vimanas mentioned in vimana shasthra in which one of the vimana is Shakuna vimana which has wings, rudder, and propeller, exactly similar to the Kitty Hawk. Many researchers now digging out the lost books of Talpade, which are also found in foreign countries. He wrote books on Vimanas, propulsion, flight and many more things in his mother tongue language. The available books are being translated as of now, which will be soon coming out.

Talpade based his design of the Marutsakha on descriptions in the Vedic texts, similar to Tripura vimana. Tripura vimana could travel three ways, which is on and in water, on land, and in air. Marutsakha used propulsion of Tripura vimana which could travel in the air. The Marutsakha was flown near beach area where there are also possibilities to make the craft go in the water and on land. When we see the photo of the Marutsakha vimana, we can understand that it is similar to an airship design with rudder and tail and a propulsion device at the top of vimana, inside a dome.

Ancient civilization was advanced than modern because they lived for more centuries compared to our modern human civilization till now. And then it's obvious that they had more chance to explore the world in all dimensions including spiritual accessed by the third eye which is not much known by a modern human. The sages used this third eye very effectively, Ayurveda, vimana shasthra, and many such amazing texts reveal useful high advanced science and technology.

I have done Re-Translations of Vimana shasthra Shlokas, and I have unearthed many descriptions from vimana shasthra and it is vastly advanced aerospace technology which is more advanced than any modern aeronautical subjects. It reveals advanced descriptions of subjects such as Manufacturing, Tactical Air Defense Aerial Wars, UFO's, Aliens, Aerodynamics, structures, Propulsion, space mechanics, Antigravity technology and many more things. I have deeply worked and presented in this book. All my discoveries, I am trying to implement in a practical manner.

Talpade made Marutsakha vimana by referring to Tripura vimana from vimana shasthra because when I examined the pictures, both are similar in devices used and the shape too is similar. As per my investigation on his history, I found proof that Talpade was looking for funds, and he was not in a position to implement high costly materials and he was not able to access the lab too for deep propulsion study. In such situation, he chose to make Tripura vimana. And I interpret that he also used antigravity device which had less power and made the craft fly for 30 seconds.

Both Marutsakha vimana and Tripura vimana have propeller, rudder, and the propulsion devices locations are similar. Both are in UFO shaped design.

3D modeling the Tripura vimana was very difficult as the vimana is too complicated, sophisticated design. The dimensions mentioned in vimana shasthra I referred, but due to the various features of vimana such as folding links and overlapping floors, the diagrams are drawn by Ellappa in various forms in each view, and this confused me a lot at the beginning but later when I referred to the texts, then I found the exact way out and I was able to finish the design. The vimana is three-way usable and it has multiple features expressed in vimana shasthra diagrams. So, I designed two type of Tripura Vimanas, one which has all the floors and the other which has been designed in the shape of vimana after folding links overlapping the floor. The special features of vimana are really impressing and I feel there is a strong need to work more on such advanced mechanisms and materials to use a single craft for multipurpose.

The shape of Tripura vimana is obviously aerodynamic. Some of the Vimana diagrams are shown in cut sections, which is of course not aerodynamic shape. But there is dome on top of the propulsion devices, which is already understood by Talpade, and I too interpreted the same and designed it. The vimana is aerodynamic. Its propulsion system is an advanced one that we currently are not capable of reproducing. But working more on it may give a way to understand the high technology of vimana propulsion.

It is obviously remarkable that Sanskrit texts record a significant amount of information of flight, much of which was not rediscovered until thousands of years later. We see the ancient Vedic civilization was most ancient and there are 500 principles of flight which include space flight too, describing interplanetary space vehicles. Vimana shasthra have solutions for most of the problems we face in modern day technology. Vimana shasthra explains technology and propulsion which has no pollution and no space debris. We need to refer to the vedic texts to solve the modern human problems. Decade by decade as modern humans, start approaching their ancient advanced technology, then we will be able to realize how much more advanced it is. Descriptions of Highly efficient propulsion devices, High energy extraction devices, and many such advanced technology descriptions are unearthed for the first time in history through retranslations in this book.

Mercury is an amazing metal and used by ancient ancestors in various forms. Mercury was used in vortex propulsion and ion propulsion too. Solar mercury boiler at the crafts' center will heat the mercury and accelerate it. The setup is in circular airframe and thus the mercury is set into the vortex, and this highly accelerating mercury is connected to the propulsion system, making the craft to travel a great distance in a marvelous manner. Mercury is magnetic, and I interpret the ancient ancestors used this property to connect mercury for propulsion and have used electromagnetic antigravity too. Centrifugal force creates a lift of the craft and the high speeds of whirling mercury could create high centrifugal force creating a great amount of lift.

In The 2nd International Space Conference held during 08-09 January 2015 in Noida, India, I presented a paper "Mercury ion engine of ancient aeronautics". The Sanskrit texts describe in detail a mercury vortex engine which is the forerunner of the ion engines being made today by NASA. As per my research, there have been various herbs which purify the mercury so that it doesn't give rise to impurity problems in ion propulsion. We can also understand that the ancient civilization invented some herbs specifically with specific properties which they wanted, to use them for space technology materials and propulsion fuels.

NASA tested mercury ion engine but failed due to impurities. The Vedic technology needs to be followed such a way that everything mentioned in it, should be implemented. But when the mercury cannot be purified, it leads to impurities. Mercury gives high efficiency compared to xenon ion engine. A text called Rasa Shasthra mentions detailed procedure to purify mercury which we can use for propulsion without any failure. The exact herbs need to be mixed properly in exact proportion leading to the perfect mixture of purified mercury. Herbs are in fact magical, I have been shown a leaf which when eaten the tongue cannot taste any sweetness and there are results in blood sugar level decrease too when taken in larger quantities. Ayurveda exposes more magical herbs which we modern humans should understand the importance of protecting the environment.

Captain Anand J. Bodas said truth at the Indian Science Congress in 2015 regarding descriptions of aviation in ancient Sanskrit texts. It was controversial because there are people who cannot understand the much advanced technology which he explained. Many modern humans limit the minds to the borders of their own proudness, and they forget that there is much more useful intelligent advanced technology which has already been worked by vedic civilization. Most of them are not ready to accept the truths because they are not ready to know or explore about them. We cannot conclude on something which we don't know, in fact we need to explore, learn, work on it and then try to understand it. Why would a discussion on much advanced technology be an insult to modern science??

We need to discuss the advanced ancient technology and work on it to make the modern into a better one than the existing low quality technology which pollute the space and earth environment. I too faced similar incident in my speech at international space conference, My paper was on Mercury ion engine of ancient aeronautics. There were scientists and audience who cannot understand highly advanced vedic technology, Similar to Bodas, I too felt difficult to explain, because, the people have limited their minds and are not ready to absorb any high technology information. I completely agree to the statement which Bodas claimed, that engineers need to take the ancient texts more seriously as technical manuals, as it would lead to improvements in today's aircraft. That is 100% correct statement.

Vimana shasthra is much advanced than modern technology. My observation on ancient technology is that it is advanced and more than 100% efficient, free energy using, and which never harm or disturb nature in any manner. Ancient ancestors had 12 strand DNA, they used all parts of their brain, hence had more intelligence than modern humans. Sanskrit documents contain advanced science and technology in them, which are documented by ancient ancestors. In the process of giving their valuable information to the next generations of human race, Maharshi Bharadwaja and several other ancient scientists or Rishis provided us Texts like Vimana shasthra.

I am happy that I am one among the researchers who did attempt to understand Vimana shasthra in a scientific manner for the first time in history.

The Sanskrit shlokas, I have retranslated, there are multiple meanings for single words in this divine language, and I choose technical meaning as per the aeronautical engineering. An assembly of all technical meanings of the words in shloka results in a wonderful message which reveals treasures of vast ancient advanced aerospace technology.

Pilot instructions are very interesting and describe something out of the world, as it is too advanced technology difficult to be understood till modern human civilization tries to approach it by decoding the mysteries. The re-translations, interpretations, 3d modeling design, and analysis I did on Vimana shasthra are performed as per my aeronautical engineering knowledge. As the works are leading to many discoveries, I am continuing to decode the vimana shasthra, and also trying to implement discovered things in practical manner with the help of scientists.

The vimana shasthra descriptions I worked on, and found that it describes wide variety of choices for a person to make his vimana, there are many yantras or devices mentioned, and the person can choose which device he needs in his vimana as per his needs of mission. Some of the interesting choices in devices is that There are time travelling devices, devices which observe humans from a certain point in space, and some devices descriptions are similar to the UFOs which we see in modern days.

The other Vimanas described in detail are Rukma vimana, Sundara vimana, and Shakuna vimana. Each vimana has its special features. Rukma vimana has 8 propellers, electromagnets, mica pillars, and mercury is used in its manufacturing material RajaLoha. Shakuna vimana has expanding wings flapping wing, rudder, and propeller and looks similar to a huge bird. Sundara Vimana is much like a Space Shuttle with solid rocket boosters. Shakuna vimana and Sundara vimana are also made of RajaLoha. RajaLoha is used to these vimana for overcoming problems faced in space travel.
RajaLoha protects from all the space environment problems. Tripura vimana is three way usable vehicle, a hybrid of submarine, terrestrial vehicle, airship and spaceship. Tripura vimana is made of Trinetra loha, and it is flexible material made of three metals mainly. This material helps the vimana operate inside water too. As a conclusion, all Vimanas are interplanetary vedic space vehicles using advanced propulsion among the 500 principles of flight.

We should stop judging vimana shasthra by its cover. We need to go deep by decoding shlokas, and by understanding the principles it describes. Relevant information are contained in the Vedic and ancient Sanskrit texts such as Ayurveda - Charaka samhitha, Susrutha Samhitha, Aviation -

Vaimanika Shastra and Yantra sarvaswa of Bharadwaja Maharishi and Atomic science also.

Many Ancient Sanskrit texts are connected to one another. Mainly Medicine, aviation and atomic science. The vimana war mechanisms, poisonous chemicals etc, describe atomic science and medicine.

I also interpret that humans are a programmed machines made by vedic scientists. In humans, all organs are devices and have specific program to make and control these. As per vimana shasthra, the one who make the vimana successfully, will reach the god. And in another page, its written that the vimana is made in reference to humans. So, I interpret that the one who decodes the program of human devices will also be able to make vimana devices with guidance of vimana shasthra, and reaching the god means that we reach the creator, which might be an advanced extraterrestrial divine being or energy. The God, or source of energy which is within ourselves, the soul.

The Sanskrit texts itself is like a code language revealing that the Vedic texts are unique amongst world scriptures in the volume, depth, and sophistication of their understanding of math, science, and technology, etc. much of this knowledge is preserved in the Vedic texts that is still relevant today.

Vimana shasthra describes human evolution, there are 4 yugas, and now we are in the final yuga as per their descriptions.

The first Yuga of humans had super human abilities, the program made by the creator/god was working fine, and perfect. They can fly by themselves without machines. The later Yuga humans need machines to fly. The second yuga was a disturbed one, due to many negative thoughts of brain device inside human. And the program was showing some errors which make the human lose some super human abilities. The third Yuga was a lot more disturbed program and lost most of abilities. The final or fourth Yuga is almost zero control with the program and near to the destruction of the human species among themselves.

In vimana shasthra, the main reason for these changes is described as righteousness. Human brain consciousness wisdom, peace and care towards nature, and universe matters a lot when we consider technology too. We can see some humans and sages who fly without machines in modern days, it means that their program is undisturbed and pure perfect. All humans need not be perfect, But at the same time there are majority of humans who fly with the machines which pollute the environment and harm the nature.

We need to realize this fact and start working on vimana shasthra, an amazing guide by ancient ancestors to make our technology eco friendly and to save human species by breathing pure air to stop mind pollution and also should stop human program in activating self-destruction.

Vedic texts can help us to rediscover ancient knowledge that will lead to future advancements in technology. Maharshi Bharadwaja's vimana shasthra is a very important book and hint for modern humans to advance our space technology and to protect our human race from being attacked by aliens. Vishwakarma is a Hindu god, a Principal Architect of the Universe. He is the one responsible for vimana dimensions and humans also said to have specific dimension design as per this god. I interpret this god as an energy of natural dimensional design. I am lucky and blessed to able to design 3D models with divine dimensions of vimana, given by Vedic civilization people.
With the help of scientists I am trying to implement vimana shasthra project in practical manner which may lead to highly advanced space technology inventions from the Vedic texts.

Appendix – 1

New Sanskrit Translations in Vimana shasthra:

Aerial Routes

- Bodhananda = bodha+nanda= knowledge+prosperity. Vritti=profession/job
- Shownaka = su+unaka= easy/rightly+inferior/removing,
- Rekhaapathha = Rekha+apatha = range/line+roadless/pathless/deviation,
- Mandala = circular/Globe, Kakshya=zone or Orbit of planet, Shakti, and Kendra = first/center/tenth lunar mansion.
- Bhooloka= Earth world; Buva=atmosphere,
- Suvarlokasuvar=the sun/light,
- Maholoka = great meteor/great firebrand/lightning,
- Janoloka= living being/creature/race/ Knowledge meteor, Tapoloka; tapa=indra and
- Satyaloka. Satya=pure/genuine/divine

Aeroplane parts
- Vishwakriyaadarpana = universal functioning mirror
- ChaayaapurushaShaastra"subject of learning from trees/nature

Manufacturing

Mantrika

- Maantrika: Chanting sounds
- Chhinnamasta: To Remove Obstacles
- Bhairavee: achieving abilities
- Veginee: rapid
- Siddhaamba: production power
- Ghutikaa: basic things
- paadukaa:remaining parts
- Bhuvaneswaree Mantra: attraction for materialistic things

Tantrika

- Mahaamaaya: complications
- Shambara:Great power
- taantric : Technique
- endow: provide

Kriraka

- Kritaka: Brought from natural parents (Inspired by nature) or artificial
- Architects: Engineers
- Vishwakarma: vishwameans- hurtful, mischieveous, injurious. It also means universe or universal. karmameans action , performance

- Chhaayaaparusha: Chaya means reflection, parusha means trees
- Mann: Thought process, brain electromagnetic waves. Heart means supply of blood fuel. Head means monitor
- Maya: Tricks illusions traps

Flight Mechanics, Tactical Air Defense and Propulsion
- Antaraala: intermediate space
- Goodha: deep mystery
- 'Vaayutatva: air/gas facts truths
- Yaasaa: air travel
- Viyaasaa: escape velocity travel
- Prayaasaa: space travel
- Vishwa-Kriyaa-darapana: universal operation mirror
- Adrishya: invisible
- Shaktitantra: shakthi means Energy/ power. tantra means having wires
- Vynarathya Vikarana: Vyna means various elements in atmospheric layers.
  rathya means taking advantage for flying chariot. Vikarana means modification/ utilization.
- Ethereal: Unearthly
- balaahaavikarana Shakti: balaha means thick / dense. vikarana means modification. Shakthi means force/power/energy
- Paroksha = Invisible/ absent
- Meghotpatthi-prakarana = energy from clouds study material
- Aparoksha = not invisible
- Shakthi tantra = electricity through wires
- Rohinee = Brightest
- Yantraangopasamhaara = Yantra + anga + samhara = machine/device + part + Manual
- Vistrita = Elaborate/Spread
- Akaashatantra = aakasha + Tantra = Sky/ether + technique
- Akaashatantra = Ether technology
- Ganita = mathematics
- Dhooma = smoke/vapour/fume
- Prakarana = subject
- padmaka + chakra = lotus flower like + disk/wheel
- bhyravee =   frightful / horrible / terrible
- Vyroopa = different/unique
- Roopantara = Transforming shape
- Tyla = oil
- Suroopa = beautiful shape
- Karaka = hailstones/ water force , it also means Operator
- pushpinee-pinjula = flowers-stalks

- suragha = strong/powerful/robust
- Samga = intercourse; naana= various/ different
- Jyotirbhaava = Glowing property
- amshubodhinee = atomic theory,
- mayookha = flame/ray of light/brightness
- ethereal = unearthly
- guhaa-garbha mirror = Concave mirror
- Tamomaya = Darkness trap
- Thamo-Yantra = Darkness force capturing device
- Vimukha = Facing opposite party vimanas
- Hridaya = heart
- Taara = Radiance
- Mahaashabda = Ultra-sound
- Vimohana = art of confusing
- Shabda prakaashikaa = Book of sound and light
- Langhana = Defense/offense
- Vaayu tattva Prakarana = Air properties subject
- baadaba = friendly or very close
- Saarpa = Snake-like
- Gamana = movement
- Dandavaktra = Elephant-like
- Chaapala = Wavering/unsteady
- Sarvatho = all sides
- Mukha = Facing
- Para = Foreign/alien vimana
- shabda = sound
- Graahaka = observing
- Sowdaaminee = listening
- Kalaa = Art
- Yantra = device
- Roopa = Visual view
- Akarshana = attraction
- Kriyaa = Action
- Grahana = observing
- Thrisheersha = Three-Headed
- Dik = Direction
- Pradarshana = display
- dishaampati yantra = direction monitor device
- Aakaasha = Sky like
- Akaara = appearance
- Aakaasha-tantra = Sky-techniques or sky-devices
- Bhoonaaga = earth-snake
- Jalada = Cloud like

- Roopa = appearance
- Bilva = wood apple
- Granthika = resin of Guggul tree
- Gugul = Balm
- Stabdhaka = Stunner or making immobile
- Apasmaara = confusion of mind
- stambhana-yantra = device which makes immobile
- Karshana = Destruction
- Jwaalinee shakit = fire power
- Vyshwaanara-naala = injurious tube

References

Main sources:

1. Brihad Vimana shasthra by Maharshi Bharadwaja
2. G.R. Josyer, "Vimana shasthra English translation" book, 1973.
3. Sanskrit Dictionary
4. Rukma Vimana Design and Analysis, Proceedings of IRF, International Conference, ISBN: 978-93-85973-93-2
5. Rukma Vimana Prototype, IAARHIES 12th International Conference on Engineering, Technology and Computer Science ICETCS – 2016. ISBN: 978-81-925978-6-7
6. Puratana Aakasha-Yantrika Nirmana Sadhanavasthu (Ancient Aero-mechanical manufacturing materials), International Conference on Advancements in Aeromechanical Materials for Manufacturing (ICAAMM-2016). Elsevier
7. Mercury Ion Engine of Ancient Aeronautics, 2nd International Space Conference, 2015. ISBN: 978-81-8011-2232
8. Research on Rukma Vimana, 2nd International Conference on Recent Advances in Design,development and operation of micro air vehicles (ICRAMAV) ISBN No: 9789351071693
9. Study of Parallel Universe Time Travel and Missing Aircrafts, International Journal of Engineering and Innovative Technology (IJEIT) Volume 4, Issue 4, October 2014 ISSN: 2277-3754
10. A Study & Brief Research on Rukma Vimana by using Deciphered Materials, Innovations in Aeronautical Engineering - NCIAE-2015
11. SWASTIK Team website: http://www.swastik9.org/kavya-vaddadi.html

3D and 4D Printing chapter:

1) Additive Manufacturing Technologies: Rapid Prototyping to Direct Digital Manufacturing by Ian Gibson, Brent Stucker and David W.Rosen, Springer, 2015.
2) https://image-store.slidesharecdn.com/bd09fbfa-3a42-471f-ab4f-b2ec8c053066-large.jpeg.
3) http://processindustryforum.com/wp-content/uploads/2014/04/CustomPartNet.png
4) http://www.stratasys.com/materials/fdm/absplus
5) http://www.lboro.ac.uk/research/amrg/about/the7categoriesofadditivemanufacturing/
6) https://www.3printr.com/file/2015/03/handbag.jpg
7) http://www.molecule.ink/img/thumbs/mjp-multijet-printing.jpg
8) http://www.star-prototype.com/blog/top-7-methods-for-making-3D-rapid-prototypes/
9) http://www.cczwei.de/index.php?id=news&newsid=2815
10) https://www.youtube.com/watch?v=wRXymDoYoWQ
11) https://www.utwente.nl/ctw/opm/research/design_engineering/rm/additive%20manufacturing/overview-of-additive-manufacturing-processes/
12) https://www.youtube.com/watch?v=7kbMyt3IM3Y

Antigravity chapter:

1) Aristotle. *Physics.* Translated by R.P. Hardie and R.K. Gaye, (350 BCE). http://classics.mit.edu//Aristotle/physics.html.

2) Maudlin, Tim. *Philosophy of Physics: Space and Time: Space and Time* (Princeton Foundations of Contemporary Philosophy). Princeton University Press. Kindle Edition. (2012-07-22).

3) Newton, Isaac, *Philosophiae Naturalis Principia Mathematica* ("*Mathematical Principles of Natural Philosophy*"), London, 1687; Cambridge, 1713; London, 1726. (Pirated versions of the 1713 edition were also published in Amsterdam in 1714 and 1723.)

4) Linton, Christopher M. *From Eudoxus to Einstein—A History of Mathematical Astronomy.* Cambridge: Cambridge University Press. p. 225, (2004). ISBN 978-0-521-82750-8.

5) Hofmann-Wellenhof, B.; Moritz, H. *Physical Geodesy* (2nd ed.). Springer, (2006). ISBN 978-3-211-33544-4.

6) List, R. J. editor. Acceleration of Gravity, *Smithsonian Meteorological Tables*, Sixth Ed. Smithsonian Institution, Washington, D.C., p. 68, (1968).

7) *U.S. Standard Atmosphere*, 1976, U.S. Government Printing Office, Washington, D.C., (1976).

8) http://instituteforenergyresearch.org/topics/encyclopedia/fossil-fuels/

9) Characteristics of Petroleum Products Stored and Dispensed (PDF), Petroleum Products Division – GN.

10) http://www.esa.int/Education/Solid_and_liquid_fuel_rockets

11) https://www.nasa.gov/mission_pages/newhorizons/main/index.html

12) http://www.world-nuclear.org/information-library/nuclear-fuel-cycle/nuclear-power-reactors/nuclear-power-reactors.aspx

13) http://www.forbes.com/sites/energysource/2012/02/29/thorium-nuclear-power-a-lesson-from-norway/#78119737187f

14) https://www.euronuclear.org/info/encyclopedia/alphaparticle.htm

15) Briggs, G. H. *The Relative Velocities of the Alpha-Particles from Thorium X and Its Products and from Radium.* 3 March, (1933).

16) The Institution of Engineering and Technology. *The Timeline of Radioactive Decay for Uranium^{238} (U^{238}).* The IET 2008

www.theiet.org/factfiles.

17) https://www.euronuclear.org/info/encyclopedia/betaradiation.htm

18) Rutherford, E. *Comptes Rendus*. Uranium Radiation and the Electrical Conduction produced by it. Philosophical Magazine Series 5. Vol. 47, Iss. 284, (1899).

19) http://www.alpharubicon.com/basicnbc/article16radiological71.htm

20) https://www.euronuclear.org/info/encyclopedia/g/gammaradiation.htm

21) http://www.alpharubicon.com/basicnbc/RadiationPenetration.htm

22) Tajmar, M. and de Matos, C.J. *Gravitomagnetic Field of a Rotating Superconductor and of a Rotating Superfluid*. PACS: 04.80.Cc, 04.25.Nx, 74.90.+n

23) Znidarsic, F. *The control of natural forces*. Infinite Energy. Sept/Oct. Issue 87, (2009).

24) AleinScientist. The Transitional Quantum State of Matter. Nov 9, (2010).

25) de Broglie, Louis Duc. PhD thesis, "Recherce sur la Theorie des Quanta", U. of Paris, (1924).

26) Episode - Voices of the Gods, Ancient Aliens show, History Channel.

HISTORICAL PART

Historical Part

By Enrico Baccarini

Introduction

The history of human civilization is a wide territory still completely unexplored, a space in which archaeology, science and research have attempted unveil to the eyes of men the origins and the events that are related to the evolution of the various civilizations.

This path, although we have tried to make it linear and coherent, it is actually a boundless territory dotted with gaps and certainties, safeties and doubts.

History is generally defined as the past that affects the present and influences the future, a past that has yet to be understood and to be built and in which the truths of today may be substituted, sometimes, and new objective realities are able to change the paradigm previously established.

Officially we use this word considering the period of time that begins with the appearance of the writing and the first written sources are placed about around 3500 BC, while the prior period is arbitrarily defined 'prehistory'. However, there is a forgotten past where lost knowledge are residing, a truth that could profoundly and radically change the concept of time which has gone before us, discoveries that open up to new scenarios and a cognitive interpretation of the civilization of our planet.

Prehistory was absolutely not a dark age in which man survived away from the ideas and civilization. The 'cavemen', in a constant struggle for survival, formed a glimpse of this period but before them, there were existed places like the settlement of Gobekli Tepe (Turkey), which they were dated back to 12,000 years, and they showed a stunning level of evolution.

It is becoming wider and wider a totally different scenario compared to what theorized until today where, next to the simplicity of the so-called caveman, islands of progress and civilization and highly evolved technology were joined, which constituted the primary cause of our progress and development but due to some disastrous climate change were destroyed and fell into total oblivion of history.

However, not everything was lost and, according to the most ancient legends of our planet, the survivors were able to re-create new realities in many places which would become the first known civilizations in the history, some people who had shared the memory of the disaster and the floods which had decimated their race.

In modern times, the concept of legend or myth has become a reality almost unintelligible, often filtered through a fantastic vision that has totally detached each objective and historical adherence from its original meaning projecting their real matrices into a limbo of confused abstraction. All this when the man of the XX century stood like as the largest creator of the myths of the history through the cinema, the fiction or the science fiction.

The modern legends form a screen through which we can alienate ourselves from the surrounding world, a way to escape to retreat when we are scared or bored. For the peoples of antiquity the myths formed, however, a fundamental point of reference, in many cases the very foundation of their inner world and their civilization.

Myths acquitted not only social or educational but also documental or historical roles and working as a tool to preserve knowledge and celebrate past events.

Undoubtedly, even if in the past time the legends suffered a transposition that re-drew their boundaries and structures, their bases were originated by the need to ensure the memories of past events for long time.

With the word 'history' we have identified a modern perception of a gradual evolution of the events, a linearity that had its origins in the ideas of Heraclitus (V century BC) and Giambattista Vico (XVII century AD) and that it was taken, with their formulation of historical courses and recourses, as inviolable model by modern historiography. In philosophy, the linear thinking is characterized by complex thought because it attempts to explain the world according to predetermined logic related to the principles of causality, of non-contradiction, of hierarchy and order, and in which every event is reduced, simplified and categorized.

Our entire modern conception of science and history is based on linear thinking with theories that are anchored to the processes of cause and effect and in which all, or nearly so, proceed logically. The most modern scientific formulations, such as quantum theory, or the historical and archaeological evidence arise in recent decades like the site of Gobekli Tepe, however, they have begun to undermine this established order by showing how some anomalies exist, and have existed, reality that strongly disrupt many of our certainties by projecting the knowledge gained to date within a new vision of reality.

Human history is part of a lineage still fully to discover and decipher.

Chapter I – Vimanas, God's flying cars

'Our learning *is nothing else than to remember'*

Platone, Fedone, 72 E

The myth is a narration that is invested by a sacral connotation. On the different traditions it is often referred to the origin of the world or to the way how the living people had reached the present form. Every people had developed his myths trough some descriptions where the most important figures were or divinities or heroes.

Often the stories narrated have their beginning in a time before the written story, during a legendary time, also representing some instruments of foundation of a specific culture or of a nation.

The myth own also an humanizing function, where the sacral dimension contained in it is transformed into a fundamental moment of the religious experience made to satisfy the need to give an explanation to the natural phenomenon or to the questions of the human existence.

The thunderbolt is transformed into Zeus's arm, the seaquake into Poseidon's fury while the fertilizing floods of the Nile into the demonstration of the benevolence of the Gods toward the people that had propitiated their magnanimity.

In every culture and in every tradition the myth take these connotations, characterizing as the different cultural forms as also as the "evolutions" made by the story itself. India is all that but also something more and something different. Inside his epic sagas about love and about wars, about conquests and about discoveries, it seems to come out some references to a superior knowledge, to a hidden reality, joined inside them as an element of union between a forgotten history and a present that we have still to discover.

The Vimana seem to absolve to this roll, almost always forming a corollary to the narrated events but also as a fundamental roll into the understanding of the myths themselves.

The term Vimana derives from the Sanskrit and it means a mythological flying object often identified as ' Gods's car' used by the divinities to move into the sky and to fight their battles.

The Vimana are described on several old Indian religious texts or on epic texts as the Mahabharata and the Ramayana.

From an etymological analysis the word is deriving from the union of the Sanskrit words *vi* (literally " bird", or generically something that is flying) and the word *mana* (that means something of artificial and generally that is lived)[1].

Another more modern etymology[2], reads in the word Vimana the meaning of "to across" that in a broad sense it is transforming into "*to across the sky*", being for definition an object that is moving upon the word.

Among the many interesting elements find into the holy texts, and into the myths, that mark the Vimana we find their manifold capacity to fly into the air, into the spaces, such as to plunge under the water.

Their mythological dimension is changed and translittered into the centuries their meaning, the original word was used in a religious acceptation generally ascribing to a "*delimited area assigned to holy purposes*", the sancta sanctorum of an Hindu temple, the *place* where the divinity was used to manifest.

Vimana, in the Tamil Nadu, a region on the South of India, is also the term with which is called during the centuries the summit of a temple, that one closer to the sky and to the divinities, the Deva, the point of conjunction among the spiritual and the material. This term in the modern language has taken many other meanings. Usually it define a palace of an emperor or of a monarch, as also the place of the entrance to a temple. This word also assume the general feature of a temple, or of a sanctuary with a particular shape known as *Vimanam's Tower*.

From this last exception, namely "the palace of a God", into the Ramayanait it is mentioned the Pushpaka Vimana or Ravana's flying palace/temple. In his original meaning the term Vimana is used as a '*cart of the Gods*', implying every closed way that is moving in the air (sometime as a throne, others as a house or a palace, others as a real cart moved by some animals).

In the ambit of the Indus medical this word had been also used to indicate the "*science of the measurement of the right proportion*", or the right equilibrium of the liquids into the body. Also if it is not well known the Vimana are not an unique property of the Hindu but are also named on the Buddhism.

The *Vimanavatthu* is a text used as a source of inspiration for the sermons where we talk about the flying residences of the Gods and how a right behavior could make the pupils to visit them. In some modern Indian dialects the word *Vimana* or *viman* means "aero-mobile"; some air companies are existed with this name while a suburb of the town is called *Vimanapura* and indicates the buildings and the areas near the airport! As we can see in all his different, and often modern, attributions the term Vimana implying always a

---

[1] As explained by D.W. Davenport in 1979, in his book *2000 BC: atomic destruction*, (SugarCo) where he provides the same interpretation from the English translation of the Vaimanika Shastra.

[2] Monier-Williams, *Sanskrit-English Dictionary*, version 0.1a_12

flying object or 'something' that let us to relate or to support to the divinities and to the possibility to move on the sky.
However the destructive passion of the modern skepticism aims to demystify such limits and researches, succeeding only superficially, in reality we are in front of some contexts extremely complex whose origin is losing on the most remote antiquity of the Indian civilization.

Ratha and Vimana, the old flying cars

The founding element from which to start our analysis is first of all the research of the old texts, to study the sources and to inquire on those elements that seem overbearingly to come out from their pages. On this case we find us in front of a very delicate contest where it seems to permeate and to amalgamate several elements tied so much to the history and to the science of the old India as to his legends and to his epic myths.
The researching of an historical adherence, or of a possible factual reality, will be the following step of our way. If from one side a remind appear through tangibly to something old, to a sensation of glamour and surprise that caught during several years the scholars, to understand all that has surely to pass through a rigorous search of objective basis to which we may appeal and to which we may recall. Fundamentally in the modern time there are three studies that started to the discovery of this mystery of the Indian past.
We saw the first study on a text dedicated to the art of the Indian war made by an Oxford teacher, V.R.R. Dikshitar, issued on 1.944. The second in a work appeared on February 1.952 and published by the Indian Institute of Culture, signed by V. Raghavan, while the third in a book of the 1.985 by the teacher Dileep Kumar Kanjilal.
Three academic teachers that, counter current and publically, manifested their own opinions on the forgotten past of their country. Their example was followed by the following generations, also if in the occident such thematic and the studies made have received almost always the indifference by the scientific world as much as by the simply curious people.
Shri V.R.R. Dikshitar publishes[3] on 1.944 an essential text to the understanding of the Indian past, *War in Ancient India*. Into the text is clearly demonstrated the knowledge of a big wisdom and cultural heritage inside the oldest Indian operas but dispersed and fragmented along the time as also filtered and distillate through a religious and epic vision.

---

[3] Shri V.R.R. Dikshitar, *War in Ancient India*, Macmillan, Londra, India, 1944, p. 284.

Dikshitar went over the simple transcription of the texts, he translated and analyzed the Vedic content in a literal way discovering, for example, as the described flying cars were able to reach the Surya-mandala, or the Sun's Region, or to arrive till the Naksatra-mandala, Star's Region.

"*No question could be today more interesting than the Indian contribution to the aeronautic sciences. There exist several explications inside our wide epic and Puranic literature that shows how wonderfully the old Indians had won the air. With unconstraint what we may find in this literature had been characterized as imaginary and briefly liquidated as unreal, a way used by the occidental and oriental scholars till a few years ago. In fact such arguments had been ridiculed and the people durst to state that it was physically impossible for the men to use the flying cars. But today (the temporal references are at 1.944, ndr) a big change has happened to our way of thinking thanks to the use of the hot-air balloon, airplanes and some others flying cars*". The scholar proceed asserting "*...the flying Rama's or Ravana Vimana has been considered as a dream from the mythographers up to the airplanes and the dirigibles of this century haven't see the light of the day*".

Dikshitar was one of the first to evidence how in the Vedic religious literature, into the Brahmanas[4], the concept of ' ships that were sailing toward the sky" was present. "*The ship is the Agniliotra which fire Ahavaniya and Garhapatya are representing the two faces tied to the sky, and the steersman is Agnihotrin offering some milk to the three Agnis. Also, in the first part of the Rg Veda samhita, we read that the Asvini freed Bhujya in safety thanks to some winged ships. This last datum could be referring to the aerial navigation in the ancient times*"[5].

Instead commenting a text of architecture of the XIII century, on which we will back afterwards, Dikshitar asserts: "*Into Bhoja's Samarangana Sutradhara, a whole chapter of about 230 rooms is dedicated to the principle constructive on which are based the several flying cars as well as to the motors used for military use or for something else. The advantages of using these machineries, particularly those flying, are noted in a very rich way. A special mention is done because of their visible attacks so as the use of invisible objects... of their movements without interruptions, of their resistance and their duration, in conclusion about their capacity to do on the air all that is happening on the earth*".

These are only a few of the starting points that the studious developed in their analysis about the old Indian knowledge, a first effort to recover a forgotten patrimony but always present and silent from too many centuries.

The studious V. Raghavan[6] went on Dikshitar's trail, he was the responsible of the Sanskrit Department of Madras's University. On his text *Yantras or*

---

[4] The Brahmanas constitute the second part of the Vedas and represent a vast literature of content almost solely focused on the ritualistic sacrifice regarded as the supreme end. The main Brahmanas are the Aitareyabrahmana and Satapathabrahmana.

[5] Agniliotra is also nominated in the section CXXXVI del Mahabharata vol. 3.

[6] V. Raghavan (M.A., Ph.D.), *Yantras or Mechanical Contrivances in Ancient India*, Transaction No. 10, The Indian Institute of Culture, Basavangudi, Bangalore, Febraury 1952.

Mechanical Contrivances in Ancient India he made to come out some elements of inestimable value, contextualizing a long time unexplored reality and surely in antithesis to the stereotyped vision existed till that time.

Through a detailed inquiry on the historical and literally sources Raghavan made a first study about the *Yantra*, or some technological machines used in the antiquity as for military uses as for amusing functions.

The reality that emerges, shows to us the existence of a technology able to create some apparatus of war highly sophisticated in which for their use it resulted fundamental the presence of a substance that had till to be invented in China, the powder of shot.

As much interesting is the reference to a Ratha-*musala,* a fighting truck that was moving *'without using the horses'* that it was sent against the enemies to rout their army. Among the other references recovered from Raghavan we also find the existence of some machines created by the wise Lalitalaya, some 'mechanical men' used during the public exhibitions to enjoy the nobles and the people. Also in this case the Vimana deserve a place of honor because they are described and analyzed into their different typologies and peculiarities.

A much more detailed, systematic and rigorous study made on the Vimana did not show up till the 1.985 when the text *Vimana in Ancient India* by the Professor Dileep Kumar Kanjilal[7] was published.

The interest of the studious for this thematic born when Kanjilal started to took an interest on the history of the Indian idolatry, he analyzed two fundamental texts the Kausitaki and the Satapatha Brahmana, datable some decennia before the 500 B.C.

The two texts in their two different parts relate us, among several things, about the physical form and about the performance of the Indians divinities showing them, out from every mythological and religious transposition, as some corporeal and physical being.

Being a matter from mythological and literary sources it was natural to consider these references as own to a metaphoric physique.

Kanjilal going on with his researches was always facing to something different, still more tangible, there were some quotations about aircrafts that the divinities themselves used to move and to displace in the atmosphere, in these references seemed to be present some concrete comparisons with some realistic technologies and not with a mere mythographyc inventions.

Kanjilal consulting the oldest texts observed as into the *Yujurveda* it was clearly spoken about some flying cars used as *Asvini*, the two celestial twins. Moving from the Ramayana to the Mahabharata and to the Bhagavata Purana, just to mention a few, Kanjilal found several references and descriptions that seemed to own a their expositive and technical coherence.

---

[7] Dileep Kumar Kanjilal, *Vimana dell'India antica*, Bhandar Pustak, Calcutta, 1985. A relation on the same subject was held in 1979 at the VI° Congress of the Ancient Astronaut Society (AAS) in Monaco.

As occidentals we are used to see represented in our mythologies some divinities using their power to move without needs of flying 'carts' or flying 'machines'.

Instead into the Indian literature the Gods needed very often these flying cars. Kanjilal made curious by this anomaly deepened this theme, he found himself in front of a huge field of information that had been unnoticed from the occidental scholars as from a good number of Indians. In this case, the will to find an historical reality concerning the facts up said, was considered from the majority of the Hindu world, till recent times, as a real blaspheme.

A very suggestive scenery it is delineating from such facts where it is necessary a research from all the available sources that are mentioning the 'divine chariots'.

We have found not less than sixteen old indian texts that describe, with a degree more or less elevate, the mythical Vimana. There are some texts entirely dedicated to this thematic as some entire chapters of encyclopedic works till to arrive to some short paragraphs or quotations that talk or describe these celestial carts.

There is here a list:

- Vaimanika Shastra
- Samarangana Sutradhara
- Yuktikalpatani di Bohja
- Rig Veda
- Yajur Veda
- Atharva Veda
- Ramayana
- Mahabharata
- Puranas
- Bhagaravata
- Avadhana
- Kathasaritsagara
- Raghuvamsa
- Abhijnanasakuntalam of Kalidasa
- Abimaraka of Bhasa
- Jatalas

Historically and literally the first reference to an aircraft to be able to move through the air, with inside some living or divine beings, is found on the oldest religious text of the humanity, the RigVeda[8].

---

[8] Rv 1.25.6

At the beginning such aircraft is not called with the name of Vimana but with the less known word *Ratha*, literally 'cart'. It is related in this text how the Rbhus had built a celestial cart for the twins Asvini, doctors among the Gods[9]. Instead on another section[10] he told us about an aircraft able to shake the houses, to pull out small plants and to cause a big wind at its passage[11].

The term *Vimana* appear for the first time in the Yajur Veda (YV 17.59) and it is surely referring to an aircraft to be able to move in the air. There are several texts where are named these flying aircrafts, in all the time they are referring to some aircrafts able to fly into the sky, to bring passengers, to lead some battles, to make long travels, and in every cases they were characterized by different but specific peculiarity and shapes[12].

Yajur Veda's other passage is more explicit to circumscribe a technology and some figures extremely specific, *"…O royal skilled engineer, construct sea-boats, propelled on water by our experts, and airplanes, moving and flying upward, after the clouds that reside in the mid-region, that fly as the boats move on the sea, that fly high over and below the watery clouds. Be thou, thereby, prosperous in this world created by the Omnipresent God, and flier in both air and lightning"*[13].

The fundamental difference that seems to submit between the use of the term Ratha and those one of Vimana it is in its conformation. With the term Ratha we usually state the 'flying cars' used only by the Gods and usually without wings instead the term Vimana is referring to some aircrafts quite always with wings and used or by some divinities but above all by the human beings[14].

Afterwards we will back on this substantial difference. Some referring to these terms have been found inside one of the oldest dictionaries on Sanskrit language, some of them are made on a time before the I century B.C., where these objects are described as *'flying cars'* or also *'some buildings projected as the specimen of the flying cars'*.

In the systematic analysis of the Indian texts we may find a substantial discrepancy towards all the mythologies on our world.

Gods's flying cars are not identified as a generic means of transportation through which it was possible to move themselves, but they are subdivide

---

[9] Rv 1.111.1

[10] Rv 1.166.4-5

[11] Rv 1.116.3, Rv 7.4.68

[12] TS 4.63.3; MS 2.10.5 e 137.11; KS 18.3; CB 9.28.17.

[13] Yajur Veda, 10.19.

[14] Vd. Kanjilal, op. cit. p.13

concerning the specific feelings and categories, functions and purposes. In this connection, in the Rig Veda we may find some unique references on its kind enough enlightening.

The *Jalayan* is described as a car projected to move or in the air or in the water[15].
The *Kaara* is a car projected to move whether on the earth or in the water[16].
The *Tritala* is a car projected to move into the three elements (earth, air and water)[17].
The *Trichakra Ratha* instead is a car with three motors projected to move only into the air[18].
The *Vaayu Ratha* is a car moved by a motor by air[19].
The *Vidyut Ratha* after all is a car described as it is drive by an extremely powerful engine[20].
These references in their apparent simplicity show us a lower description extremely more complicated and articulated.
It is interesting to notice, also if it not well known, as a divinity had existed since the most remote Indian past and he was pointed out as the first and most important figure chosen to build the Vimana for the gods, Visvakarma the celestial architect, the divine engineer and artisan.
The Adi Parva, or Mahabharata's first book, describe him with these words "...*Viswakarma, the founder of all the arts. He was the founder of thousand arts, the engineer of the immortals, the founder of all the types of ornaments and the first among the artists. The one who has built the celestial cars of the gods, and the humanity can live thanks to the inventions of this illustrious man and for this he is adored by the men*"[21].
Also the old texts assert that its 'tradition', or knowledge, descended among the people[22] letting them 'to imitate' his inventions.

---

[15] Rig Veda 6.58.3

[16] Rig Veda 9.14.1

[17] Rig Veda 3.14.1

[18] Rig Veda 4.36.1

[19] Rig Veda 5.41.6

[20] Rig Veda 3.14.1

[21] Adiparva, Ch.66, vss. 26-31.

[22] Adiparva, Ch.66, ib.

A further reference is coming from the epic test Brihatkatha, of the VI century A.C.[23], where the Vimana had been called *Akasa-yantras,* or 'the car of the ether' instead their builders are identified as the Yavanas, this term indicated generically some people coming from the West on which real origin and nature had never been throw light on that subject[24].

In the text are narrated several episodes ascribed to some mythical reports where the Vimanas are as the most important element: where it was used a Akasa-yantra they ordered the maximum of secrecy about its characteristic and its components, the one who had revealed the secret was murdered.

We will remember such warning as an inalienable corollary in every test where is related about the flying cars (that it was epic, religious, or technical), everyone who had known their secret could have preserved at the cost of their life. Of the same opinion was the Samarangana Sutradhara, which after dedicating 230 verses and some instructions how to built these aircrafts, peremptory imparted the warning do not divulge that underlining as in the same test it had been exposed only a part of the constructive ways because they wanted to avoid that they could fell in the wrong hands.

Following our research we have also found that also Pali's test titled *Paramattha-jjhayotika,* dated VI century A.C., has some references how to built a wooden car which intents were also 'the aerial transport' for supplying with food and articles of clothing from a town near India's heart[25].

The traditional interpretation, also in this last case, is that is about some mythos created by the human being but it is curious to see how their uniqueness to any other known tradition, what it had been described seems incredibly to look like to the modern cargo aircrafts for the transport of some goods. If it is a myth, it is a *unicum* that has surely preceded the times!

As said before, in the *Rig Veda* the word Vimana does not appear but its presence is indicated through the oldest word of *Ratha* or flying chariots. We find in Veda's poetic transliteration others terms that will describe the Vimana,

---

[23] Maurice Winternitz, Moriz Winternitz, Subhadra Jha, *A History of Indian Literature,* Volume 3, Motilal Banarsidass Publ., 1986.

[24] The word "Yavanas", in the ancient Indian texts and especially Tamil, identified peoples or ethnic groups coming from the West. The use and contacts led initially to identify the term with the Greeks, 1probably from the corruption of the term "Ionia." This identification, however, was late and left unknown the true origins of an highly civilized population during pre-historic times that had deep contacts and influenced the civilization of the Vedas. A similar phoneme to indicate the Westerners, is also used in other Asian languages: Assiro = Iawanu; Persiano = Yanua ; Arabo e Turco = Junan ; Cinese Han = Ta-Juan.

[25] These references were identified by Professor Devaprasad Guha, an authority in the Pali language texts which had held the chair of Sanskrit literature at the Oriental Conference University of Santiniketan.

words that in their etymological essence tried to give a deeper meaning of what they were trying to describe, or some objects able to fly.

The *Purunishshidvana Dadhika* is an example among the several, literally *'aircraft that transport its occupants through the sky and among the planets'* showing us the purpose and the functionality of such cart. Instead into the *Budhaswmin Brihat Katha Shlokasamgraha* is described the landing of Vidyaharas's king, Naravashanadotta, in Uijayani's town near the Gange river, with a cart presided by his "celestial" to find Ipploha, who had raped the princess Surasamanjari.

On the I° book, hymn 164, of the Rig Veda we instead read: "*Obscure are the origin, these mechanical birds by the golden color, fly till the sky hidden into the water. Then they descend again from the "residence of the order"* (the sky, ndr) *and all the earth is soaked by their "capacity to fertilize"… Twelve are the segments of the wheel rims, but the wheel is only one; three are the nave. Which man could understand their working? The 360 rays* (of the wheel., ndr) *are regulated together, in a way that they could not be absolutely loosened*"[26].

On the VI° Book, hymn 58 devoted to Pusan, we read "*O Pusan, with those yours golden ships that are traveling on the ocean and on the air*", instead on the IV, hymn 36 entitled Rbhus we find written "*The car that was flying on the sky was not made for the horses or reins, but it was made by three wheels, deserving to be commended*".

Ayyavazhi's mythology speaks us about the *Vitthakalai* Kali's "*golden flying coach*".

In all these cases happen something of exceptional, the described cart does not own the valence of a mythic object but his description, and the same attribution that had been done, seems to correspond to something of concrete but especially that the old editor seems to have seen with his eyes and he is trying to understand and to transcribe the peculiarities and the astonishments to the posterities.

The most disconcerting thing it is that we are facing, at last in two cases, to semi-legendary wisemen, or Rishi, of the old India who handed down their own knowledge to realize these aircrafts.

The two wisemen are Agastya and Bharadwaja.

Other references to flying carts

Analyzing the *Rig Veda* we may find at least 20 passages, or at least 1.028 hymns to the gods, where they clearly quote the flying chariots utilized by the Asvini. This aircraft is represented "*with three floors, triangular and with three wheels*". According to the tradition it may bring at last three passengers and was made up of gold, silver and iron and had two wings. Through this aircraft the

---

[26] *Rig Veda*, translated by Ralph T.H. Griffith, 1896.

Asvini saved the king Bhujyu while he was missing into the sea in a dangerous situation.

We find another reference of big interest into the epic text Ramayana (Yuddha Kanda, chapter 125) where it is mentioned the *Pushpaka Vimana* and the modalities that preside its 'take off', "... *after Rama's order, the Vimana raised up with a big noise…*".

We clearly understand in this description how this 'mythic object' owned some unusual peculiarities that quaintly it seems to a modern jet plane. Following the reading of the book, Rama escorts on the Pushpaka Vimana his wife Sita to lead her flying towards Ayodhya's town.

It is surprising how Rama during his travel describe to his wife the flown over the lands by the Pushpaka identifying the places and the localities with an incredible accuracy.

Also it seems banal to us, this passage show us something of extremely important. From a terrestrial point of view it is enough simple to individuate a specific place, instead when we take an airplane and we observe the view under us everything seems to vanish and to homologate in a confused repeating of similar lands.

Only some pilots trained to fly, or some people accustomed to move into the sky, are able to sharpen their own perception and to recognize with no problems the flied over places.

In this case, Rama seems to show such reality with a so detailed level as the one who had become the owner of this art but more surprising is the fact that the places that he describe are corresponding exactly to that one that we can see doing the same way from an enough high quote.

Dileep Kanjilal has further on analyzed[27] the passages up said on the Ramayana observing them by a technical and scientific point of view. First of all Pushpaka Vimana seems to be able to house at last 12 people. His living from Lanka in the morning and his arrival to Ayodhya in the afternoon, with only two stops in Kiskindhya and Vasisthasrama, let us to calculate the total distance or 1.800 km in 9 hours. From a simple calculation we may estimate an average speed of 200 km/h observing the map of the way, and considering that the event described on the epic test happened before the wintertime, we see as the Pushpaka Vimana had followed a way that let him to avoid the monsoon clouds of the South of India.

It is a strange event if it was treated only of a myth where some so specific details would have surely resulted needless! Reading the Mahabharata that is considered from the Hinduists as the literally transposition about some facts really happened, we find some interesting references to the Vimana and to

---

[27] Kanjilal, op.cit., p. 16.

their incredible peculiarities. It is an example the acquisition of one of these aircrafts by Uparicara for Indra's intercession[28].

Into the fifth book of the Mahabharata, titled *Vimanapala* or ' the guardian of the aircraft'[29], it is indicated a high specialized person and deputed to take care of the Vimana. Of the same nature it is the reference to an old ' science of the arms for the war', and it was inclusive of some specific techniques to use into the four elements, and it was divided into ten branches and four *padas*[30].

More interesting it is the description of the travel made by Arjuna to the highest celestial spheres. The wise man during his travel was accompanied by Matali, he proceeds through several celestial lands, he goes beyond a portion of the firmament where there are the stars and he find himself in front hundreds of Vimana[31].

Some of these aircrafts are described while they are moving others instead are stopped and some others are taking off[32].

Here it is a brief extract *"Arjuna, the killer of the heroes after having greeted the mountains, he was shining as the sun, he raised on the celestial cart and the prince Kuru, gifted by a great intelligence, with a happy heart, he ran across the firmament on that celestial cart shining as the sun and full of extraordinary conquests. And after to have became invisible to the mortals of the Earth, he saw thousands of Carts of an extraordinary beauty. And on that region there was no sun, no moon, no fire to give the light, but it blazed into a light generated by his own ascetic merits. And those sparkling regions that we see from the Earth under the form of stars, as alight lamps (in the sky) – so small consequently of their distance, also if they are very big, Pandu's son saw them".*

On Vana Parva's same paragraph he spoke of Arjiuna's ascent to the celestial spheres, we find another reference very explicative about the Vimana from the god Indra.

"...Vaisampayana said: After that the Lokapala went away, Arjuna – the killer of all the enemies – started to think, or Monarch, to Indra's cart!.. the cart wrapped into of a big splendor and driven by Matali, arrived separating the clouds and shining in the sky and filling the whole celestial vault with his groan so dip as the thunder of the powerful masses of the clouds. Swords, and missiles from a terrible shape and maces from a horrible description, winged darts with a celestial splendor and lightning flash, and darts, and helix provided of wheels, and (the flying cart) *worked with the expansion of the atmosphere*

---

[28] Rv 1.161.3 e Rv 1.163.2

[29] Udyoga Parva, Ch. 118, vss. 19-21

[30] Mahabharata, B.O.R.I., Salyaparvan, Ch. 7

[31] Vana Parva, Ch. 42, vss 30-34.

[32] Mahabharata, B.O.R.I., Ed. Vanaparva, Cp. 43, vs 8-10, 28.

making some loud sound as the thunder of big mass of clouds, all of these was into that cart"33.

Every comment seems superfluous! Inside the text known as Sabha-parvan, or the second book of the Mahabharata, we may find instead some important references regarding some celestial beings generally called Deva. It is said explicitly that in the Krita Yuga the Gods usually used to go down to the earth with human features, "*eager to contemplate what could be seen here*" and they had shown to be so much curious to go around on our planet freely[34].

The recall is clear enough and needs no further explanation.

The reports that are describing the Vimana and are lingering on their technical characteristics could go on for many pages again, but we think that ' this brief hint' to the Indian literature is extremely exhaustive and that it could itself show the great differences from such stories if related to the occidental myths as to show the reader the total depth that these traditions seems to own.

Among the most recent texts that have analyzed Vimana's and Vymanika Shastra's mystery we find instead a book written by the Captain Anand Jayram Bodas, one of the most famous pilot of the Indian subcontinent with to his active more than 30 years as a pilot for the government, for some private lines, as a military pilot and as an instructor on some schools of flying.

During his carrier he also received several recognitions for his work of research and for the historical divulgation on the Indian aviation, among which the C.D. Deshmukh, an high decoration given to him by the government of Maharashtra's state for his book "*Vimangatha*"[35].

The text closer to our argument is the following book titled "*Prachin Bhartiya Vimanshastra*"[36], or "*The aeronautics of the old India*". Jayram Bodas taking the cue from the content of Shastry's text, has made some others analysis focalizing his attention on the technical and aviation aspects presented on his pages.

The result has been the confirmation of these elements but much more Bodas has confirmed as the technique of flying described into the Vymanika Shastra are extremely similar to those used on the modern times and also as the elements and technological knowledge described are really the product of a scientific and technological evolution and not the result of a literal invention. Something of extremely deep seems really to be on the bases of these very old traditions.

---

[33] Vana Parva, Ch. 42, op. cit.

[34] Ib., Sabhaparva, Cap. 11, vss.1-4.

[35] Literally '*On the various aspects of aviation*', Ananda Jayarama Bodasa, *Vimangatha*, Edition Navina avrtti, Imprint Mumbai Manorama Prakasana, 2004, DKMAR-1588.

[36] Literally '*On the history of aviation in India*', Ananda Jayarama Bodasa, *Pracina Bharatiya vimana sastra*, Edition Prathamavrtti, Imprint Mumbai: Manorama Prakasana, 2004, DKMAR-1582.

Not of a minor valor is the fleeting or sibylline technical hints that could be recovered and extrapolate from the texts themselves.

In their classical meaning we should interpret these histories as some narrations made to exalt the history of a people and to teach to their own generations the value and the virtue to go on in their own life. In this sense we will be leaded to think that every possible rational explication should be impugned from the legendary constructions of which they are parts and objects.

If the myth is ending where the reason is advancing, all that is coming out from the study of these tests will have necessarily to lead to a new taking up to a conscience on the heritage preserved inside these very old traditions.

The necessary elements to understand this historical enigma are already in our disposal, and following Marcel Proust's words perhaps *"the true travel of the discovering is not to find new lands but to have new ayes"* to discover and to understand them.

A Vimana for every era

On our researches and analyses of the tests that were treating about the Vimana, an opera among all seems to have interested the European studious in the last decenniums, a test written on the first years of the twentieth-century but it had arrived to the honors of the occidental world only on 1.973 when it had been made a partial translation from the Sanskrit to the English language.

The Vimanika Shastra, literally *Aeronautics Science's*, it is controversial, problematic, obscure, criticized book and source of strenuous debates and contestations it seems to have not yet pacified the eternal discussions that have inflamed it since its first apparition.

The test was dictated between the 1.918 and the 1.923 by the Pandit, from Sanskrit *Teacher*, Subbaraya Shastry was a person extremely respected and venerated who lived between the XIX and XX century.

His contents have been source of infinite disputes where on its pages were described some techniques of construction or of pilotage of the mythical Vimana.

We will deepen this subject on a special chapter but since now we feel important to clear that the reliability of this test is sure, at least on hits genuineness of the one who was the first bearer and especially because of some contests that happened from the Occidental word that never had really and seriously deepened the contests and the notions described into its pages limiting themselves to contest the truth only because it was not conceptually and ideologically possible to think that on the old India it could have been possible to built some flying cars.

During these years we have made some analysis, some researches, we have kept some epistolary contacts with scholars and Indian university teachers that have open to us some doors of a worlds till then unknown.
It is thanks to the help of the friend and films maker Diego D'Innocenzo, with who we have realized some documentaries for the RAI (National Italian Television), to let a further and fundamental jump to the acquisition and to the verification of the materials that otherwise should had remained in some dusty Indian desk.
The searching of Vimana's references and Vimanika Shastra's analysis[37] bring us to a further, curious, discovery.

In Bharadwaja's test, the author make a distinction on these flying carts for typology fixing them on an historical time temporally different and ascribing them some "propulsion systems" different for each time they had been built.

This "classification" and subdivision do not constitute a *unicum* and it could be found also on some others tests or old traditions, as for example that one of the wise man Lalitalaya[38].

Such situation testify the different use and also the evolution that these aircrafts could have had during the centuries and the millenniums and the different traditions that of this specification they become bearers.

In Vymanika Shastra's case it treat of a strange division that could be mainly justified by the idea that the old Indians had, or on the same faculty, and on the grade of progress, that on the respective described Era the Veda ascribed to the human evolution.

It is a different way to conceive our history, and the human evolution, all-different from the occidental vision, so linear as much limitative.

In the Indian conception of the time some other époques had existed before us, called *Yuga,* where the men lived and created a flourishing civilization that had been destroyed at the end of every cycle from the *Pralaya,* a catastrophe

---

[37] Maharshi Bharadwaaja, *Vymaanika-Shaastra Aeronautics*, International Academy Of Sanskrit Research, p. 87.

[38] V. Raghavan (M.A., Ph.D.), *Yantras or Mechanical Contrivances in Ancient India*, Transaction No. 10, The Indian Institute of Culture, Basavangudi, Bangalore, Febraury 1952, p.8.

that should have decimated every time the human being leading them to start once again.

Also if this idea could seems incredible, the existence of one or more antediluvian civilizations remind in Occident on several myths among which the most known are the platonic Atlantis as Aztlan's south American legend, that was the original country of the Aztecs people made "*on the middle of the ocean*".

All over the world had been identified, not less than 650 myths or traditions about a Flood, happened on some extremely remote times, which should have destroyed a civilization existed before. Bharadwaja describes three big classes of aircraft on his Vimanika Shastra; the Mantrika, the Tantrika and the Kritaka Vimana.

An interesting element is also the presence of a description of the different propulsive typologies that the Vimana should have used.

In the Kritayuga, or "*Yuga of the truth*", lasting as much as 1.728.000 years, it is said on the old tests that the Dharma wrapped up every action and thought of the man.

It is the mythic Golden Age where everything is living and looking forward a beginning towards equilibrium, a peace, and where the harmony and the propensity to the divine could explicate without the presence of some destructive and harmful elements.

The men of this time should have been able to reach any place through their Ashta Siddhis, the eight enhancements obtained by Hatha Yoga's practice. During the Tretayuga, lasted 1.266.000 of years, the Vimana had been called Mantrikavimana and flew through the 'power of the hymns' or the mantra. According to the Vymanika Shastra and the tests cited in it, twenty-five varieties of aircrafts, included the Pushpaka Vimana, belonged to this era.

According to the Veda the human being succeed on this Yuga to understand the 'divine magnetism' that is the origin of the several electrical strengths.

In this case the bearing element of the system of working of these aircrafts seems to be the vibrations, the frequencies, today one of the systems among the principal candidates for the future discoveries and conquests on the scientific field.

Instead the Vimana used on the Dwaparayuga, lasting till 864.000 years, were called Tantrikavimana and flew, always according to Bharadwaja, through tantra's power. According the Veda on this time the human being was able to understand the several electric power and the relative property that are the fundamental of the rudiments to create the material word.

Fifty-six kinds of aircrafts, included the Bhairava and the Nandaka Vimana, belonged to this Yuga. The aircrafts used in the Kaliyuga, lasting on the whole 432.000 years, or the actual era so on our most remote past, were called Kritakavimana and flew thanks to the power of the motors.

In this case Bharadwaja speaks of about twenty-five kinds of Vimana, included the Sundara, the Shukana and the Rukma.

The general idea that spring from the description of the mystic Bharadwaja is that on the precedent times of our planet some civilizations had been existed that developed some techniques and technologies based on some principles all different from those modern, but everyone able to produce and to be able to fly some particular types of Vimana.

The existence of some other civilizations in the past of our planet is a concept deeply rooted inside the Indian culture as a part of their myths and of their same religiosity.

Inside this point of view, and flanking to the cyclical idea of the time, every Yuga saw the birth and the decline of a civilization with the consequently blooming and the fall of a progress and of a spiritual and technological evolution that decree the uniqueness respect the precedents.

We are going to analyze the details of the data provided by Bharadwaja.

The Mantra Vimana

The men in the Treta Yuga should have used the power of the Mantra "*to fly from a place to another*". Bharadwaja cites the mystic Shaonik Rishi that for him there should have been 25 types of Mantra Vimana during this time.

The Treta Yuga into the Hindu region constitutes the second of the four eras in which the life has developed and it is the time during which the human being succeeded to understand the divine magnetism that could be to the origin of the several electrical strengths.

According to the Hindu mystic the magnetism is in a close correlation with the existence of the entire universe. The Treta Yuga has a duration of 3.600 divine years, corresponding about 1.296.000 years.

The Vimana ascribed to this time are: Pushpak, Ajmukh, Bhraj, Swjyotirmukha, Kaushik, Bhishn, Shesh, Wajrang, Daivat, Jwal, Kolahal, Aachirsh, Bhushnu, Somank, Panchban, Mayur, Shankar, Tripur, Wasuhar, Panchanan, Ambrish, Trinetra, Bherund.

The Tantra Vimana

For Bharadwaja the Tantra Vimana should have been built using the elements of the nature and through the power of the mantra. Bharadwaja cite the wise men Maharshi Shaonak and Maharshi Gautam, together to some tests ascribed to them the "*Shaonak Sutra*" and the "*ManiBhadrakarik*", where there should be described 56 types of Tantra Vimana used during the Dwapar Yuga.

This time in the Hindu religion is the third of the four eras where the life has evolved and is the time during which the human being can understand "*the several electrical strength and the relative propriety that are the basement of the beginnings that created the material word*". The Dvapara Yuga lasted 2.400 divine years, about 864.000 human years.

The Vimana ascribed to this time are: Bhairav, Nandan, Watuk, Wirinch, Tumbar, Vaintey, Bherund, Makardhwaj, Shrunagarak, Ambarish, Sheshasya, Saintik, Matruk, Bhraj, Paingal, Tititabh, Pratham, Bhushrni, Champak, Dronik, Rukmapunkha, Bhramani, Kakubh, Kalbhairav, Jambuk, Girish, Garudasya, Gajayasya, Wasudev, Shursen, Virbahu, Vrusund, Gandak, Shuktund, Kumud, Kronchik, Ajgar, Panchdal, Chumbuk, Dudunbhi, Ambarasya, Mayurasya, Bhiru, Nalik, Kampaal, Gandrksha, Pariyaatra, Shakunt, Ravimandan, Vyaghramukh, Vishnurath, Sauvarnik, Mrud, Dambholi, Bruhtkunj, Mahanat.

The Krutaka Vimana

The Vimana built with the elements of the nature without using the mantra are the Krutak, or Krutaka, Vimana. According to Bharadwaja these Vimana had been built during the end of the Dwapar Yuga and used at the beginning of the present Kali Yuga[39].

This time, as the interpretation of a great number of the Hindu Holy Writings, among them the Veda, it is the last one of the four Yuga. It was an obscure era, characterized by several conflicts and by a diffused spiritual ignorance. His beginning started with the physical death of Krishna[40], his length should be of 432.000 human years and will ended on 428.899 A.C.

On this remote date will appear Kalki, the tenth and the last one Vishnu's avatar *'he was riding a white steed and he was with a blazing sword with which he will dissipated the wickedness'*. After this time there will be a new genesis, a new Satya Yuga and a new *Golden Age*.

The new beginning will bring the end of the world as we know, the History will fall into the oblivion, the Earth will be back as a heaven. Maharshi Shaonik and Gautam mention 25 Krutak Vimanas's types for this time:

Shakun, Sundar, Rukma, Mandal, Vakratund, Bhadrak, Ruchik, Vairaj, Bhaskar, Gaj, Aavart, Paushikal, Viranchi, Nandak, Kumud, Mandar, Hamsa, Shukasya, Som, Kronchik, Sainhik, Panchban, Oryayan, Pushkar, Kodand. What we said till now it is presented as a conception of the world extremely complex and variegated and not the fruit of a creation of Subbaraya Shastry as someone superficially and lightness affirmed in the past. If the venerable Pandit should had created the Vaymanika Shastra, ascribing it to Bharadwaja to confer an aura of sacredness, he should have had an incredible creative work, of enormous proportions and unique on his type, finding himself parallel to give existence and truthfulness to an infinity quantity of technical, scientific and philosophical concepts, without ever to incur in some known or thin contradictions or mistakes, done to not let to discover the possible false.

We add to all that the style and the prose with which the test had been written is corresponding exactly to that one used on Bharadwaja's time, so around the VIII century B.C., quite incorporating some inflections and terms not more used since a millennium.

We have to add that some recent new analysis made by some Indian Universities had let us to verify the genuineness of some formulas and chemical processes described on the test recognizing in its inside a scientific and technological level extremely advanced, also for the modern time, such as to ask for some knowledge and some instrumentations that we can dispose only since a few decades.

---

[39] Literally *Kali*= Black, *Yuga*= era, would correspond to the Iron Age in Greek myths.

[40] Occurred, according to the Surya Siddhanta the astronomical treatise that forms the basis of the Hindu calendar, at midnight of 18 February, 3102 BC.

So what it is described in the Vymanika Shastra seems to delineate a unique scene, but also under certain point of view, not too different from what is described on others mythologies.

In the specific case of the Vimana 'that they are different for every time', it is as in our civilization based on the technology and the progress, there was following another one based on the natural value and that developed one of its own evolution and technological form based on these principles.

The use of the crafts able to fly should not have changed but on the second case it could treat on some different methods or principles from that one that we are using or that we known today, only because of this civilization should have moved in the respect of the nature and the use of its instruments.

Vimana, the confirmation inside the edicts of Ashoka

The historiography analysis on the Vimana made us to discover one more element of a big interest, or the edicts made by the emperor Ashoka Moriya the Great[41] to legislate on the huge dominions conquered.

After several military campaigns Moriya reigned on a territory that included a big part of the Indian subcontinent, the present Afghanistan, a part of Persia (the present Iran) Bengal (today divided among India and Bangladesh), and Assam. The history remind him not only for his big conquests but also for his religious conversion happened after his innumerable battles, a meeting with the Buddhism that transformed him in a man all different from the belligerent leader that he had been.

His edicts inscribed on the stone constituted some unique epigraphs through which the sovereign entrusted his legislative and moralizing wills toward the subdued people .

These edicts[42] had been engraved between the 256 and the 237 B.C. and form an archives one of the most interesting that the history had transmitted us, but also it is an element extremely important to analyze the presence of the Vimana in old India's skies. Inside them the term that it is used to designate the Vimana is, as usually happen, the term 'celestial chariot' but curiously it is quite often flanked by the term 'divine beings', so connecting and fastening indissolubly the two realities.

Particularly on an edict[43], referring to the Darma and the cosmic low, it is told *"...the flying carts, the elephants and the other divine beings"* had to be submitted to

---

[41] Asoka or Ashoka (Pataliputra, 304 BC - Pataliputra, 232 BC), Sanskrit word which means 'without suffering', was ruler of the Mauryan empire the largest that India has ever known.

[42] Kalsi R.E. IV.9, Mansehra R.E. III.13, Shahazgarh R.E. IV.8, Girnar R.E. IV.3, Dhauli R.E. IV.2.

[43] See the Kalsi edict, R.E. IV.9.

them. So everything was regulated by some superior and cosmic laws and nothing or nobody could oppose to them.

Kanjilal think that the relation between the two terms could lead us to identify exclusively *"something over the human being or celestial"* excluding any other interpretation on the conjunction of the two word.

In our opinion all that does not imply necessarily that at the time of the imperator Ashoka still existed some working Vimana, but surely the echo of their presence was still alive into the popular traditions as into the holy tests from that their use for the edicts as some images and some metaphors of the same concepts contained.

God Salva's Vimana and his terrible arms

After having analyzed and discovered some old tests containing some descriptions and some secrets ascribed to the Vimana, it is arrived the time to leave directly the word to the Indian traditions to see the roll and the power that these ' legendary' celestial carts infused on these people permeating their legends.

Among the most fascinating descriptions that we had been able to find, that one on the *Bhagavata Purana* is one of the most emblematic and meaningful. It is considered one of the holiest tests of the Hindu tradition, the fulcrum of its story rebound Vishnu's attitude considered the most sublime among the Gods. The Work narrates about the events, the works, and the God in his several incarnations, or Avatar, in a particular way that one of Krishna.

In his twelve songs in particularly one attracted our attention, the chapter 76, where it is described the war provoked by the diabolic God Salva against Krishna and his holy town, Dwarka.

Through his pages the carts described are moving in the air apparently without wings or some motors instead the arms that they used incredibly seem to refer to some modern contrivances wars.

For most of the Hindu translators, had been simple to transpose the Sanskrit word with the modern term 'aircraft' because, since the most remote times, the same term has always indicated an aircraft made to travel in the sky and in the space.

It is through the direct words of the Bhagavata Purana that we can understand the complexity and the depth of what is transposed by his pages.

"Salva chose a Vimana that could not be destroyed by the Deva, by the Asura, by the human beings, by Gandharva, by Uragas nor Rakshasa, which could travel everywhere wished to go and terrorized the Varishnis…then Shiva said ' So be it. Under his order, Maya Danava, who had conquered some towns from his friends, built an iron flying vehicle called Saubha, and he presented it to Salva".

Proud of his new and powerful means, that in its name remember the mythical flying towns of the gods, Salva gather his army and he prepared himself to fight to the earthly Krishna's town.

"This unassailable vehicle was full of darkness and it could move everywhere. To obtain it, Salva, reminded Varishnis's hostility toward him, so he proceeded toward Dwarka's town. Salva besieged the town with a great army, Bharata's best one (the old name of the India, ndr) *destroying the suburban parks and the gardens, the buildings with their observatories, towering doors and walls and also the public recreative areas"*.

Quickly the fortune changed and in a few minutes Salva's army was decimated. The king was furious and used his new aircraft to attach the town with every means possible.

"From his excellent Vimana he threw a stream of bullets…a savage vortex raised up and covered the all area with some floating dusts"[44].

Suddenly Krishna appeared, on his shining celestial cart, to challenger on fight the king Salva. With the appearance of the God on the fighting field Salva revealed a big and powerful weapon that *"it fly in the sky with a roaring sound as a big meteor"*.

The text describes it so shining *"to light up the all sky"*. As soon as Krishna starts his offensive, Salva engaged the special powers of his Vimana with an obsolete effort to avoid to be destroyed.

"The aircraft used by Salva was very mysterious. It was so extraordinary that sometime it seems to be many aircrafts in the sky and some other time nobody. Some time the aircraft was visible some others was invisible, and the warriors of Yadu's dynasty were perplexed on where could be the aircraft. Sometime they sow the aircraft on the floor, some other times flying on the sky, other times was laid on the top of a hill and sometime was floating on the water. The fantastic aircraft was flying in the sky as a whirling firebrand and did not stopped neither for a moment"[45].

It is a description enough 'strange' and, also if it is coming from a recent translation is completely true and specular to the original test in each of its meanings. First of all Krishna tried to destroy Salva's "big weapon" throwing one of his weapon described as *"clear as the sun in the sky"*.

Afterwards he made Salva's Vimana completely inoffensive releasing a destructive 'shower of arrows' such as *"the fantastic Salva's Vimana exploded in several pieces and fell into the sea"*. Salva at the last moment miraculously escaped to the destruction of his Vimana and walking, and violently, he throws himself against his hated enemy.

This one, *"shining as the sun rising on the mountains"*, throws his final shot using a "shining disk". Usually the scholars of Sanskrit read these histories as they are

---

[44] Bhavagata Purana, 10.76.

[45] Traslation of Bhaktivedanta, Swami Pradhupada, A.C., in *Krishna*, The Bhaktivedanta Book Trust, Los Angeles, 1986.

nothing else that a pure mythological imagination depriving such stories from every base of reality.

There is no doubt that along the centuries they had been enriched by religious and literary shades to extend their capacity and their meanings but, as recited an old proverb, the truth is hiding behind the details and the meanings and those that we may find in these epic poems are escaping to whatever possible imaginative creation receiving inside them some notations and descriptions that remain difficult, if not impossible, to explain if only calling in cause one real technological reality.

As the *Vishnu Purana,* Dwarka's town was submerged by the sea a little after Krishna's physical death. Such event happened in the 3.102 B.C., fixing both the end of the Dvapara Yuga, the third era of the world, and the beginning of the Kali Yuga or the actual era.

Just till a few time ago we thought that such legend was only a mythological heritage and nothing more.

What if Dwarka had been really discovered under the waves of the sea? Will change something?

On Mai 19 of the 2.001 the Indian Minster for the Science and the Technology, Murly Manohar Joshi, announced the discovery of some submerged ruins on Khambhat's Gulf found by the National Institute of Ocean Technology (NIOT) on December 2.000.

On November 2.001 some more inquiries including the dredging of the sea bottom to recover some possible manufactured, they returned to the first confirmations about an old human installation today submerged.

Among the 2.003 and the 2.004 the NIOT made some new inspections recovering some samples of what they thought to be ceramic.

The founds had been sent to Oxford's and Hannover's laboratories as to several Indian institutions to demonstrate the authenticity and to value a date[46].

The results had been so much astonishing, the found objects and the identified submarine traces had been made by the man and they constituted the rest of an old town today submerged from the devastating fury of the waters. By the way, the disposition of the urban installations showed a strong affinity with the towns planning of Harappa's culture letting to leak out a possible connection between the two realities.

The legendary Dwarka said in the Mahabharata was returned to the light. Most difficult was to date it.

The recovered samples showed how such installation had been active since the 1.900 B.C.; some precedent centuries or perhaps some millenniums could have sheltered a population now disappeared under the waves of the sea.

---

[46] S. Kathiroli, *Recent Marine Archaeological Finds in Khambhat, Gujarat,* Journal of Indian Ocean Archaeology 2004: 141-149.

It is not surprising that the reaction of the Occidental archeologists had been to ascribe some strange submarine formations to a natural origin[47] also if no one of them could had made some inspections on the place to see the genuineness or could had directly analyzed the recovered samples.

To confute such assertions the date of some wooden objects arrived of a clear artificial nature recovered from Alok Tripathi, a subaqueous archeologist and a superintendent of India's Survey, some wares that had been submitted to carbon 14 furnished some exceptional results.

Some of the founds had been dated[48] to the 7.500 B.C., instead the study of the raising of the oceanic levels in the following time to the last glaciations showed how between the 8.000 and the 2.500 B.C. Dwarka's coast was at the light of the sun and only afterwards had been submerged by the actual 36 meters of water, with a maximum margin of raising, in some other Indian borders, of 13 meters.

On January 19 of 2.001 the BBC News opened his evening edition announcing: '*Lost town can write again the history*' instead the National Geographic, May 28 2.002, recovered the news titled from his portal '*New subaqueous discoveries raise some questions about the myth of the Flood*'.

The mythic Dwarka, Krishna's holy town, after several millenniums of oblivion was raising from the waves of the sea where the tradition had always placed, conferring an historical veracity to the thousand legends transmitted into the *Mahabharata*.

"Chapter 76"

Here it is an extract of Bhagavata's Purana Chapter 76 which contents surely will be explicative as referring to the Vimana as also on their differences to what there is on the occidental traditions.

"Sri Krishna realized that all the town was heavily menaced. He himself ordered to His charioteer Daruka: *"Quick, lead me to Salva! You have to known that he is a very powerful and mysterious fellow; you have not to be scare of him"*.

So Daruka, to a high speed, drove the Master to Salva. Krishna's cart brings a standard with Garuda's image, and as soon as the warriors of Yadu's dynasty sow that flag they understand that the Master was on the battlefield. Almost all Salva's men had been already killed, but Salva, who had noted Krishna's pres-

---

[47] Witzel, Micheal, *Rama's realm: Indocentric rewritings of early South Asian archaeology and history in Fagan*, G. G., ed., Archaeological Fantasies, Routledge Taylor, and Francis Group, New York, 2006.

[48] Sundaresh, S. Gaur & Tripati, Sila, *An ancient harbour at Dwarka: Study based on the recent underwater explorations*, Current Science, Indian Academy of Sciences, May 10, 2004.

ence, throw a terrible weapon that, as a big meteoroid, furrow the sky with a deafening crash.

His dazzle was so alive that all the sky was lightened. But Krishna, arriving, with only one arrow reduced on thousands of small pieces that terrible weapon. Then with sixteen arrows he hit Salva, and wrapped his vessel with a rain of arrows as the brightness of the sun with his several molecules he wrap the immensity of the sky without clouds.

Salva answer with a terrible stroke to his left side, where Krishna bring his arcs Sarnga that slip from his hand.

What astonishment! The big personages and the *deva* were observing the battle and they were extremely shaken and said: "Alas! Alas!". Salva thought to be winning and he turned to Krishna with a roar: "Krishna!

What a heel you are! You raped Rukmini with the strength, and in front of us! You humbled my friend Sisula wedding Rukmini. Then, in the big meeting of the sacrifice Maharaja's Yudhisthira *raja-suya,* you have profited of a momentary inattention of this friend of mine and you have killed him. Everyone think that you are a big hero and no one can win you.

But now you have to show your power! I think that if you stay in front of me a little more, I shall sent you with my sharp arrows in a place from where you will never come back". Sri Krishna answered: "How stupid you are, Salva! Your words are all idiocies! You do not know that the death is already on your head!

The true heroes *do* not get drunk with the words, but they try their value with the facts!". In his great anger, Krishna wields his mace and hit Salva at the clavicle with a stroke so violent to provoke an internal hemorrhage. Salva start to shake, as he was near to enfeeble, broken by an iced cold. But before that Krishna could hit him again, Salva becomes invisible thanks to his supernatural powers.

After a few second, a mysterious stranger presents himself in front of Sri Krishna. Shaken by violent hiccups, he bended to the foot of the lotus of the Master and told him: "As You are the dearest son of Your father Vasudeva, your mother Devaki has sent me to inform you of these misfortune: Your father had been taken prisoner by Salva and brought far away with the force. Salva had taken possession of him as a butcher grasp an animal with no pity". At these news Sri Krishna was very upset, as a normal man. On his face appeared some signs of pain, and start to cry in a moving way: "How could it happen? My brother Balarama is there, and no one can win him.

I entrusted him for defending Dwarka, and I know that he is always on the look -out. How had been possible for Salva to enter into the town and to grasp my father in this way? How big is his power, Salva has his limits too; how could win Balarama's power and to rape my father, as says this man? Alas! After all the destiny is very powerful!".

While Sri Krishna was giving himself up to these thinking Salva made to appear in front of the Master a prisoner perfectly like to his father Vasudeva. But this apparition, as that one of the mysterious stranger, was a Salva's trick, the result of his supernatural powers. Again Salva addressed to Krishna: " You are only a heel, Krishna! Look! Here there is your father that gave you the life and thanks to him you are still alive! Look now how I am Going to kill him! And if you have some power, try to save him!" Pronouncing these words, the smart wizard cut the head to the false Vasudeva. Then, with no hesitation, he brought the body and jumped on board of his aircraft.

Sri Krishna is God…but since he plays the part of a human being, here he is for a second in a way of completely humiliation, as he could have really lost his father; but it is only a moment, because Krishna understands that such arrest and such killing are only the fruit of the magic that Salva has learned from Maya's *asura*.

Krishna, coming back to his normal conscience could see that there was no messenger nor his father's head, but simply Salva who had left away with his aircraft in the middle of the sky… Salva thinking that Krishna had been deviated by his magic, felt encouraged and attached him with a big energy and power, throwing against him a stream of arrows.

But Salva's enthusiasm is like the rush of the insects that throw themselves into the fire. Answering, Sri Krishna bombard him with his arrows threw with an unimaginable strength. Salva is wounded and his armature, his arc and his helmet spangled with precious stones fly on thousands of pieces. Then with a terrible stroke of his mace, Krishna make to burst on thousands of pieces the aircraft, that suddenly sink into the ocean.

But Salva was very prudent and, instead to crash with his aircraft, he throw outside and with a jump he was on the dry land and again with impetus he throw himself towards Krishna, with his mace on his hand. But he cut at a blow that hand, that fall on the ground, together with the mace that he was tightening.

Once forever he decided to finish with Salva, Krishna grasps his fantastic disk, radiant as the Sun at moment of the dissolution of the material creation. Krishna, standing up with the disk in his hand, appear reddish as the rising Sun that up on the mountain and cuts its head, that roll on the ground with the earring and the cask. Salva died in the same way as died Vrtrasura by Indra, the king of the celestial planets. Salva is dead, and all his warriors started to scream: "Allas! Allas!".

And while Salva's men were complaining, the *Deva* from the celestial planets were throwing a rain of flowers on Krishna and were announcing his victory beating on their drums and blowing on their trumpets".

Chapter II – Lost Technologies

Technological Hints on the Vimana on the Holy literature

"...*With no doubt it was lived before every other zone and could also have been the place of all the creation and of all the sciences .Certainly as we know, the Indian culture is coming from Tibet, so as all our arts as the agriculture, the numbers, the game of chess, etc.., seems to come from India.*"

Immanuel Kant

It is necessary, after having analyzed some fundamental parts of the old Indian history, of its territory and of its flying carts known with the term Vimana, to identify some fundamental points, to extrapolate some basic elements that could show us explicitly how these knowledge, kept by a small caste of wise men, simply do not were a rich mythology but sank their roots on something older, of extremely remote.

The electricity into Vedic's literature

The wise man (Siddhar) Agastya into the text *'Agastya Samhita'* gives us the description of two different typologies able to move into the air. The first is a *"Chchatra"*, from a shape of an umbrella or a balloon, which the principal characteristic is to be completely filled by hydrogen. In this case the process for the extraction of the hydrogen from the water is described on detail and it should use the electricity to obtain the most pure and stable shape.

The *Chchatra* is described as a typology of a primitive airplane that mainly was useful to escape during a state of emergency, from which afterwards took the name of 'Agniyana'.

The second type of aircraft was similar to a parachute and, according to the Agastya Samhita, they said that it could be opened by using some strings. This 'aircraft' is described as a *"vimanadvigunam"* that is of an inferior level respect to a Vimana or a 'normal aircraft'.

We have to consider that Agastya, the author of the text, is one of the Indian Sapta Rishi or those wise men that after the last Pralaya, or the Indian flood, repopulate the world with their knowledge.

Agastya Samhita himself had been made in writing about around the 500 B.C.[49], a time when the old religiousness and the Vedic knowledge were on their decline and where many old wisdoms passed from an oral tradition to a writing form.

It is interesting one more reference inside this very old text where seems to be held the instructions to produce the electric energy. *"You have to put a well clean dish, or a plate over a ceramic pot; you have to cover it with a stratum of copper and fill up the rest with wet sawing. You have to put a sheet of zinc covered by an amalgam of mercury on the sawing. If you will have care to leave to stick out from the pot a piece of copper joined to the plaque, among the piece and the sheet of zinc it will produce a stratum of energy called Mitra-Varuna with which we can divide the water into Pranavayu and Udanavayu* (Hydrogen and Oxygen, ndr). *You must be careful that the pieces of iron do not touch the sheet of zinc; on this case the effect will disappear. If you will made a chain of these pots one after one, you will have a big energy"*.

We tried to reproduce the proceeding described by the Agastya Samhita on a laboratory and the result had been of an open circuit to 1.138 Volts and a current of a short circuit to 23 mA. The first historical testing on the electrical energy had been on the '700, but only with Alessandro Volta on 1.801 that the Royal Society had announced to the world the invention of the electrical pile.

The unique precedent known had been Baghdad's piles, a ware dated back to Parti's dynasty (150 B.C. - 226 P.C.) which its historical uniqueness had showed the possibility that some technologies could be used in the most past of our planet.

The invention of this incredible instrument, in Agastya Samhita case we have to draw back of all last four centuries, but we should go back before where this text had been written after centuries, if not millenniums, of oral tradition.

On the treaty of architecture called Shilpa Sutra[50], ascribed to the legendary Visvakarma[51] an Hindu divinity patron of the artisans and of the architects as well as the builder of the Pushpaka Vimana, we find one more reference rather surprising.

---

[49] One of the oldest versions of this brahmanic manuscript is preserved in the Library of Ujjain in Madhya Pradesh.

[50] The Shilpa Sutra is contained in the Shilpa Shastra, a series of Hindu texts that describe manual arts, the rules for religious iconography, architectural requirements, the proportions of a sculptured figure, as well as general rules of Hindu architecture. It can be considered a forerunner of Samarangana Suthradara written by King Bhoja. See P.K. Acharya, *Indian Architecture according to Functional the Manasara Shilpa Shastra*, London, 1927.

[51] It is considered by Hindus as the "Principal Architect of the Universe" the man who built and designed the divine architecture of the universe, the Lord of Creation. The term *Vishwakarma*, also known as Vishwabrahmin, also describes a caste of priests, engineers, architects, sculptors, temple builders and Indian artists. The term is applied to five sub-castes of blacksmiths, carpenters, coppersmiths, goldsmiths and sculptors.

"*If you take a terracotta pot, you put a cooper sheet and the shikhigreeva. Then you spread it with wet sawing, mercury and zinc. Then if you will join the wires it will give you an energy (Tejas) called Mitravaruma. This will bring to the disintegration of the water into the Praan vayu and Udaan vayu. They say that we need a chain of pots to obtain a very active and efficient strength. The Vaayu Udaan created could be entrapped in a cloth to airtight with a certain way. If that will be realized, thanks to Udaan's vayu push, it will be possible to built a structure able to fly in the air*".

In this case it is also evident the reference to the creation and to the application of some techniques for the production of electrical energy and also to the electrolytic decomposition that in a gassy form produces Hydrogen (H) and Oxygen (O). From an analysis of such passage there come out some fundamental points that could show us an advanced knowledge on determinate branches of the technical sciences.

Here there are the most important points:

1. The energy is produced by this particular mixture of chemical products and transmitted thanks to the use of a wire of an unknown nature. It is specified the double nature of the energy.
2. It is expressed the concept of the electrolysis of the water. A current at a low voltage which across the water and forms gassy oxygen to the anode and gassy hydrogen to the cathode.
3. Through the scission of the water, are made the two gases up said.
4. The energy may be modified if is in possession of "pots" placed on succession.
5. It is used a material, or tissue, waterproof to avoid to the air to go out.
6. The Udaan vaayu produced after the scission of the water could be entrapped into an hermetic container.
7. Udaan vaayu is a light gas.
8. The Udaan vaayu could be exploit to built a structure able to fly on the air.

As it has been underscored by the Agastya Samhita, the *Chchatra* was a kind of primitive aircraft filled by hydrogen and especially proper to escape during the state of emergency.

The junction of the techniques contained on these two treaties should had let to built a rudimental aircraft but with the characteristics up showed, such aims that in the official history should had been reached only among the 1.700 and the 1.800.

One more analysis of the test just cited could be made on the term Mitra-Vaarun, or Mitra Varuna, aggregation of two divinities in charge together to preserve the order and the low and had been translated into the Shilpa Sutra as the form of energy generated by the described mixture.

In the Indian pantheon Mitra and Vaarun coexist as a couple and are two twins divinities similar to Ashini-Kumar.

The energy that has been generated from the blend of the cited elements seems to call in cause the negative and positive terminal of certain cells used with the consequently generation of charge of an electrical field. Positive and negative cannot exist one without the other so as Mitra and Varuna.

The verse examined seems to point at, on his apparently simplicity, a wisdom spring highly evolved so as its antiquity could seem to testify a remote heritage preserved during the times by the Indian wise men.

According to the oldest traditions the Shilpa Sutra ascended absolutely to the 2.000 B.C., also if the language and style used seem to belong to a time immediately after to the reformer of the Sanskrit grammar Panini[52] (about in the 500 B.C.). Also if this temporal gap is not indifferent it is possible that on his first transcription the old knowledge had been filtered by some words and codes dictated by the Panini himself.

In this connection also the widest Vastu Shastra[53], or the old science of the Indian architecture, constitute a very old discipline, which had conglobated inside not only some constructive and architectural knowledge, but also a bigger spectrum of techniques and scientific competences, a patrimony of the old Vedic people. So as in the Samarangana Sutradhara are present some notions of all strange to the architecture as such as in the chapter XXXI also dedicated to the Vimana, so also in the Vastu Shastra as in the Shilpa Sutra could be found some knowledge belonged to a wider heritage of the scientific knowledge.

The scholar Smita Gupta in her thesis of doctorate, of the Arizona State University, has traced a brief historical sight "... *the Vastu Shastra is the old Indian science that state the guide lines to fix the summary and the delimitation of an area for several activities. The Vastu Shilpa Shastra find his origins in the Sthapatya Veda (Science of the Architecture – a Veda's part written by the Aryan in the following times to the civilization of the Hindu Valley). We say that Maya, the great scientist, architect and urbanist was the responsible for the basis of the Vastu Shastra, and that ascended to a time between the 10.000 B.C. and the 5.000 B.C.*"[54].

---

[52] Panini (c. 600-500 BC) was an Indian grammarian famous for having formulated the 3,959 rules of Sanskrit morphology. Panini's grammar conventionally marks the end of the Vedic Sanskrit and the beginning of Classical Sanskrit.

[53] D. N. Shukla, *Vastu-Sastra: Hindu Science of Architecture*, Munshiram Manoharlal Publishers Pvt. Ltd., 1993.

[54] Smita Gupta, Arizona State University, School of Architecture, *Vastu Shilpa Shastra: The Ancient Indian, bioclimatically responsive science of building*, 2005.

It is a tradition that for the same Indians had been lost in the old times placing itself in a time outside from every date officially known. As we sow it is not the only one!

The Saubha, the town on the space

If the Indian divinities seem to tie deeply to our planet and to the "other inhabited worlds" it is also interesting to observe how their residences, during their stay on the Earth, often there were some real flying towns. Such buildings, definite *Sabha or Saubha,* were of huge proportions and seem to "be a bigger kind" of the above-mentioned Vimana.
Often these one were described as integrant elements of the same Sabha where they get in and get out for the most disparate motivations. The tests speak about flying towns belonged to some divinities as Indra, Brahma, Rudra, Yama, Kurvera and Varuna and many others.
These towns were moving in the sky and are described as shining as the silver, full of foods, drinks and water and all the possible pleasantness as also of weapons and munitions for their protection.
Some of these Sabha are described in detail, specifying that they were able to move not only on the sky but also in the space[55].
In this connection it is interesting Maya's Danava history, the master of the illusions, a big builder and a divinity to chief of Danava's kingdom placed in the ' sphere', or could be a planet, called Talatala as well as the builder of the Vimana belonged to the God Salva.
Maya is also considered among the mythical authors, or inspiring, of the Surya Siddhanta a astronomical treaty of highly understanding taken on a written way about the 600 B.C. and precursory of many discoveries and concepts that in Europe should have seen the light only in the following centuries and millenniums.
During his permanence on the earth, Maya is described as a big king who leaded the races of the Asura, of the Daitya and of the Rakshasa. His incredible wisdom made him to built three flying towns known together with the name of *Tripura,* made with gold, silver and iron.
Into the Sabhaparva we find a description of the Tripura where it is clearly said that they could not only to move in the sky, but also to entertain in what today we could define as an geostationary orbit around the Earth and also they were owing some ' doors' able to make to enter some 'small flying carts'[56].

---

[55] Mahabharata, III° book Vanaparva, ch. 9 vss. 25-61, ch. 3 vss. 168-170 e ch. 181 vss. 33- 38; ch. 200 vss 50-60 e ch. 207 vss. 6-8; Adiparva ch. 33 vss. 8-10, 28, ch. 134 f.n. 301; ch. 57, vss. 13-14 etc.

[56] Sabhaparva, Chs. 3, vss. 6-10.

In their greatness and strength the three towns let Maya to dominate the world for a long time, but because of their impious nature Shiva, the destructor of the worlds, shot them down making them to crash down into the occidental seas as big torch of fire and making a 'strong noise'[57].

A Sabhaparva's passage gives us one more start to understand the nature and the purposes of these flying fortresses. "*...While Arjuna was fighting against Indra, Krishna saw a demon who, among the flames, was finding safety in flight. He had just throw the disk Sudarshana to kill him, when the danava, sawing to approach the infallible Master's arm, run towards Arjuna asking him help desperately. The Pandava touched by his prayers turned to his friend...*". You have saved my life, so I would do something for you. Tell me how can I pay off?

I am Mayadanava, the Asura's architect, I could built for you the most beautiful towns... The Pandana do not have a Sabba to the grandeur of their fame and you are one of the less all over the universe able to built it. If you really want to show your gratitude built a Sabha so beautiful as ever we had seen all over all worlds...After a few weeks they started to built it. He was helped by several rakshasa with an extraordinary strength, Mayadanava went to Kailasha, from where he came back with unimaginable wealth of gold, diamantes and other very precious metals that in the past time had been used for some rites extremely expensive. Coming back from Himalaya, besides the danava made a present to Bhima a powerful and very heavy mace and to Arjuna a cart of war with some magic powers. Helped by the Asura, the sabha was finished very soon... After a few weeks many kings and princes started to arrive from all over the world to Indraprastha. The place most visited of the fantastic town was the royal palace built by Mayadanava, where everything was shining... Everyone, coming in, was thinking that the sun was posing every time its rays on the walls of the several rooms. Millions of people were admired and astonished commended the fantastic Pandava's sabha"[58].

The genesis of this story is dating from Krsnayajurveda's time (6.2.2) that, as the Indian tradition, we fell that had been set before the 3.000 B.C. The number of the flying towns do not exhaust with the epical sagas or with the religious Indian text. Often we read about Hyraniapura, the golden town, built by Brahma for the demon Puloma and Kalaka. This town was invincible and inexpugnable, as many of its sisters, to the point that the demons, which were leading that, had been able to refute every attack from the other divinities.

Matali, trying to end the dominion of the demons and to destroy their flying cart, asked to the brave Arjuna, one of Mahabharata's principal characters, to destroy the celestial fortress. So the brave leader addressed himself towards the town flying his cart, a Ratha, kindling the fury of the demons and causing a terrible battle that ended with Hyraniapura destruction.

---

[57] Matsyapurana, Ch 129, vss. 20-21, 30-34 e Ch. 140, vss. 40-44.

[58] Sabha Parva, paragraph 36.

We find on the third book of Mahabharata, called Vanaparva, some other descriptions of the flying towns able to move in the sky and in the space and some specific names are assigned to them as Vaihayasi, Gaganacara or Khecara[59].

The Professor D. K. Kajilal comment Saubha's stories contained in the Matsya Purana with the following words: "*Behind the legend there is a scientific truth that three aerodrome basis had been built and used by some strange beings. One was on a stationary orbit, another was mobile on the sky and the third was permanently placed on the terrestrial surface. These were similar to the modern space stations reachable at a particular hour from a prefixed latitude and longitude. Shiva's arrow* (about what is talking the text), *is referring to a burning missile threw from a satellite in the orbit which is going to hurt a spatial ship making that to fall down into the Indian Ocean. The traces of an old healthy civilization, destroyed by battles, come out through these old stories*"[60].

From these words come out several points of remarkable interest, not only the reference to a civilization extremely evolved, disappeared because of several wars, but also to a technology similar, if not superior, to our and the reference to the same artificer of these marvels identified as '*strange beings*'.

It is natural to see on these traditions a modernity typical of our times, it should be anomalous the contrary, but it is also undeniable that their uniqueness place them outside to every contest, myth or tradition previously known and especially guessed.

The classical concept, in these traditions, of " myth" meet with an evident abnormality chiefly suggested by the presence of a real technological and scientific references, not made by a pure poetical invention but adherent to an objective reality to have been, on several cases, recreated inside of university laboratories!

The Saubhikas

Sometimes some particular figures, not for that less important and not less interesting, are called and are quoted from the old Indian texts. Reading the text called *Arthasastra* ascribed to the wise man *Kautilya* and written around the III century B.C., we find mentioned the *Saubhikas*[61] definite as '*pilots that drive carts on the skies*'[62].

---

[59] Mahabharata, III libro Vanaparva, Chs. 168, 169 e 173.

[60] Dileep Kumar Kanjilal, *Vimana in Ancient India*, Sanskrit Pustak Bhandar, 1985.

[61] A.S., chapter 48 and 135 from the beginning.

[62] A.S., chapter 135 from the beginning.

Kautilya, a contemporaneous of Aristotle, was the minster and counselor of the semi-legendary king Chandragupta Maurya, Ashoka's The Big grandfather instead his *Arthasastra* has always been considered the most exhaustive and futurist treaty of Politic Sciences on the Indian History, inside which had been enclosed several scientific knowledge.

In the old times, the term Saubha identified king Harishchandra's flying towns, he was the 36° sovereign of the Solar Dynasty, instead the term *Saubika*[63] identified *'those who are flying or those who know the art how to make flying an aerial town'*.

On his treaty, Kautilya also use another word extremely interesting, *Akasa Yodhinah,* that could be translated as *'the person that had been trained to fight in the sky'*[64].

According with the Indian scholar Deelep Kanjilal we feel that the complex of informations described by Kautilya three centuries before Christ, that we have exposed in minimal part, are the testimonial of a previous knowledge of the techniques and traditions concerning *'the fighting in the air with flying carts'*.

The *Arthasastra* is a elaborate text of a sophisticated intellective complexity where had been codified some refined knowledge regarding the art of command and of the power, a real sapiential code which possession made the king Chandragupta himself one of the most powerful monarch of the Indian history.

Inside that, Kautilya codified the old knowledge acquired in several fields as how to treat the metals for war material, the distillation of the mercury whether for scientific use or mechanical use, the use of different types of powders to produce fire (as the *tejana curna*) or several types of vegetal and mineral combustible oils able to exhale flames also if used inside the water, just like the Greek fire.

Always in the Arthashastra, chapter XVII titled 'how to do and how to transgress the Peace', we state "*Dancers, actors, singers, musicians, clowns and court-poets, swimmers and pilots (Saubhikas) previously under the control of the enemy, may now to follow their occupation under the command of the prince*".

The *Arthashastra* mentions also two typologies of fighting carts: statics and dynamics. On one case Kautilya speaks about a machine able to throw water

---

[63] In the VI century A.D. in India the term Saubhikas took over a different meaning by identifying singers, religious, nomads, from house to house recounted the epic stories of these people through some banners variously painted. Nine centuries before Kautilya this term identified, however, very different figures. It is possible that over time the 'drivers' of the Vimana that moved in the heavens had been transliterated into singers who donated religious knowledge by moving from place to place.

[64] Look at A.S., chapter 108 from beginning and also Balaram Chakravarti, *The Indians and the Amerindians*, Volume 1, Self-Employment Bureau Publication, 1992, p. 144.

with such pressure to force the enemies to escape to do not be killed by his power[65].

A following commentator of Kautilya's work, called Bhataswamin and Pratipada's Panchika author, asserts that these 'automatic carts' used three fundamental powers: the pressure, the movement and the weight.

It is with us a verse of the Yajur Veda (Verse 10.19) where he says *"Or specialized enginery, you that are planning oceanic ships pushed by water engines as those used by ours aircrafts, that give the possibility to rise on vertical over the clouds and to travel all over the state. You have to be prosperous on this world and fly through the air and through the light"*.

Parallelly *the* term *Kathasaritsagara* indicated, on the old texts, some work men highly specialized who could be some *Rajyadhara*, experts in mechanic and able to built some oceanic ships, or some *Pranadhara*, experts to built some flying carts able to transport over 1.000 passengers and to cover in a few minutes very long distances. All that shows us a scientific knowledge extremely advanced and necessarily before the third century before Christ.

It is not possible to speak about some myths or legendary heritages, Kautilya's work picked up the most refined and advanced knowledge of his time about the art of the war and of the command transposing them in a text that could contained every technique, tactic, weapon and unknown instrument.

In a similar way during the XII A.C. century the Samarangana Sutradhara of the king Bhoja constitute his work on the architecture of the old India and also on this case, on the Chapter XXXI, the Vimana made their appearance because of they had been considered the models on which had been built the Hindu temples. Always the king Bhoja was the author of the *Yuktikalpataru*, a treaty about the construction of the ships, their classification and the differentiation in the construction for those that had to sail on the sea instead that one on the rivers.

The text presents some references about some aerial carts on the verses 48 – 50.

Besides the text describe several ways of transport used in the past by the Kings asserting that those aircrafts that were flying in the sky were called Vimana or *Vyommayana*[66].

A copy of the original manuscript is still today at the Sanskrit Calcutta College and is dated from the 1.870. Taking as an example these three classical works we find inside them some clear and right references to the 'flying carts', to some knowledge and technologies extremely advanced, to a technological and scientific patrimony unique on its kind.

---

[65] A.S., chapter II° 108 from beginning.

[66] Y.K.T., Chapter I, verse 48-52.

All that could find only one rational explication, the heritage of an old knowledge already arrived to his historical epilogue and of which they continued to transmit during the centuries some poor fragments and memories.

1. The Sanskrit and the artificial intelligence

In the last twenty years we had spent a lot of resources, time and cash to a plan a clear representation of the natural languages to made them more accessible to the elaboration through the computers. The natural language, or that one spoken by the human being in everyday life against the languages of informatics programming, shows several ambiguities especially of a semantic type and it lend easily to misunderstanding especially in the case it should be understand by a computer.

According to several scholars the natural language should not be proper to connect correctly to the rigorous and mathematic logic of a computer, anyway there exit at least an exception: the Sanskrit.

Actually the Sanskrit is a "dead language", is not spoken by any people but is used by the Hindu for the celebration of their religious ceremonies or for reading the holy texts.

What nobody could imagine is to discover that the Sanskrit seems to be a real and true language for the computers.

The *Forbes* magazine on July 1.987 sheltered an article titled *"The Sanskrit is the ideal language to programming software for computers"* but also two years before a NASA's searcher come to the some conclusions much more astonishing.

On 1985 Rick Briggs a scholar of Artificial Intelligence of the Roacs NASA Ames Research Center of Moffet Field, Ca., published a scientific article titled *"Sanskrit and Artificial Intelligence – NASA Knowledge Representation in Sanskrit and Artificial Intelligence"*[67].

In his abstract Briggs asserted, *"In the old India the intention to individuate the Truth became so worn that, in this process, we discover perhaps the most perfect instrument to satisfy such research that the world had ever known, the Sanskrit language... More than some works of a literary value, existed a long philosophical and grammatical tradition that had existed with an unchanged vigor till the present century. Among the realizations of these grammarians, we think that they have discovered a language, the Sanskrit, that is identical not only in its substance but on its shape with their current work in the artificial intelligence. This article shows that a natural language could also to serve as an artificial language and several studies on the artificial intelligence are nothing else that an unconscious appropriate again of some much older studies"*.

---

[67] Rick Briggs, *Sanskrit and Artificial Intelligence — NASA Knowledge Representation in Sanskrit and Artificial Intelligence*, AI Magazine, Vol. 6, n.1, Primavera 1985.

Briggs analyzed the semantic nets, in the first '80 of recent invention, and he found a striking resemblance with the Sanskrit language. Starting from the 500 B.C. several Indian grammarians, and specially the reformer Panini, had realized a tight codification of the language from the oldest Vedic Sanskrit till would be a useful element to transmit some logical concepts with an extreme accuracy. During the time, from such reform it was created a more concise and "condensed" and rigorous Sanskrit similar to a modern language of informatics programming.

Already, in its original form of Vedic Sanskrit, the reformed Sanskrit kept those peculiarities that should have been introduced and renewed only on the VI century B.C. The discovery of a connection between the Sanskrit and the Artificial Intelligence is of a monumental importance.

A language so old of about 6.000 years seems to be in any case a so perfect idiom to let to interfacing with the modern informatics systems. Such peculiarity could seems casual but in the reality of the facts it shows itself as the highest representation of that concept of perfection that constituted the *animus* of the Hindu thought. The uniqueness that the Sanskrit language own respect towards the others known, seems to express itself also in the etymology of the word itself, *sams-kr-ta* that means *"perfection"*.

It is natural to ask who is that one that had 'created' this language and to which levels of 'perfection' its inventor had reached to make it so unique and accomplished, peculiar and incomparable respect to every others.

With regard to that it is necessary an historical incised that came out from some recent archeological studies. The Daily Mirror on the 9 of July 2.012 published a disconcerting news.

From some studies leaded it resulted that the Kannada, a dialect spoken on the south India, it is the oldest language in the world and it ascend at about 10.000 years ago, inexorably connected to the legend of the three Sangam and of the submerged continent of Kumari Kandam mentioned on the first chapter.

In front of these new evidences we observe, as a new element seems to add and to impose on that longue list of anomalies and peculiarities that characterize India and its old civilization.

The gunpowder

The research of the sources and of the documentations regarding the lost knowledge that had existed in the old India has leaded us to a discovery extremely interesting, an evidence that have passed under the eyes of several scholars without to have received the right attentions.

In the huge work titled *The holy Hindu's books* (1.923), a kind of encyclopedia where they had tried to collect all the holy texts of this people, on its thirteenth book titled Sukra Niti[68] dedicated to the economy and to the politic, there is a treaty of military science from the unimaginable contents.

Mindful of the fact that on the most unsuspicious books had been found some referring and descriptions on the Vimana, as for example on the treaty of architecture called Samarangana Sutradhara, at the page 235 of the Sukra Niti we have been attracted by a paragraph where had been described several kind of "fire arms".

The understanding of these knowledge can let us first of all to probe the high level acquired in the remote Indian past from its civilization as also to contextualize, and perhaps also to explain, several of the incredible arms described into the holy texts or into the epic sagas.

Perhaps it is through the Sukra Niti that the legendary epics could now really come out to the history not anymore in their mythological form but as a poetical transposition of some facts really happened.

First of all the Sukra Niti is ascribed to the wise man Sukracharya, a legendary figure also named in the Mahabharata.

Therefore the text was made in writing only in a recent time, several centuries ago, probably through that passage from an oral tradition, called Sruti, to a writing knowing, that has involved several texts in this country. Here it is the passage extract by the text:

"Astra is what has been thrown or made to be fallen through some spells, carts and fire. Sastra is every other arm as swords, daggers, kunta etc...".

"Astra is of two types, magic or tubular. The king wishing to win, has to use the tubular type, where the magic one doesn't exist, together to the Sastra".

The experts of military instruments know the several types of Astra and of Sastra that change with the form or dimension.

The *Nalika Astra* (tubular or cylindrical) is of two kinds: the small one and the big one.

The short, or small, *Nalika* is an instrument used by the infantry and by the cavalry. It has a big horizontal hole and another one perpendicular on the base. It is long about five *Vitastie*[69] (about 114 cm, N.d.A.). It has a point on the mouth and another one on the base that had been used to aim the object. The fire is made by a mechanism. Inside that there is some gunpowder and it has a good handle of hard wood.

The big Nalika is inside a wooden frame, is transported on a truck and has on its basis a wedge, moving that can be pointed toward the target. If it is well used, will bring to the victory.

---

[68] Major B.D. Basu, *Sukra-niti* (Hindu Economics and Politics), *The Sacred Books of the Hindus*, Vol. xiii, Allahabad : The Panini Office Bahadurganj, 1921-1923, versi da 381 a 409.

[69] 1 vitasti (palm: distance between the tip of the thumb and the tip of the little finger when the hand is open) = 188 mm to 250 mm.

Five *palas* of *suvarachi* salt (a salt similar to the alum, N.d.A), a shovel of sulfur and a shovel of coal of arka's wood, snuhi and some other burned tree in a way to avoid the coming out of the smoke, they have to be purified, pulverized and mixed together, so they have to dissolved on disnuhi's juice, arka and garlic, dried with a moderate heat and at last pulverized as sugar.

This substance is the gunpowder!

The balls are made of iron, with or without any other substances inside. For the smallest *nalas* the balls are made on lead or on some other metals. The *nala astras* can be made on iron or of some other metals and must be cleaned every day and watched by armed men. The experts made the gunpowder on several ways and on several colors, in consequence of the ingredients used: coal, sulfur, *suvarachi* stones, *harital*, lead, *hindu*, iron filings, camphor, *jatu*, *indico*, *juice* of the tree *sarala* etc.

The balls introduced into the instrument are thrown toward a target because of the touch of the fire.

The instrument has to be cleaned before, then we introduce the gunpowder, and we press lightly through a stick.

So we have to introduce the ball, then some other power had to be placed on the ear of the weapon.

Finally we have to approach the fire to this powder and the ball is thrown against the goal.

Certainly the term 'gunpowder' is used by the translator, where in some other texts such mixture was called with the name of *Agnicurna*. The Major Basu commented *"We found ourselves in front of the description of some rifles and guns very similar to those modern, surely it is a later interpolation"*.

Unfortunately this comment does not give the truth to the history and it is the consequence of such colonialism that saw the Hindu a population of barbarians from the picturesque habits and mores.

To face the reality should have meant to remove the technological supremacy to England and, with such annotation free of every counterpart to support that, they cleared quickly the all matter.

Unluckily there are some other Hindu texts that tell us a different story.

The type of the firearms just described concerning some riffles and guns with a muzzle loading is of a kind simple enough. Anyway it is undoubted that such technology could give the predominance on a battlefield and, if refined, could have let to have some real and proper weapons of destruction!

Such science could not bet cited only by the Sukra-Niti, so our examination has expanded making us able to discover the Agneta Astra, from the term Agni that on Sanskrit literally means *fire*.

The wise man Kautalya, in his Arthashastra, describes a weapon called Agni Bana[70] and mention three types of ' recipes' to get it, the agni-dhrana, the ksepyo-agni-yoga and the visvasaghati.

The Visvasaghati was composed '...*by some powders of all the red metals as the fire or by a blend of power of kumbhi, lead, zinc, mixed with coal and with wax and oil of turpentine'*. The compound that derived from that is highly flammable and it is difficult to be extinct.

On 1.932 the Indian scholar Chandra Ray stated that *"The Visvasaghati-Agniyoga practically was an explosive bomb which its metallic scraps were thrown far away. The Agni-bana had been the precursor of the gun"*[71].

Instead Sir A.M. Eliot remember how the Arab had learned how to make the gunpowder from the India and how before their commercial relationship the Islamic used some arrows soaked with naphtha.

However the Persia owned some saltpeter in abundance, the studies that had been made tent to indicate that the original source of the gunpowder was just the India.

Also the German Gustav Opper the translator of the Sukraniti was of the same opinion[72].

During his studies he found many evidences of a knowledge, and utilization, of these typologies of weapons from the old Indians[73].

On his text *Political Maxims of the Ancient Hindus,* Oppert states that the term *Sataghni* used in the Sundara Kanda of the Ramayana could not less than referring to some real and true cannons[74].

Oppert's contribution goes over showing, as a half dozen of temples on South India owned some finding of the use of the fire weapons in antiquity.

Palni's temple, in the district of Madura, own on his outside part of a *mantapa* some carved scenes on the stones with some soldier that take in their hands small firearms as described on the Sukranitisara.

On Sarnagapani's temple of Kumbakonam, on the fifth history of the principal portal it is represented a seated king on a truck drawn by horses and

---

[70] Ganga Ram Garg, *Encyclopaedia of the Hindu World*, Concept Publishing Company, 1992.

[71] Rai Bahadur Jogesh Chandra Ray, *Fire-Arms in Ancient India*, I.H.Q. (Indian Historical Quarterly), vol. VII, 1932, p. 586-88.

[72] Gustav Oppert (1836-1908) born in Hamburg, Germany, taught Sanskrit and comparative linguistics at the Presidency College in Madras for 21 years. He was the Telugu translator for the government and the curator of the Government Oriental Manuscript Library.

[73] Valentine Stache-Rosen, *German Indologists: Biographies of Scholars in Indian Studies writing in German*, p.81.

[74] Dewan Bahadur e K.S. Ramaswami Shastri, *Hindu Culture and The Modern Age*, Annamalai University 1956 p. 127.

surrounded by a group of soldiers. Before the truck two Sepoys seem to keep on their hands the small *Nalika* otherwise said *Nalastra*.

In Conjeevarn's temple in the Nurrukkal Mantapan, we find instead on the Nord side of the mandapa, a pillar where had been represented in relief a soldier that seems to held a firearm against an enemy. Similar scenes had been seen also on Tanjore's temple and on Perur's temple on Coibatore's district.

Canda Baradayi, who lived between the 1.126 – 1,192, speaks about small firearm similar to the Nalika but called Tupak[75].

The encyclopedic test Manasollasa, made by Sri Sureshwaracharya on 1.130 A.C., speaks about the installation of the war machines (*yantras*), with the same characteristics that we have described till now, on the walls of the forts.

The architecture text Samarangana Sutradhara, made up by the king Bhoja around the 1.050 A.C., speaks about the use of the *Sataghni* (literally ' the killer of hundreds of men') as an instrument made to defend the forts.

To the light of the documentations reported till now it is possible to trace an evolution in the use of the gun powder and of the fire gun on the old India, an instrument that as we said before may let us also to understand under a different shape the incredible destructive weapons used by the divinities during their epic fighting.

The Vedic hymns are rich of compliment towards Agni, the fire, able to defeat the enemies. The Atharva Veda, one of the four Samitha, or holy texts of the Vedic religion, made on the 1.200 and the 900 B.C., describes the use of the firearms with some lead shots.

In his absence the Agnicurna, or gunpowder, is composed by 3 til 6 elements that are:
- Saltpeter
- A part of sulfur
- A part of arka's coal, sruhi and some others trees that have been burned in a pit and reduced on powder.

According the Indian historian Ramachandra Dikshitar[76], the described composition on the Sukra Niti could be probably dated at least to the time pre-Gupta, an empire that reigned in northern India among the 240 and the 550 A.C., but the references present on the Atharva Veda seem to anticipate of about thousands years the use of this explosive compound.

We are in front of some technologies that go back dawn of time and its use could be possibly changed into some more allegorical and epical forms as to describe some weapons of a divine nature.

---

[75] G. Kuppuram, *India Through The Ages: History, Art Culture and Religion*, p. 512-513.

[76] V.R. Ramachandra Dikshitar, *War in Ancient India*, 1944, Motilal Banarsidass Pub., 1987, p. 103 -105.

Perhaps, this possibility could explain the events happened in the town of Mohenjo Daro, today in Pakistan, but perhaps do not explain totally those that are the descriptions presents on some technical texts as the Amsu Bodhini or the Vimanika Shastra or as some Astras's descriptions reported in some old religious texts.

The border between these two realities is very thin but certainly the knowledge inside the Sutra Niti could constitute the tangible test that something of extremely more complex and dipper has existed in our planet.

Amsu Bodhini

The Vimana Shastra is not the only forgotten pearl in the Indian past. Less famous, but as many rare and fascinating, is another text with the contents at least incredible, a treaty of which we have[77] only the first chapter[78], but that, on his known extension of 187 pages, it shows us some contents extremely amazing.

First of all, to known the Amsu Bodhini it is basic to rediscover his origin that is attributed to the mystical Bharadvaja the author of the Vimanika Shastra.

The transposition of the Amsu Bodhini from the oral tradition to a written text had been made, also on this case, by Subbaraya Shastry, the Pandit or wise man who had brought the light the several time cited Vimanika Shastra. In this case more than in others it seems to strengthen the hypothesis that a wisdom tradition extremely evolved and old was preserved orally during the millenniums through the Sruti.

After a personal research lasted several years we have been able to recover[79] two copies of this treat, the first one on his original version[80], dated 1.931, and the second one[81] on its edition commented by the technical – scientific side issued on 2011 made by the Sanskritist Narayan Gopal Dongre and from the physicist of the particles Shankar Gopal Nene.

"The evolution of the earth into the galaxy start from the sun, on the same way when the creation of the galaxies took shape from the primordial atom".

---

[77] The Amsu Bodhini consists of 12 chapters and more than a thousand sections.

[78] Titled *'Srstyadhikarah'* or *'The Evolution of the Universe'*.

[79] Kept in the library of the Oriental Institute, Vadodara, Baroda.

[80] Maharshi Bharadvaja, *Amsu Bodhini Shastra*, Edited by Pandit Subbaray Shashtri, Printed by V.B. Soobbiah of Bangalore, Published by F.K. Dadacharji and R.R. Mody of Mumbai, 1931, 187 pagine.

[81] Narayan Gopal Dongre e Shankar Gopal Nene, *Bharadvaja's Amshubodhini: Ancient Indian Treatise on Cosmology and Physics of Nuclear Particles*, 2011, 156 pagine.

These are the words with which the Amsuboodhini starts, text that after the analysis made by Nene and Dongre has resulted to be a treat of cosmology and physical of the nuclear particles.

The verses of the book describe in a very elaborated way the evolution of the cosmos from the Big Bang from the creation of the universe till to that one of the solar system and it is interesting to note how all the definitions given to all these phases are comparable, and also on certain case speculate, with the modern description[82] given by Hans Stephani on his book *'The general relativity'*.

The chapter available today also presents a Bodhananda's commentary the same historical of the antiquity to which we have had the explanatory notes presents on the Vymanika Shastra.

It should be complex to analyze in detail this text but trough the words of the two scholars it is possible to retrace the most important elements as well as the incredible range of its contents. The Amsu Bodhini is fundamentally a cosmological treaty that describes the evolution of the universe. Through his pages it is explained how the cosmos had been originated by the Bindu vishput/Maha vishput[83] a term with which today we could identify the Big Bang that brought to the composition of the solar system and of the Sun[84].

The text, ascribed to Maharshi Bharadwaja, was edited and dictated by Pandit Subbaraya Shastry on 1931, therefore after the dictation of the Vimanika Shastra[85].

The treat is described by the two scholars as a technical manual focused on the *"Crystals, the mirrors and the solar energy of the old India"* and, curiously, these same topics seem to fall fully inside the technologies described by the Vimanika Shastra.

The introduction to the book had been prepared by Subbaraya Shastry himself and it result exceedingly illuminating, as clarifying, on several points till today remained completely obscures.

Shastry described into the detail the time calculation and the units of measure of the old Indian science, starting from the Nimisha, namely from a 'blink human' till to reach the time of one year.

---

[82] Hans Stephani, *General Relativity: an introduction to the theory of the gravitational field*, Cambridge University Press, 1982.

[83] The Maha vishput is also mentioned in the Rig Veda.

[84] Dr. A.S. Nene, Former Professor of Civil Engineering, VNIT, Nagpur, *Introduction to Amshubodhini Shastra by Maharshi Bharadwaja*, Bouddhik Sampada, 2012.

[85] The first edition of this rare text was given to the press by two Indian scholars and devotees of Shastry, V.B. Subbiah of Bangalore and curated by F.K. Dadacharji and by R.R. Mody of Mumbai.

He also described the extension of the time in all the different Yuga arriving to the worth of 4.320.000 years, definite Maya Yuga[86] and cited on 24 old scientific texts not more available to this days.

The Pandit also mention an old text that today had been lost, written by the wise man Agastya (one of the seven Indian wise men) and titled *"Shakti Tantra"*, literally 'Treaty of the Energies', the book is mentioned also by the Vimanika Shastra[87].

Agastya describe 32 types of different kinds of electricity as well as the corresponding machines that could be built using them. Shastry underline as *"All the creations take place thanks to the rays of the sun and so the text speaks of the property of the star"*, but Shri G. Venkatachala Sarma, Pandit's adopted son and the translator from the original Sanskrit text to the English, to make available to the great public the book mentioning on the introduction itself the big limitations that had become evident into the transposition of the book, some links that belonged to a perfect language as the Sanskrit, which the literal meaning is 'perfection', in relation to some languages more poor as for example the English language. Sarma remind us also as the old wise Indian men had focused their own asceticism first of all through a spiritual quest without however to set aside the physical and material part, structured partly into Yoga's practices to control their body through their mind. Such a reality, goes on Sarna, had not delineated a net and right demarcation with the other sciences that were resulted to be interconnected between them and in which the religious contest could melt with their material side with no contradiction.

The Asmu Bodhini contains 50 sutras, or aphorisms; among the many points the most significant, and perhaps also incredible among the pages 60 and 77[88], they speak about a Spectrometer. A Spectrometer is an instrument that let us to measure the spectrum of the electromagnetic radiations namely the property of the light in function of its wavelength.

The majority of the spectrometers exploit the principle of the interference to decompose the light radiation on its wavelength and to measure the intensity

---

[86] Kalpa: Unit of time Indian, having a total length of 8.64 billion years, which is equivalent to 2000 Great Yuga (v.) Or Maha-yuga, each of which is 4.32 million years. This result can also be obtained considering that the time of Brahma is divided into four Yuga (age), which together add up to 12,000 years (Maha-yuga) which, multiplied by 360, or the number of days in a normal year, would result in the 4.32 million years. Seventyone Maha-yuga form a Manvantara, and 14 of these constitute a Kalpa of Brahma. So, in summary, 4,320,000 x 71 x 14 = 4,294,080,000 years of our present calendar. The Kalpa is also defined as "a day and a night of Brahma".

[87] In VS is written that *"In Shakti Tantra is written, that with a projection of the light beam Rohine, the things in front of the Vimana are made visible even in darkness"*. Rohine means red, the meaning of the phrase may therefore indicate an infrared device.

[88] In his commentary on the text, Bodhanada recalls how the description of this machine, called Chhayaprakasharna Yantra, appears in the 17th aphorism.

of them. Bharadvaja's Spectroscope absolves to all these functions structuring instead in a different way from those instruments that we own today.

In certain cases the elements from which are composed are different in their shape from those modern, as the prism to divide the bright spectrum, but the result is the same.

Basing on the description contained by the Amsu Bodhini, Shri Dongre developed this complex instrument and presented his work to the international scientific community in two articles that had been published on the Indian Journal of History of Science[89].

Bharadvaja's spectroscope, called '*Dhavantapramapaka-yantram*' or '*Tamapramapaka-y ant ram*', is a device made by five yantra able to measure the spectrums of the five optical regions, infrared, visible and ultraviolet. Among the fundamental points emerged by Dongre's researches the scholar stated that: "*The nature and the proprieties of the ultraviolet, of the visible and the infrared radiation were well known by the old India. They knew the properties and the methods to realize these materials useful to study these radiations and supposedly, they knew how to use and theoretically how worked and also these and the four kind of spectroscopes*".

The described instruments are of a new typology of spectrometers not diffused in the modern scientific world. On parallel, thanks to the annotations contained on Amshubodhini's 24°aphorism, the scholar Shri Didolkar together with his team of the VNIT (Visvesvaraya National Institute of Technology) of Nagpur, developed[90] a glass infrared transparent ceramic.

This new material had been tested by the researchers of Kolkata's CG&CRI (Central Glass and Ceramics Research Institute) and stated the authenticity and concluded that such ceramic could be used for the most disparate uses.

The Amshubodhini's 24° aphorism describe the five basic elements, called Panch Mahabhutas, to realize the ceramic alloy made by Didolkar; subsequently the text cites some others substances that are useful to realize the compound: the 10 Aakash's powers (the Sky), the 13 Vayu's powers (the Air), the 8 Agni's power (the Fire), the 10 Aap's powers (the Water) and the 5 Pruthwi's powers (the Earth).

Instead on the 8° Amshubodhini's aphorism are listed some materials, in relation to some old texts where we deepened their use, useful to manufacture the glasses, prisms and mirrors. Their use diversified from the manufacturing of instruments as Bharadwaja's spectrometer and some machines, Yantra, that should be mounted afterwards on the Vimana.

---

[89] The first, N.G.Dongre, *Dhvantapramapaka Yantra of Maharshi Bharadwaja*", Indian Journal of History of Science, 29(4), (1994), pp. 611-627. The second, N.G. Dongre, *Spectroscopy in ancient India, an application of Spectroscopy to Astrophysics*, Indian Journal of History of Science, 33(3), (1998), pp. 229-238.

[90] V.K. Didolkar, *Infrared transparent glass ceramic as per ancient Indian text "Amshubodhini"*, Jr. of Bharatiya Boudhik Sampada, May 2012, pp. 28-32.

The next table shows the materials listed on the 8° aphorism.

Table 1- Substances used for the preparation of the glasses, prisms and mirrors etc.

Sn	Material	Type	Reference text
1	Oils	64	Unknown
2	Water	100	Apatwaprakasika
3	Miror (Darpan)	1914	Darpana Prakaran
4	Glasse(Kach)	407	Kach KalaKoushalya
5	Kristals (Mani)	6	Manikalpataru
6	Stone	12	Shilodaya Prakaran
7	Metals	7	Loharatna of Shakatayana
8	Trees	12	Agatatwa Lahari of Aswalayana
9	Vegetable wax (Lakh)	3	
10	Leaves	6	
11	Flowers	7	
12	Fruits	8	
13	Pearls	8	Vajrakalpa
14	Precious stones	9	Ratnakara
15	Cloth	12	Patakalpa Mahadadhi
16	Hides	5	Charma Shastra

| 1 | Mica (mineral) | 3 | Abhraka kalpa |

7

As we said before, it is highly possible that the old people knew the listed substances into its most pure form, or into their chemical constituents, and they knew how to purify and to work them to reach to realize the substance or the machinery that they wished.

It is interesting to notice as Bodhananda, commenting the Amsubodhini, states that the description of Bharadvaja's Spectrometer contained in the text, is the same present[91] on *Dhvanta-vijhanabhààskara* a book made by Sarikanatha among the 780 and the 825 A.C., in its turn based on the *Yantrasarvasva*[92] by Maharshi Bharadvaja.

This fact let us to see how, over the centuries, some knowledge had remained inside a small cultural elite and occasionally emerged to the honor of the history.

The same phenomenon is found also on Samarangana's Sutradhara pages compound by the king Bhoja during XIII century A.C.

The Professor S.N. Takufr, a physician of the University of Benares, trying to reproduce this machinery and following the instructions of the Amshubodhini, had been able to reconstruct the *Dhavanta pramapaka yantra*, Bharadvaja's Spectrometer able to analyzer the light coming from the stars. The producer Diego D'Innocenzo, on 2010, interviewed the Professor Takur for the documentary that was realizing with the title *"The Vimana, the flying God's Charriots"* (broadcasted on RAI 2, National Italian Television, within the transmission Voyager).

The physician said, *"...the peculiarity of this spectrometer is in fact a crystal prism. A conical prism, very different from those that we use today. The prism and the convex lenses made a spectrum that is projected on a screen in the form of bright rings with different colors that are corresponding to the respective colors present on the light source"*.

Much more explicit was his position on the authors of this instrument. *"... perhaps, thousands of years ago, our ancestors studied the circles of light as those one coming from the Sun and from the other stars... and so they had been able to categorize the celestial corps in different groups, all this happened much before of the modern classification of the stars"*. This statement could allow us to understand the reason why the historical India, since its most remote past, had been elected as unmoved

---

[91] Prakarana Pancikâ, Ban Aras Hindu University, pp. 56-57, 1961.

[92] In the Yantra Sarvasva, Maharshi Bharadwaja described the spectrometer as the 109° instrument of the text, a yantra composed of 32 auxiliary components that is able to analyze the radiation in general. Furthermore, according Sârikânàtha, of the 32 components from which it is already composed only 13 are sufficient to analyze 'Tama', ultraviolet rays, UV.

motor on the evolution of the science offering to the world some of the most excelled minds as well as of the discoveries and of the most sublime knowledge that the mankind had managed to conquer. Instead, the fundamental point, had not yet found an answer.

There is no doubt that in India the knowledge and the scientific thought had had a fundamental role since its ancient past.

Some texts as the Amsu Bodhini, the Vimanika Shastra, the Surya Siddhanta, to name just a few, are a condensed of notions and scientific information of immense value and depth.

The same Rig Veda or the Purana, contain within them some notions extremely high of historical type, anthropological, scientific that only in recent times the scholars had started to understand and to decode.

This immense heritage, the old Indian science, at a certain moment started to take a role increasingly marginal and slowly disappeared on the fogs of the history.

The professor S.G. Nene, a Sanskrit of the University Of Pune, said "*…it was among the IX and the X century B.C. that started the decline of the knowledge. The rational thought started to decline and was completely replaced by the intuition and by the mysticism that are deriving from the meditation*"[93].

From the Vedism, on the VII –VI century B.C., developed the Brahmanism and after, around the VII-VI century B.C., the Hinduism as we know today.

It is through these historical and religious passages that we witness to the decline of the old Indian knowledge and to its substitution with a tight religiosity based on the rituality and on the mysticism.

"*But the heritage of the scientific knowledge did not gone completely lost and survived largely through the oral tradition. Still today there are some Sanskrit schools where the knowledge is given orally, word of mouth. A way that had let to a part of this old knowledge to arrive till us*", state the Professor Nene.

Cosmology of the old India

The Amsu Bodhini is a fragment of a knowledge extremely wide and elaborated and it belonged to a remote past, a knowledge that passed into the epic texts as into the oldest Indian religious texts. The Vedic universe, or the concept of the cosmos and its rules that the old Indians owned, is something of extremely complex and articulate. The astronomy, the cosmology and the astrology become a unique case.

In this vision the totality of the visible universe, the macrocosm, is composed by a microcosm of infinitesimal particles.

---

[93] During an interview with the director Diego D'Innocenzo.

Only if we consider such conception, we are catapulted into the theories of the modern physics but if we look at the Indian past times we will discover as on the VII century B.C. Achrya Kanad, or Kanada, had been the true and the first postulator of the atomic theories, two centuries before the Democritus and Leucippus formulations.

Acharya Kanad was also the founder of one of the six most important philosophical Indian schools, the Vaisheshik Darsan. Kanad classified all the objects of creation into nine elements:

> The earth, the water, the light, the wind, the ether, the time, the space, the mind and the soul

On the VII century before Christ he stated, *"every objects of the creation is composed by atoms* (Anu in the original formulation, ndr) *that are in their turn in touch between them forming the molecules"*[94].

Kanad described also the dimension and the movement of the atoms and their mutual chemical reaction.

His formulation anticipated to the world John Dalton's atomic theory, postulated 2.500 years after. The eminent historical T.N. Colebrook stated, *"Kanad and some other Indian scholars respect to European scientists, had been the global teachers on this field"*.

The wise men of the old India measured and learned every imaginable phenomenon, from the orbits of the planets to the fluctuation of the subatomic particles.

Such peculiarity was codified in an advanced system that stretched between some valor of 10^{-8} and of 10^{421}, a system used and mentioned by Buddha himself and where to every strengths was attributed a name.

Much before Acharya Kanad another great wise man, which was celebrated as the founder of the Samkhya philosophy, had left to the world a heritage of knowledge of exceptional value.

Acharya Kapi (also called Kapila) was the father of the Indian cosmology and the traditions[95] want that he lived around 3.000 B.C., during the Harappa's civilization.

In the old India the numbers were not some abstract concepts, so the question naturally arise of what they could measure with the value of 10^{421}.

---

[94] Taken from *The Indian Encyclopaedia*, Volume I, edited by Subodh Kapoor, under Vaisesika Sutra and Acharya Kanad, Cosmo Publications, 2002.

[95] Dasgupta, Surendranath, *A history of Indian philosophy, IV: Indian pluralism*, Cambridge University Press, 1949.

On the *Srimad Bhagvatam,* on the chapter 11, it is told that the division of the time had calculate on the basis of the range that the Sun employs among the smallest and the biggest dimensions of the matter[96].
These divisions are:

- Paramanu = primeval atom
- 2 Paramanu = Anu
- 6Anu = Trasarenu
- 3Trasarenus = Truti
- 100 Truti = Vedha
- 3 Vedha = Lava
- 3 Lava = Nimesha
- 3 Nimesha = Ksana
- 5 Ksana = Laghu
- 15 Laghu = Nadika
- 2 Nadika = Muhuruta

A Muhutra defines 48 minutes, anyway a Paramanu should amount to 1.3×10^6. The time scale continue with:

- Divasa = 24 hours
- Saptaha = 7 days
- Masa = 4 weeks
- Varsa = 12 months
- Satabda = 100 years
- Sahasrabda 10 centuries
- Deva yuga = 12.000 Deva's years or 4.32 millions of human years.
- Manvantra (day and night) = 71 Deva Yuga or 306.72 millions of years or 613.44 millions of years (in its entirety).
- 7 Manvantra = 4.3 billions of years, the age of the Earth
- 1 Kapa (corresponding to 14 Mantravantra) 8.64 billions of years, the duration of the life of the Earth.
- 100 Brahma's = 311 trillions of years, the estimate duration of the life expectancy of our universe.

At the present moment we are living on the 7° Mantravantra, this means that the age of the Earth according to the Indian calendar and chronology, should be 4.3 billion of years.

---

[96] For the following definitions see Gītā Śloka (8.17) and especially Burgess, Ebenezer, *Translation of the Sûrya-Siddhânta: A text-book of Hindu astronomy, with notes and an appendix,* originally published: *Journal of the American Oriental Society,* 6 (1860), 141–498.

On the 5 A.C. the Indian mathematics[97] calculated that the age of our planet was about 4.35 billions of years.

In modern times we could calculate, with the most advanced technologies, that the age of the Earth is around the 4.3 and the 4.5 billions of years!

It is completely incomprehensible how we could reach to some so accurate valuations without the modern scientific instruments and we feel that it is not a coincidence.

Perhaps this classification could result to some readers a heavy exercise of transcription of old texts or perhaps we could believe that such measures are completely arbitraries and valueless.

In reality these estimations and classifications are corresponding almost exactly to the most recent scientific discriminations, they show us exactly how this people, in this case the ancestors of the today Indians, the Veda and perhaps the same Harappa's civilization, had been the custodians of a knowledge extremely high and a product not only of a logical speculation but more probably of previous scientific achievements acquired or accrued by an unknown source.

To get an idea of the importance of this simple as incredible knowledge, we think as Aristotle considered that the Earth was eternal, for the Chinese Han the Earth existed in cycles from 23 million of years, the count of Buffon on the 1.799 estimated an age of 75.000 years instead John Phillips calculated, from the analysis of some stones, that the Earth had about 96 millions of years.

William Thompson on 1.862 published some calculations according to which our planet had an age between the 20 and the 400 millions of years instead the German physicist Hermann von Helmholtz and the American astronomer Simon Newcomb, estimated an age of 100 million of years.

At the beginning of the XX century Bertram and Rutherford, using the first systems of carbon dating, estimated an age of 1.6 billion of years. How could the Indian mathematical on the 5 A.C. to be able to achieve estimation so precise to be almost completely right to those calculated on the last decennium?

We were speaking of Acharya Kanad's formulations on the VII century B.C., if we observe such theory we will be surprised to note its total modernity.

According to Kanad there exist five principal elements:

- *Pritvhi:* atomic elements
- *Vayu:* the attraction and the repulsion strengths
- *Tejas:* the electromagnetic energy and the photons
  Apas: the fluids

---

[97] Nancy Wilson Ross, *Three Ways of Asian Wisdom: Hinduism, Buddhism, Zen and Their Significance for the West*, Simon & Schuster, 1978.

- *Akasha:* the vacuum energy

Kanad seems to have codified all these concepts but before him there is a text that seems to give us some oldest and circumscribed concepts on the existence of the atom.

The Bhagavat-puranam, known also as Paramahamsa Samhita (III/XI[98]), according to the tradition was asserted by Sri Sukadeva Goswami[99], 30 years after the beginning of the Kali Yuga, the current era, exactly 3.072 B.C.

We have extrapolated some verses, transposing them on this book, without altering any form.

Their contents are much more interesting because shows us how certain scientific knowledge had been preserved during the millennia codifying their contents under a religious way.

So it is definite a concept already known in modern times, the "atomic time" definite on the third Skanda (book 3), chapter 11, verses 1, of the Bhagavata Purana.

Here it is the translation of some verses:

- *Verse 1:* (the great wise man) Maitreya said: "The last particle of the material manifestation, which had not been combined to any other similar particle, is called Paramanu. Paramanu always exists both in the been dormant that in those manifested of the material existence. It is the combine of more of one Paramanu that give rise to the illusory concept of unity (material)".
- *Verse 2:* (Paramanu) it is the all manifested material existence, regarded as a whole not specified, and before to come back to a way not manifest (dormant), is definite as a the biggest compound (material).
- *Verse 3:* We are able to understand the long or short dimensions of the time (material), as a power of supreme God's strength that all pervades and that, under the shape of the Sun, pass through the big and small dimensions of the things (materials)
- *Verse 5:* The combination of two paramanu is an Anu (atom) and three anu (atoms) are a trasarenu. The Trasarenu are visible [to the naked eyes] if are seen fluctuating in the air upwards while we are watching through the rays of the sun that are entering into a room through a window.
- *Verse 6:* Three Trasarenus are called *Truti* (8/13.500 part of a second), that is the measure of the time used (by the Sun) to cross three Trasarenu. The combination of hundred trutis is called *vedha* (8/135 parts of a second), and three Veda all together are known as *lava* (8/45 parts of a second).

---

[98] Srimad Bhagavat Maha-purana, Chant III, Chapter XI.

[99] Son of the legendary sage Vyasa and a man who according to tradition had put in writing the Vedas.

Summarizing some of the described measures, we find that a Truti, the union of 6 anu/atoms, is corresponding to the 1.687,5 part of a second.

Starting from this base we find the corresponding time to the other measures or a *Kastha is* corresponding to 8 seconds, a *Laghu* is corresponding to 2 minutes and a *Danda* is corresponding to 30 minutes. Keeping on we observe as 2 Danda are corresponding to an hour and 6 Danda to 3 hours (a quarter of the day).

The fundamental answer is how they could made such measure so precise, and corresponding to the physical reality, if not measuring them directly.

Is it credible to think that they had obtained them only through the logical-deductive processes? How could they have realized such precise measurements without instrumentations and only with the use of the logic?

To get to define a Truti as the 1.687,5 part of a second is already perceived as something of upsetting! To read this book made us in front of some questions that is not possible to liquidate generically and it give us a vision of the history and of the human 'Knowledge' entirely different from that one present into the myths and into the holy texts of the western tradition.

It is immediately clear a net differentiation among these two worlds, an abyss that become stronger in front of a double message, that one religious tended to explain the created as a manifestation of a superior will flanked by the rationalization the same that the wise men of the Indian past had and that own in its inside a real scientific base.

Now we do not known what is the origin of this knowledge, there are some hypothesis but no one such to exclude completely the others.

It's up to us to reconstruct this intricate enigma with the certitude that the final vision that we could obtain will certainly give us a new understanding not only of ourselves but especially of our history.

Atomic philosophies

Showing our attention from the Indian cosmology to Acharya Kanad's 'atomic philosophies', we may observe how the level of complexity reached by this people is becoming exponentially more complex.

There are two lines of thought, two schools that illustrate us the concept of the Vedic physic and its big secret.

The universe is called Prakriti and is manifested by a vibration of the Svra, a current of vital force called Parabrahman and Purush.

This makes it that vibrating, origins the five thin elements; every differentiated vibration originates the Pancha-Mahabhttas, the five physical elements.

These elements follow on another in the following order of aggregation:

Akasha (ether), Teja (energy), Vayu (strengths/fields), Pritvhi (atomic elements) and Apa (fluids)

Except in the case of ether, all the elements are compound, for the Vedic conception, by distinct and indivisible particles, the Paramanu. These concepts are contained on the *Vaisesika Sutra,* a treaty ascribed to the mentioned Kanada.

The text in its modernity shows us some concepts rather astonishing. Here there is a synthesis of the part concerning the origins of the universe.

The ether has the highest level of vibration and was the first element to be formed and from it the other forms had been originated. After that, the great breath blown on these five principles from which had originated the physical elements that made the space-time.

Through Vayu, the strength of attraction and repulsion that after shaped the Brahmand an universe with the shape of an egg in expansion, a sphere or a primordial seed called Hiranya Garbha (the origin of all the energy) in which is contained the sum of the Prakriti in the universe and from which started an explosion (Bindu Vishput) and the third element, the Teja, the electromagnetic energy.

The Teja expanded and formed the fourth element said Pritvhi from which had formed Apa, the fluids, completing the creation. To be a 'myth' of the origins we have to recognize that it is at least elaborated.

If we compare the concepts just exposed with those of some others cultures, we immediately can see a clear line of demarcation that project the Vedic conceptions into the most modern scientific theories.

The *Vaisesika Sutra* continues by stating that all the material of the universe is composed by five Tattvas, five vibratory modes of the vital force. All the Prana consisted by several points belonging to several classes depending to the vibratory state of the Tattva.

Every point of the Prana is a microcosm of the universe and is inclusive of all the five Maha Tatavas in varying proportions[100].

The cosmic seed, or the state in which is located the Universe to his origins, is described on the Rig Veda[101] and on the Atharva Veda[102] with the term *Hiranya Garbha* that, literally means ' the seed or the lap of the whole energy'.

---

[100] These same concepts are also expressed, in the simplest way, even in the Bhagavad Gita and the Upanishads Vedas.

[101] RV 10.121.1

[102] AV 4.2.7

The Hindus spirituality exposes several elaborated concepts on the creation and everyone agree to assert that every universe, at the end, dissolves itself to create another one.

This transformation, the dissolution for a new generation is the Bindu Visphot that occurs periodically in a rhythmic process of death and rebirth. The modern theories called them Big Bang or Big Crunch.

A hymn of the Rig Veda[103], dated on the 3.000 B.C., said that at the beginning of the creation there was neither to be nor not-to be, only the dark.

From the cosmic void was originated the heat, from the heat the desire and from the desire was originated the spirit. It was the time where the Gods did not existed and the hymn goes on saying that some Wise Seers tried to join the emptiness with no form into a primary reality.

Another hymn[104] say that the '*Lord of the Hymns*' initiated the creation with a cosmic explosion from which originated the Deva, the Gods, and with them the existence of the non-existence.

This proceeding of dissolution and of generation always repeat itself; according the Rig Veda[105] an universe has existed before that one that we are living now and there will be another after the current one.

Also the Shiva Purana presents the same creative myths, underlining how all the existence is "*a perpetual oscillation of the current of the vital wave*".

The Vedic creationistic model, if it read under a scientific lens, cannot be understand calling into question only one scientific theory but it seems to combine inside the most recent postulation of the physic. If we would try to identify the correlations among the modern scientific theories and the concepts present on the Veda we will find several disconcerting similarities.

The theory of the Super Strings says that the matter is a vibration of the vital force. The theory of the *inflated and symmetric universe* states that at the beginnings of the creation there was a symmetry of the Four fundamental forces (electromagnetic, weak nuclear, strong nuclear and gravity) and those were unified into only one force that existed as a singularity.

After a period of the following time to the zero moment of the Big Bang, equal to 10,43 seconds and definite Plank's time, the gravity was split as the first fundamental force breaking that symmetry that had formed the singularity with the other forces and starting the cosmic symmetry or a rapid expansion that would have created the universe.

Instead the Vedic theorization says that the ether has been the first macroscopic element of the universe, the reality from which had originated the

---

[103] RV X.129

[104] RV X.72

[105] Rig Veda, 10-190-3 and alsoGita, 9-7.

Vayu, the force of attraction and repulsion, that formed the primordial seed where there was contained all the material, the energy and the space (the singularity).

The Vayu burst into the Tejas, the electromagnetic energy, that condensed forming the Pritvhi and the life was created after Apas's generation, the fluid. The *theory of the quantum vacuum* say that the universe is in a constant state of boiling of some couples of virtual particles that steadily born and cancel themselves producing a dynamic balance among the material and the antimatter.

Because of some unknown effects it was originated a singularity. The Vedic word states that the first elements to be created were the Thin Ethers that manifested, afterwards, as a primordial seed.

It is to be noted that the shape of the Universe is described as spherical, however this concept would seem to contradict the Vedic postulation of an infinite multiverse.

We could bypass this problem if we should insert a unique constant, the hyperspace a space that has a superior numbers of geometric dimensions superior to the 3 of the traditional physical space.

The Veda's speed of light

A reflection had to be made on the reality with which the modern world daily lives, the speed of the light. Its official discovery was made on 1.675 by the Danish astronomer Ole Romer, but it was not completely accepted until James Bradley's studies, on 1.727 confirmed the validity and the value of 300.000 km/s.

To read the Rig Veda it seems to reserve some surprises extremely interesting. The hymn 1.50 dedicated to Surya, the Sun, reads textually "…[Oh Sun] *I bow to you, You that are coming to us in 2.202 yojana in the middle of nimesha"*.

A passage extremely short but full of deep meanings. In its inside the Dr. Padmakar Vishnu Vartak[106] has evidenced a puzzling matter, an element that had been more manifest when it has been supported to an old commentary drawn up by Sayana (c. 1.315 – 1.387), a minister in the south of India at the court of the King Bukka I during Vijayanagar's[107] empire.

As the Vedic text the light of the Sun employs 2.202 Yojana in half Nimesha. The Yojana is an old unit length, a value that Arthasastra defined as equal to 8.000 dhanus, or to 9.09 miles (14.63 kilometers).

---

[106] Vartak, P.V., 1995. *Scientific Knowledge in the Vedas*. Nag Publishers, Delhi.

[107] Friedrich Max Müller, *The Hymns of the Rigveda, with Sayana's commentary*, London, 1849– 75, 6 vols., 2nd ed. 4 vols., Oxford, 1890–92.

A nimesha corresponds as a unity of time to a blink of eyes, equal to 16/75 of a second (0.2.133 seconds).

Kautilya in the Arthashastra, about 320 B.C., defines the nimesha as a1/360.000 of a day and of a night, so about 0.24 seconds. So, knowing these values we know that *"2202 yojana in the middle nimeshā"* are corresponding to about 189.547 miles per second, namely 304.981 km/sec.

The resulting value is extremely close to that one of the speed of the light that, in its modern estimate, is equal to 186.281,7 miles per second namely 299.792,458 km/sec[108].

How is it possible that a value so accurate is present in one of the oldest religious texts of the human history?

How was possible to obtain it and especially which use had been made with it?

According to the foregoing in the Amsu Bodhini, or into the technical details contained to built an optical spectroscope, we should believe in antiquity of the Indian subcontinent had really existed the necessary instruments to study and to probe the cosmos and its mysteries and that their old wise men, the Rishi, had held a knowledge, in certain areas, comparable to that one of the modern time.

The piece of the Rig Veda that Sayana commented is not an isolated reference, the same value and the same peculiarity are cited by another precedent commentator who lived on the X° century A.C., a scholar called Bhatta Bhaskara who had reported to the ' speed made by the Sun' in his text Taittiriya Brahmana.

Bhaskara point up as the value of '2.*202 yojana in half nimesha*' constituted a holy number and that it was the heritage of an oldest tradition preserved by Rishi's wise men.

We are not in front of a mathematic fortuity, as someone said, the extensive and accurate calculation of the values presented by two ancient commentators made us to dispel any doubt on their reliability and the fact that there was referring, although in different terms and with a minimum waste of error, to a value corresponding to the speed of the light.

According to what shown it is not possible to deny that the knowledge and the right evaluation of the speed of the light are present in the first hymns of the Rig Veda, one of the oldest religious texts of the human history and officially put into written form, after a long oral tradition, around the 2.000 B.C., more than 4.000 years ago!

Who could be able to obtain this accurate estimate, how is it possible in a past so remote?

In the light of the findings till now shown and preserved into the hearth of the Indian religiosity, appears to be always stronger the consciousness that

---

[108] Siddhanatha Sukla e Vijay Shankar Shukla, *The Rgveda Mandala III: A critical study of the Sayana bhasya and other interpretations of the Rgveda (3.1.1 to 3.7.3)*, Sharada Books, 2012.

something of immensely deep should be existed in our past, a knowledge whose traces are only emerging in these modern times.

Parallel universes in Hinduism

The existence of some parallel universes is one of the most interesting postulates of the modern physics, a theoretical hypothesis not yet corroborated by some scientific and secure evidences but widely supported by the quantum and relativistic theories as also by the most modern theory of the strings.
Basically the theory of the multiverse[109] hypothesizes that there exist infinite "mirror" universes of our dimension in which should be present, and should manifest themselves, some other "reality" respect that one that we are experiencing.
These alternative worlds would present, according to some scholars, a different performance that our reality could have taken instead for other people would host some conjunctures completely different and distinct from that one we know and experience.
The philosopher and psychologist William James[110] coined the term on 1895 and, over the time, was resumed and developed by several theoretical of the science as well as by philosophers and scholars of the mind.
Some analyses on its nature and plausibility have been made in physic, cosmology, astronomy, philosophy, in transpersonal psychology till to arrive to its 'practical application' in the world of the fantasy and of the science fiction.
In its most modern and scientific enunciation, the concept of the parallel universes was formulated by Hugh Everett III on 1.956 but, curiously, the all question seems to sink in oldest and circumstantial roots in the basin of the Indian Ocean or into the beginnings of the Vedic culture.
One of its formulations of scientific and religious kind can in fact be found in the complex body of the Hindu traditions present in the texts as the Bhagavata Purana.
Also on this case, several scholars have recognized this strict parallelism and have evidenced a similarity, in the concepts expressed and in the formulations, that goes well beyond the simple casual correspondence but instead it seems to constitute a legacy, filtered by the lens of the religiosity, inside which seems to have been preserved some deep and technical and scientific concepts.

---

[109] The combination of all possible parallel universes is called the multiverse.

[110] James William, *The Will to Believe*, 1895; and previously in 1895, as quoted in OED's from 2003 at voice "multiverse": "*1895 W. JAMES in Internat. Jrnl. Ethics 6 10. Visible nature is all plasticity and indifference, a multiverse, as one might call it, and not a universe*".

Here are some examples:

"*Every universe is covered/included by seven levels – earth, water, fire, air, sky, the total energy and the false ego – each of them ten times bigger of the previous. There are countless universes in addition to this, and also if they are unlimited big, they move themselves as atoms inside Yourself. Therefore you are called without limits*"[111].

Shiva said *"My dear son; I, the Lord Brahma and the other Deva, that are moving inside this universe under the misunderstanding of our magnitude, we cannot show the power to compete with the Supreme Person, for the several universes and of their people that come there and are annihilated by the simple sensitivity of the Lord*"[112].

And more "*… after separating the several universes, the big universal body of the Lord, who come out casually from the ocean, the place where appeared the first Purusca-avatara, entered in every distinct universes, wishing to lie on the transcendental water that he created*"[113].

From that is possible to learn by the Puranic literature, the number of the universes seems to be incalculable and incommensurable: "*Also if in a period of time I could count the atoms of the universe, I could not count all my opulences that I have manifested on the several universes*"[114].

In ancient texts we also find several analogies to describe them, "*What am I? a small creature that measures seven spans of my hand? I am enclosed in an universe similar to a bowl of material nature, the total energy of the material, false ego, ether, air, water and ground. And which is your glory? Illimitated universes are passing through the pores of your body as well as some dusts particles are passing through the openings of a window screen*"[115].

Or even "*Because you are unlimited, nor the Lords of the sky and not even yourself could ever to reach the end of your beauties. The several universes, everyone wrapped on their shells, are obliged by the wheel of the time to wander inside You, as some particles of dust that are blowing in the sky. The Sruti, following the method of eliminating everything that is separated from the Supreme, to reach the success, revealing to you as the final conclusion*"[116].

Instead, particularly a verse seems to call to mind the modern formulation of the String Theory.

---

[111] Bhagavata Purana 6.16.37

[112] Bhagavata Purana 9.4.56

[113] Bhagavata Purana 9.4.56

[114] Bhagavata Purana 11.16.39

[115] Bhagavata Purana 10.14.11

[116] Bhagavata Purana 10.87.41

"The stratums or elements of the universes are each of them ten time thicker than the precedent, and all the universes grouped together appear as the atoms in a huge combination"[117].

In the Brahma Vaivarta Purana, one of the eight major Purana, where between the different realities they speak about the creation of the Universe, we find an explicit reference to the Hindu awareness on the existence of the parallel universes.

The text reads, "*...who will seek through the extended infinities of the space to count the sides of the universes next to each other, each containing its Brahma, its Vishnu, its Shiva? Who can count the Indra in all of them- those Indra side by side, who are reigning at once all over the several worlds; those other have passed before them, or also the Indra, that are following one another in each line, ascending to the divine kingship, one by one, and, one by one, passing away?*"[118].

The citations could last much longer. We have tried to present to the readers the most important, those more exhaustive and less misunderstanding.

Some concepts and notions concise in their simplicity, had been preserved to posterity of an incredible value, knowledge that place themselves as an *unicum* in the Vedic/Hindu landscape but that, as we have seen till now, are surrounded by a constellation of many knowledge of an inestimable value yet to be discovered.

Continuing along this path of search and rediscovery we see as a deep analysis of Srimad Bhagavatam's V Canto had been made in *Vedic Cosmography and Astronmy's*[119] text, by the scholar Richard L. Thompson[120] where the author made a comparison between science and religion tended to show how the Vedic cosmology hide inside several scientific and technological mysteries tied to an old culture extremely evolved.

Recently, in some academic contests, some scholars of scientific and religious extraction have also tried to suggest a theistic[121] vision of this subject trying to amalgamate the two realities to understand the most hidden meanings and to find some links, and a meeting point, able to conciliate faith and science.

---

[117] Bhagavata Purana 3.11.41

[118] Mani, Vettam. *Puranic Encyclopedia*. 1st English ed. New Delhi: Motilal Banarsidass, 1975.

[119] Ricahrd L. Thompson, *Vedic Cosmography and Astronomy*, Motilal Banarsidass Publ., 2003.

[120] Richard L. Thompson born in Binghampton, New York, in 1947. In 1974 he obtained his Ph. D. in mathematics from Cornell University, specializing on probability theory and statistical mechanics.

[121] See "*A Theistic Perspective on the Multiverse Hypothesis*". This study was presented at Stanford University between 27 and 29 March 2007 in Symposium "*Universe or Multiverse?*" sponsored by the John Templeton Foundation. See also Bernard Carr, Universe Or Multiverse?, Cambridge University Press, 2007.

In front of what had been presented, only one option seems to be able to cause a point of contact objectively satisfactory, or to create a common ground where to be able to understand the high knowledge level content into the old holy Indian texts.

To find this place it means to consider these same knowledge as real and truthful, fruit of a forgotten past and belonged to a superior civilization of which only on these present days, we are slowly discovering the traces.

Which is its *origin* it is not yet know exactly, there are some hypothesis, some theories, but all of them searching of a solid ground where to validate their reality.

Some clues and evidences are slowly emerging from the dark mists of the history, forming an always more coherent picture of something that our species has forgotten but that had been preserved into the Indian and Pakistani traditions as also of some other people in the world.

During our search we stumbled on a scientific discovery that could further confirm as till now exposed or else to give a solid ground to the theory of the parallel universes and so to confirm how transcribed on the old Indian texts.

The first *'concrete evidence'* of the existence of the multiverse had been announced on 2005 but only on June of 2.013 gained strength in sufficiency to be once again supported and sustained with concrete proofs.

Studying a map of the universe obtained from the data collected by the space probe Plank, the cosmologists have identified, on the recorded tracks, some 'anomalies' that "*may have only been caused by the gravitational attraction of some others universes*".

These are the statements made by Laura Mersini-Houghton, a theoretical physicist at the University of North Caroline at Chapel Hill, and Richard Holman, a teacher at the Carnegie Mellon University.

The map obtained shows the radiation originated by the Big Bang of 13.8 billion of years ago, a radiation still detectable in the universe and known as a cosmic microwave radiation.

The scholars had predicted that it should be equally distributed, but the data collected have shown a higher concentration in the southern half of the sky and in a greater extent in a *'cold point'* that cannot be explained with the current knowledge of the physics. Mersini-Houghton said, during an interview to the British newspaper the *Daily Mail* of May 19 2013, " ...*these anomalies had been caused by the attraction of some other universes on our universe, as when had been formed during the Big Bang. This is the first concrete evidence of the existence of some other universes that we have ever seen*".

Although there is still a lot of caution, these results are a fundamental key step towards the change of our scientific horizons.

The same European Space Agency, that manages the telescope Plank that cost 515 million of pounds, has publicly stated that "...*considered that the precision of the Planck map is so high, it has let us to notice some peculiar unexplained characteristics that may require a new physic to be understand"*.

So, present and past seem to melt to only one direction leaving every skepticism away from the objectivity of the texts that are emerging.

The same theoretical formulation had been reached by a speculative point of view, and also if sometime with some different way or approaches, also from the physicist Fritjof Capra that on his best-seller *The Tao of Physic,* has proved how incontrovertibly exist a substantial harmony and correlation among the spirit of the oriental wisdom and the most recent and advanced concepts of the modern occidental science.

So, should we deduce that these correlations are the result of a simple fortuity? If the old Indian philosophies had been able to go over those limits that only the most modern theories of the quantum and relativistic physic seems to have theorized or reached, then it become logically that some 'other' historical realities had existed which for now we have not come to know but logically may and must, have existed to justify a so refined and high knowledge guarded for millennia by the Indian subcontinent.

Stealth's materials on the old Vedic texts

The Deccan Herald[122] of November 2nd 2002 published an article signed by Rajaesh Parishwad with the title "Stealth bomber from Shastra", by making known a sensational discovery that seemed to give back to life the old Vedic knowledge. *"A material similar to the glass based on a technology found in a old Sanskrit text could be used in a stealth bomber"*.

Several people thought it had been a *boutade* of the Indian newspaper born to give luster to a glorious past as also to the creation of some nationalist group in the mood for spectacular. Indeed those few lines, able to change the preconceived vision of our past, had their roots in an article published on a prestigious scientific journal, *The Indian Journal of History of Science,* where was analyzed a discovery that after a short time unfortunately it would be returned in total darkness.

The material produced in the laboratory, called *Prakasa Stambhanabhida Lauha,* had the unusual distinction of not being detected by the radar and so an aircraft covered by this same glassy substance would have been virtually invisible.

The formula to realize it was not the result of the modern technological conquest, it came instead from a text judged by the occidental academic world a fake of the beginning of the XX century while the Indian traditions attributed to it three thousand years.

---

[122] From Deccan Herald, 2 November 2002, article by Rajesh Parishwad, DH NewsService Bangalore.

If it was a fake it is astonishing at least that to the beginning of the era of the flight a poor Indian Pandit had been able to create a compound so futuristic as far away from the most extreme theories of that time, but most incredible was the fact that nothing like that, except in this specific case, had been created until then.

The compound had been rediscovered by the Professor N.G. Dongre[123] studying the Amsu Bodhini a text ascribed to the Rishi Bharadwaja, who lived about on the 800 B.C. and it was dictated on the first decade of the XX century by Subbaraya Shastry.

These names seem to recur often on our study. The article had published on 1.998 by the *Indian Journal Of History of Science* with the following wording:

Dongre, N.G., S.K. & Ramachandra Rao, *Prakasa Stambhanabhida Lauha of Maharsi Bharadvaja (A novel transparent material of range 5:000 to 4:000cm.)*, IJHS 33.4 (1.998) 273-80.

The applications of this substance had been further expanded by the studies of the Prof. M.A. Lakshmithathachar, Director of the Academy of the Sanskrit Research in Melkote near Mandya, that on the interview released to the Deccan Herald he stated that the texts that he had made on the material had confirmed that the substance was totally radar reflective.

The discovery and the interest towards Bharadwaja's text is originated from a study commissioned by the Aeronautical Research Development Council of New Delhi, a project called *"Non Conventional Approach to Aeronautics or Aviation Science in Ancient India"*.

As in some previous cases, or in some others that we will see afterwards, had been some public or governmental institutions to focus their own interest towards some old texts and on their contents considered in agreement true and even more futuristic.

Therefore through a multidisciplinary team compound by Sanskritists, physicals, mathematics and engineers, Prof Lakshmithathachar tried to give a new life to the secrets in its guarded. *"The book gives some interesting information about the Vimana, and some different type of metals and alloys, on a spectrometer and on some machinery to fly"*, Lakshmiththachar stated from Deccan Herald's pages adding how in its interior it was written "...*the metallurgic system to prepare a very light alloy that could resist at high pressure"*.

The reflecting power of this glassy substance makes it able to refract the radio waves, but not only.

---

[123] Prof. Dongre from the study of Amsu Bodhini has also published the following scientific studies: Dongre, NG, *Dhvantapramapaka Yantra of Maharsi Bharadvaja* (Spectrometer / Monochromator), IJHS 29.4 (1994) 611-27. See also Dongre, NG, Spectroscopy in Ancient India: *An Application of Spectroscopy to Astrophysics*, IJHS 33.3 (1998) 229-38.

Dongre called in his report the compound as *'an absorbent material of the visible light'* and stated: *"…in the light of the discussion it result that this material is a solution of ferrous silicates in metacalcio silicate and uniformly mixed with tricalcium phosphate, in their liquid phases, produces a clear liquid that then is cooled to a solid mass. According to the Sanskrit text* (the Amsu Bodhini, ndr) *this material would be able to absorb the visible light and to retransmit the infrared radiations through it"*[124].

You can add a few comments as just described, but it is interesting to notice how such substance was not the only fragment of a lost knowledge emerged from a forgotten past.

The Chumbakamani

Even more surprising is to discover the result of the research conducted by another Indian college teacher of chemistry, subsequently become the director of the Centre of the Nanotechnology and the Bionanotechnology in Ambernath.

These studies had let to patent five new metal alloys, and some substances, following what was indicated in the Vimanika Shastra. The professor Maheshwar Sharon[125] from Bombay whose contribution to decipher this riddle, has let us to develop some substances as the Chumbakamani, the Panchadhara-loha and the Paragrandhika-drava.

Dr. Sharon's studies have evidenced the equivalence of these old materials respect those developed by the modern science during the last decades of the XX century.

But the prodome of this search dated back to the 1983 when the Indian Institute of Technology (IIT) of Bombay, under the egis of the Council of industrial and Indian Scientific Research, had started to study and to try to develop some of the materials described on the Vymanika Shastra.

On 1989 saw the light a preliminary analysis of the same Institute signed by Dr. Maheshwar Sharon, where he stated *"…with our modest knowledge of the Sanskrit and thanks to the help received from several scholars, we are able to confirm that Chumbakamani compound and the materials prepared from this formula operate as a semiconductor with a very low range gap (0.53 ev) and shows some property of photoconductivity as theorized. Therefore it is extremely possible that this material may have been used as a detector IR* (infrared, ndr) *as it is described on the book"*.

---

[124] See Dongre, N. G., S. K. Malavia & P. Ramachandra Rao, *Prakasa Stambhanabhida Lauha of Maharsi Bharadvaja* etc., p. 6 (278).

[125] Indian Institute of Technology, Bombay and later Director dell'NSN Research Center for Nanotechnology and Bionanotechnologies to SICES College of Ambernath (NSN & Nanotechnology Research Center fir bionanotechnology, SICES College, Ambernath http://www.nsnrc.com/aboutus.php).

What to say, it is difficult to comment a statements of this kind whether we think that the book is a product of the early decades of the twentieth century, fact highly unlikely if not impossible, or much more that we setting it on the first millennium B.C.

The attention of Sharon and of the Indian Institute of Technology was originated from a section of the Vimanika Shastra where it is described an instrument called *Guhagarbha Dharshana Yantra made* by three principal components the *Chumbakamani*, the *Paragrandhika-drava* and the *Pancha-loha.*

These substances were prepared using some natural compounds as plants, bones of animals and some minerals.

When these three compounds were combined together they formed the Guhagarbha Dharshana Yantra that allowed the Vimana's pilot to obtain some information about the hidden arms under the earth, possibly as a result of their input infrared, while the same Vimana was moving over the enemy area.

In the specific the Chumbakamani was described as a solar cell that fed the same instrument.

According to Sharon, the Chumbakamani seems indeed to work as a semiconductor with a short band gap, the Pnacha-loha is resulted to give an ohmoc contact to the Chumbakamani while the Paragrandhika-Drava is a liquid highly photosensitive.

Further we may known in detail Dr. Sharon's work thanks to an interview released the documentarist Diego D'Innocenzo, a meeting where it was replicated the process to realizing the Chumbakamani.

"I found three interesting sloka (verses, ndr) that are describing an instrument that, if it is exposed to the sunlight, generates electricity and it is able to discovery all the materials hidden underground, under the airplane. The first one describe the Chumbakamani which serves to produce electricity and it is composed by 32 ingredients, the second sloka the Paragrandhika-drava or the liquid part while the third sloka describe the Pancha-loha of 6 ingredients that is the other electrode".

The surprising fact is that the first solar cells were made on 1972 by two Japanese scientists, from Prof. Keniki Honda and from at that time his student Akira Fujishima[126].

Even if we considered the Vimanika Shastra a work made at the beginning of the XX century, and not 3.000 years ago, we should still compare with this incredible technological anachronism!

Anyway, on the Vimanika Shastra there are not provided the names of the necessary elements to realize such compounds in a direct way but through those who were the forms found in nature.

Therefore, we will find some reference to 'elephant tooth' or to the 'deer antler', but according to Sharon *"… I think that in the past the scientists had a*

---

[126] Fujishima, Akira; Honda, Kenichi, "*Electrochemical Photolysis of Water at a Semiconductor Electrode*", Nature 238, (5358): 37–38, 7 July 1972.

written knowledge of the elements and the animals and they could select these materials as well as we have today the elements synthesized in the laboratories and we may do it by ourselves, they took them in the form of plants or of animals and combined them so to obtain the ingredients", and that it is what Sharon had recreated in a modern laboratory.

The surprising fact is that all these elements combine together to form the Chumbakamani at a temperature extremely high, superior to the 1.200° C.

This type of procedure is known only since 50 years and is called Flux Technology. On May 10 2013 the Italian newspaper *Corriere della Sera* reported, on its scientific pages, the news that seemed to trace faithfully the purposes and the goals of the Ghagarbha Dharshana Yantra, one of Chumbakamani's components.

The text recited "*...with the laser a new system to detect explosives. It can detect a quantity extremely small of explosive, lower to 1 milligram, to a distance of 20 meters. It is the portable device developed and texted successfully by the consortium Optix, that brings together several European partners, some industrials and academics, and has received a financing of 2,4 millions of euro from the European Commission for the project ' Optical Technologies for the identification of explosives'. The innovative instrument, useful to avoid bomb attacks, uses an advanced optics technology and, thanks to the laser, is able to identified exactly the atomic and molecular structure of the explosives and to identify the residual traces, examining rapidly and remotely some suspicious objects: for example a car or a suitcase. Particularly, the optical system of detection of Optix is based on the combined use of two technologies: the spectroscopy Libs (Laser Induced Breakdown Spectroscopy) that identify the atomic and molecular emissions generated by the rupture of the elements following the excitation by a high-energy laser and the spectroscopy Raman, that measure the changes in the state of vibration of the exited molecules with the laser, making possible to identify unequivocally their molecular structure*".

Two technologies that are so far away in the time of about 3.000 years but that seem to be both united by the same use and able to exploit the principles apparently similar to obtain the same results. What can be added in the above?

Till today, if the skepticism has denied the literary evidences it is anyway problematic to deny and obstinately such tangible evidences as those exposed just now.

The Panchloh, a corrosion-resistant material

On March 11th 2012 the newspaper Pune Mirror titled "*Eureka, two researchers of the town had made a concrete discovery*", a title of a poor recall if it had not contained inside the news of an incredible discovery.

The object of the discovery was "*a formula used in the antiquity to built airplanes and one of them, the Panchloch, will be produced for commercial purposes*".

Once more from India's past, some forgotten knowledge comes to light. The newspaper continues stating "*...in what might be a step forward in the industry, two* *researchers of the town while were studying some Sanskrit texts and some old manuscripts have incurred in a material that was resistant to the corrosion called Panchloh that may be used to built cars and airplanes".*

The authors of the discovery are two scholars of proven reliability and experience, the metallurgist Satish Kulkarni and the Professor Ram Prasad.

At the origin of all it was a text dictated by the Pandit Subbaraya Shastry and titled *Bhautik Kalanidhi,* also attributed to the mystic Bharadwaja already author of the Vimanika Shastra. Once more seemed that there was a wire to connect a Pandit of the beginnings of the nine hundred century with some knowledge extremely advanced held in antiquity. Much more disconcerting is, as for the skeptics had tried to refute and to ridicule this field of study, the most direct and concrete evidences about its authenticity seem every time to show out and to deny their own statements.

Hardly we could think of a mystification of the beginning of the nine hundred century when only recently we had been able to recreate *ex nihilo* some substances ever realized before and even more on the basis of some 'old' manuscripts and using knowledge relate to it.

The Panchloh was indicated as one of the fundamental substances to build up the Vimana, the flying carts that Shastry listed under different names and typologies corresponding to several 'models' according to their employment and use. So they are listed "... *the Shakatyudgama, the Bhutavargan, the Dhoon Vargam, the Sikhedgama, theTaramukha Vargam, the Muivhaka and the Marutsakha",* not counting the more well known Vimana mentioned on the Vimanika Shastra as the Rukma or the Shakuna. The Panchloh is basically an alloy of copper and some analysis of laboratory found as its speed of corrosion is at least 100 times lower than stainless steel.

It is a substance highly ductile and extremely resistant to moisture and to the salt water, of a yellow color. This last notation is extremely interesting especially when is compared to the old descriptions that were made about the Vimana. We have extrapolated several of them, *"Or Pusan, with those your golden ships that are travelling on the ocean and on the air..."*[127] or *"The substance of their cabs is gold; their pillars are on metal, and they are shining in the sky as some lightning..."*[128].

The Indian literature is full of references to golden Vimana able to move in the sky. Perhaps it is possible that such appearance, or their coloration, was due to the use of the Panchloh? According to what described on the Vimanika Shastra, the answer is undoubtedly yes. The discovery made by Kulkarni and Prasad is not the only on his kind, indeed it joins to the job made for more than twenty years by Dr. C.S.R. Prabhu, the Vice Director of the NIC (National Informatic Center) of Hyderabad, who has identified, and partly recreated,

---

[127] RigVeda, Book 6, Hymn LVIII, Puṣan. Rig Veda, tr. by Ralph T.H. Griffith, 1896.

[128] Rigveda Samhita, Book 5 (5.062.07).

more than 200 'materials' used on the old time and described on some texts as the Vimanika Shastra. Some of these materials are the Tamogarbh Loh, the Panchloh, the Arara Tambra, the Chapala Grahak and the Agni Sthambak. Four of these materials are solid while the Agni-Sthambak was in a liquid form. We are in presence of some physical, concrete, tangibles evidences that in the past of our planet a knowledge extremely advanced formed a cultural and scientific heritage of an heritage of a civilization located in the northern Indian continent.

Even if no archeological reference has brought, till now, brought to light the ancient testimony of this heritage, it is true that there are some texts carefully concealed to the sight of the common people and preserved with extreme caution over the millennia that are proving the reality of what has been described by the epic Indian poems. The reason for this caution is explicitly suggested in the text Samarangana Sutradhara that gives the final warning do not tell this type of knowledge to avoid an improper use if it fell in the worng hands.

Test Tube Babies

Medical science is presented in the books of Ayurveda as a separate Shastra. However the original medical science present in the Vedas, the Mahabharata and the Puranas is far more superior to the Ayurveda. All the Shastras (sacred scripture) of India have their origin in the Vedas. The Vedas are composed in the Vedic era extending from 23960 years B.C to 5560 BC. The Mahabharata is composed during 5560 BC and the Puranas still later. Animal and Human cloning is well explained the Veda. Rubhus Brothers produced a Horse from another horse and a cow from the skin of a cow. They also brought youthfulness to their old parents. Their father was old but they were cloned young to bring back the youth of aging parents. Other achievements like Parthenogesis, Test-tube-baby etc are also reported in ancient literature of India. Early in The Mahabharata, there is a story about how the hundred Kaurava brothers came into being. Gandhari became pregnant naturally from her husband Dhrutarashtra. However she did not deliver a child even after two years passed. Therefore, burning with anger, Gandhari aborted her foetus. In fact it was not a foetus, not a developed child; but a mole, a mass of living cells having no shape like a human child. As soon as the sage Vyasa heard about her abortion, he came immediately, took the aborted cell mass in possession and divided that mole. Sage Vyasa dissected the mole carefully and found out normal living cells, which he kept separately in Ghruta Kumbha. Ghruta Kumbha does not mean an ordinary pot of clay, filled with Ghee. 'Ghruta' means a nutrient medium supplying life energy, which was kept in a special Kumbha, a highly sophisticated urn. There were 101 cells and those cells were grown separately in

Ghruta Kumbhas or nutrient urns. In due course children were born, 100 Kaurawas and their sister Dusshala. (Mahabharata, Adi, 115) Sage Vyasa relied on the principle of 'Chamasa division' (Division of Living cells), laid down by Rubhus brothers in the Rigveda and worked further to give birth to 100 Kaurawas. Following Vyasa the modern scientists must work on the principles stored in the Rigveda and other ancient literature, so as to produce new miracles. Many experiments on Test-Tube babies can be found in Mahatbharata's Adi Parva. Draupadi and Dhrushtadyumna got birth from an utensil only from the semen of king Drupada. They had no mother. Drona was born from the semen of Bharadvaja, without any help of any mother. Further observations were also done in the Mahabharata period that one test tube baby Drona was married to another test tube baby Krupi and they lead normal married life producing one son Ashvatthama. Another test-tube baby Krupa was married to normal girl. This couple too behaved normally. Such an experiment has not yet been done in the modern science. Concept of test tube baby is also found in the birth of Lord Balarama. An embryo of Devaki was taken out and implanted in the uterus of Rohini. Rohini delivered a child, which was termed as Samkarshana, who later became famous as Balarama, elder brother of Krishna. (Bhagawata 10-2-13). This means that the ancient sages had the technology to implant the embryo of one woman into the womb of another woman. Kunti used a technique of Parthenogenesis. She stimulated her ovum with the help of solar energy and produced Karna.(Mahabharata Adi 111.) Such experiments are done in the modern science to produce young ones only in the lower animals and not yet in the human beings. The technique is termed as Parthenogenesis by the present day European scientists. The term Parthenogenesis contains a word 'Partha' which means 'of Prutha'. Prutha is Kunti's second name. Was this a chance happening or did they coin the term from Kunti's experience? Thus it seems that medical science was very well developed in ancient India and all that knowledge is stored in our Shastras. Ancient method of test tube baby was more efficient as they knew the way of creating babies outside the mother womb. Ayurveda is an ancient medicinal science that not only has information on many highly advanced genetics theories but also many others that is either discovered or yet to be discovered[129].

Human and Animal cloning

Cloning is the scientific process of creating an exact replica of any living being. Such a clone has the same face, same body structure and same genetic

---

[129] Material from Medical Science from ancient Indian Shastras. by Dr. P.V. Vartak.

code. This means DNA of clone is same as the DNA of the living being that was used to create a clone. Clone can be created from a single cell and thus, thousands and millions of replica can be created from a single living being by transfusing his/her living cell in an embryo of another living species. In Vedic age, cloning of animals was done by sages to clone the species of powerful horses and productive cows. Ancient Indian produced a horse from another horse and a cow from the skin of a cow. Other achievements like Parthenogesis, Test-tube-baby etc are also reported in ancient literature of India. When Dr. P.V. Vartak found mentioned in RigVeda about this miracle of human and animal cloning he told this finding to Augustus audience of scientists and doctors during May 1976, but everybody laughed, saying if the modern science can not do it, how is it possible for the Vedic people? However after 21 years animal Cloning was done successfully in 1997 when a lamb was produced from udder of a sheep, which is a part of skin. Therefore the Vedic principle of cloning appears correct. Below are some of the stories on cloning from Vedic age. In Rigveda, Rubhus are mentioned as brothers (Rubhu, Vajra and Vibhu) who brought youthfulness to their old parents. Their father was old but they were cloned young to bring back the youth of aging parents. They even managed to bring back their lost skills. (They must have used ayurvedic recipes to bring back youthfulness. So much literature about such recipes exist even today but is neglected). Additionally, they also created a cloned horse and cow. According to Dr. Vartak these experiments are done about 25000 years ago. Cloning of Rubhus is mentioned by seven different sages in seven different verses of rigveda. All of these sages were from different generations thus depicting that such a technology existed over a long period of human life and Rubhus Brothers were famous for a prolong period proving greatness of their work. The first work of Rubhus was to bring back youthfulness to their old parents. These seven sages are Kanva Medhatithi (1-20-4), Angirasa Kutsa (1-110-111), Dheerghatamas (1-161), Vishwamitra (3-60), Vamadeva (4-33), Vaisistha Maitravaruni (7-48) and Sharuna Arbhava (10-176) and the hymns they composed on Rubhas are as follows:

1-20-4: With their exclusive power they made their old parents youthful again.
1-110-8: They made their old parents full of youth again by their skill.
1-111-1: Ingenious Rubhus prepared a chariot for Indra, prepared two powerful horses (2nd being cloned from 1st), made their old parents youthful again and gave new mothers to orphaned calves or children.
1-161-7: Oh Rubhus, with the power of your intelligence you converted your old parents into youthful state.
4-33-3: Rubhus who made their old parents youthful again may come to our Yajna.
4-34-9: Rubhus gave a new life to their old parents, to Aswinau, to a cow and a horse.

4-35-5: Oh, ingenious Rubhus, you made your old parents youthful again.

4-36-3: Oh, Rubhu, Vaja and Vibhu, your workmanship of making your old parents youthful again was praised by gods.

Some of the Verses on cloning along with their translation are as follows:

Rigveda, Book 1, Hymn 20, verse 4

युवाना पितरा पुनः सत्यमन्त्रा रजूयवः | रभवो विष्ट्यक्रत

TRANSLATION: WITH THEIR EXCLUSIVE POWER THEY MADE THEIR OLD PARENTS YOUTHFUL AGAIN.

Rigveda, Book 4, Hymn 33, verse 3

पुनर्ये चक्रुः पितरा युवाना सना यूपेव जरणा शयाना | ते वाजो विभ्वां रभुर इन्द्रवन्तो मधुप्सरसो नो ऽवन्तु यज्ञम TRANSLATION: RUBHUS WHO MADE THEIR OLD PARENTS YOUTHFUL AGAIN MAY COME TO OUR YAJNA.

These eight evidence proves that hey had definitely made their old parents youthful. It show that they were expert in medical science. This miracle is not yet done by any scientist in modern world. During the age when Rubhus were born, there was shortage of milk and cloning was essential to produce high-yielding cows. So, next task of Rubhas was to create a high-yielding cow that yields copious milk (RigVeda 1-20-3). Puranas describes this method in details where skin from cow's back in taken and cells are multiplied from it to produce a new cow (named Viswaroopa) which looks alike.

Rigveda, Book 1, Hymn 110, verse 8

निश्चर्मण रभवो गामपिंशत सं वत्सेनारूजता मातरं पुनः | सौधन्वनासः सवपस्यया नरो जिग्री युवाना पितराङ्ग्णोतन TRANSLATION: OUT OF A SKIN, O RUBHUS, ONCE YE FORMED A COW, AND BROUGHT THE MOTHER CLOSE UNTO HER CALF AGAIN.

Third work of Rubhus brothers was to produce two powerful horses. First they created a horse (named Hari) and presented it to Indra. Then they produced another horse which is a clone of Hari and now both were used to yoked to his chariot (1-161-7, 4-33-10). Considering all these facts we have to admit that the creation of a horse Hari from another horse and creating a cow Vishvaroopa from a skin of a cow were experiments of cloning, just like that of producing a lamb Dolly from a sheep. During 1997 many newspapers compared the scientists, who prepared Dolly by cloning, with Gods. In the same way the Rubhus were compared with the Gods by the Vedic Sages. (1-161-6,7) Before cloning, Rubhus brothers seem to have worked on living cells and their multiplication. Sun gave Amruta in a Chamasa to Rubhus (1-110-3) and they divided it into four equal parts.

Chamasa here can be considered as a cell and amruta means Immortal thing or life. Therefore a Chamasa full of Amruta is a cell full of life, a living cell. Rig-Veda (1-110-3) calls it Chit–Chamasa. Chit Means energy or rather life energy. From this we can imagine a pot containing life energy. It is nothing else but a living cell or a living animal or a plant. The basics unit of life is a living cell. Therefore Chit-Chamasa means a cell full of life. A fertilised egg can be called as 'Chit Chamasa' because it is a living cell and has capacity to divide and redivide to form billions of cells. It was divided into four by Rubhus, which means Rubhus evolved four animals from a single zygote or fertilised ovum.

During the experiment, eldest brother (Rubu) said, "we can make two cells from the original one" (4-33-5). However, the younger(Vajra) brother said, "We can make three" , because he observed that when one cell was divided into two, there were three cells. The youngest brother(Vibhu) observed and said , "We can make four", as second cell also can be divided equally like the first. Rubhus had also made a chariot flying (Vimana) in air, so they were automobile engineers too. Next was human cloning, which is mentioned in story of King Vena. He was a great king but became evil and corrupt. So, Mother Earth (Bhoodevi) decided that she would not provide crops to humans anymore. A group of Rishis (Sages) killed King Vena out of anger. They removed all the evil from his corpse's thigh. Then they created his clone from his arm. This clone had similar body but however, mind cannot be cloned. So, a new King who was named 'Prithu' was created with pure mind. He brought back Mother Earth (Bhoodevi) to her normal state and promised be her eternal guardian. This is why Earth (Bhoodevi) is also known as Prithvi. Shrimad Bhagawata (4/15) states that the sages, later, also had cloned a girl from the left hand of king Vena. This girl later married Pruthu. Such type of experiment of producing a male and a female from a dead male is not yet done in the modern science.

There are examples of abnormal cloning, like of Nishada in Vishnu Purana. Nishada, though an abnormal clone, could live long and could produce children. His tribe was known after his name and King Nala (of the famous Nala-Damayanti story in Mahabharata) and Ekalavya were his descendants. There are many more instanced like the demon 'Rakta Beeja' or AhiRavana and Mahi-Ravana who produce Rakshasas(demons) who look like them from their blood drops in war. It might be cloning from the white blood corpuscles, which contain nuclei. Similar other clones are also reported in the Puranas. Lord Ganapati was produced from the 'Mala' (ey), which means the superficial layers of skin of Parvati. This is possible according to the modern science.

Chapter IV – Astra God's weapons

"The Indian wisdom is the most deepest that there is and the search of the psychology confirm step by step the statements contained in it. The old science of the Indian soul is expressed by the yoga that appear as the way towards the self-perfection".

(Carl Gustav Jung)

Astra, God's weapons

Between the nodal points of our analysis, a predominant role is that one relative to the weapons that the old Indian gods possessed, the Astras. In the remote antiquity of this country exists a fundamental division of the instruments used for the war, a categorization that sees two distinct classes:

- The *Shastra/Sastra*, or the weapons to hold as swords, the spears and the clubs;
- The *Astra (Astras* at the plural*)*, the category of the divine weaponry with some characteristics completely different from those used by humans, usually invoked through a mantra or through sacred hymns.

In the Hinduism the Astra are supernatural weapons, because of their nature, they are different and they totally differ from the traditional weapons. Each of them is controlled by a divinity and their uses require from the same gods a specific knowledge on how to invoke and how to use them.

The old Indian texts also speak about some specific conditions that are necessary to use this type of weapons, some conditions whose transgression would have been fatal for the divinity himself. Some testimonies of their nature and destructiveness are widely present in the Mahabharata and in the Ramayana as in the old Purana, among the oldest commentaries to study the Veda. Because of the energies and the powers involved, Astras's knowledge was handed down by the teacher to his pupil only in oral form and only when the young pupil had become mature and had understood all the consequences of their use and would have gained full power, control and utilization.

In some cases the specific knowledge could be given and handed down to the men only by the divinity who presided over the control; the simple knowledge of his Mantra would not have 'enable' it and only the divinity could give the necessary teachings for a correct use of them.

However, what had always left baffled the scholars about these old 'epic saga', and parallel has led at a very low production of an analytic literature about that, are the incredible capacity that the Astra own if compared with the common weapons of war.

The fantasy of ancient commentator or the will to deify some common weapons, it seems, cannot justify the details of expertise used to describe the divine's weapons. Even more, up to now, no other mythology, coming from other parts of the world, seems to match these weaponry characteristics.

For example, in the *Ramayana* are described some Astras that seem incredibly to coincide with the modern weapons, a peculiarity that leaves us completely astonished. We find *Lakshya,* a weapon that can be followed on its way (remote-controlled); *Modana,* the weapon of inebriation; *Moha,* that get lost the consciousness; *Prasvapana,* that causes sleep; *Pratiharatara,* that neutralizes the effect of the weapons; *Shatagni,* that kills in one time hundred of men; *Shoshana,* that drains the waters and contrasts the effects of the weapons *Varshana; Soumanva,* that controls the mind; *Vidhuta,* that vibrates strongly; *Vinidra* that causes somnolence; *Sanvarta,* the covering weapon chaired by Kala (the Time) used to destroy the worlds; *Tvashtar,* that own the power of the architect of the Gods. Examining this list, we have to keep in mind that Valmiki, the first known compiler of the Ramayana, when he wanted to speak about the common weapons usually used some appropriate terms as sword, lance, arrows, shields etc.

Which motivation necessitated of a distinction so clear between these two categories of weapons and why both differ in their nature and power in a so radical way? The simple explanation is that we wanted to make 'literarily' powerful the Astra owned by the divinities, it seems colliding hard with an objective reality made of descriptions and technical details so realistic to validate the possibility of their real existence.

As we got to see and to deepen on Chapter III, titled *The Vimana and the Lost Technologies,* sometime the simple literary fantasy seems to transform itself in a objective evidence such to leave no other explanation except that who described these weapons had observed a real object. In the following chart the details of the 24 major astras and their features.

Table of the main Astra and their characteristics

Astra	Divinity	Effect

Indraastra	Indra, the god of the time	It produces a ' shower' of arrows from the sky.
Aagneyaastra	Agni, the god of the fire d	The weapon generates inextinguishable flames with the usual instruments.
This Astra Varunaastra	Varuna, the god of the water	The weapon generates volumes torrential of water. Usually it is mentioned and used to contrast the Agneyastra.
Devastra	Devas	Equivalent to the modern missiles.
Asurastra	Asuras	Equivalent to the modern biological missiles.
Manavastra	Manu, the father of the human race.	It may exceed the supernatural defenses and to reach the beneficiary far away hundreds of kilometers. It may infuse wickedness in the humans.
Varunapasha	Varuna	It is a lace divine to which even the gods can escape.
Bhaumastra	Bhoomi, the divinity of the earth.	It may create deep tunnels in the ground and to gather gems.
Bhargavaastra	Bhargavas, the clan to which belonged Parashurama	Parashurama gaves this weapon to Karna. Through its power caused several damages to the army of the Pandava. This Astra has 'Parashurama's ability in the shooting of the bow'. It may carry with it a considerable number of the most imposing weapons and may destroy every military training and to cut the enemies defenses.
Nagaastra	Naga	It is infallible against the aims, and assumes the form of a snake and result mortal at the moment of impact with the enemy.

Nagapaasha	Naga	This weapon at the moment of the impact it binds at the objective as the loops of a poisonous snake. (In the Ramayana was used against Rama and Lakshmana by Indrajit).
Vayvayaastra	Vayu, the god of the wind.	It produces a wind able to raise the army by the land.
Suryastra	Surya, the god of the sun.	It makes a dazzling light that dispels the darkness and drains the courses of water.
Vajra	Indra, the god of the time.	The aims is hit by thousands of thunderbolts (Vajra is refering to Indra's thunderbolts)
Mohini	Mohini is Vishnu's Avatar.	He dispels any magic near it.
Twashtar	Twashtri, The Celestial builder.	When it is used against an adversary army it does take the form to its own mates of the appearance of the enemy doing that they fight between themselves.
Sammohana/ Pramohana		It causes in the army a state of trance.
Parvataastra		It causes the fall from the sky of a mountain on the enemy.
Brahmaastra	Brahma, the Creator	It is able to destroy a whole army and to contrast the greatest part of the other Astra. It is a weapon of a total destruction and is compared, in its destructive effect, to the modern nuclear bombs.
Brahmashira	Brahma, the Creator.	It is able to kill the Deva, the demigods. We think that the Brahmashira is an evolution of the Brahmastra and it is 4 times more powerful.

Narayanaastra	Vishnu, the Conservator.	It creates a rain of discs and arrows. The Power of the weapon increases with the resistance that it is opposite. This Astra can be obtained only by Vishnu and can be used only once. If you tried to use it a second time the weapon would turn against those who would have invoked and against his army.
Vaishnavaastra	Vishnu, the Conservator	It destroys completely its aim regardless from the nature of the target.
Maheshwarastra	Shiva, the Destroyer	It contains Shiva's third eyes. It generates an instantaneous beam of fire that can turn even the Deva in ash. Infallible.
Pashupatastra	Shiva, the Destroyer	It destroys completely its target regardless of the nature of target. Infallible. When the weapon is released, recalls several demons and a huge spirit that personify the weapon. Sometime it can cause a catastrophic explosion comparable in its devastation, to a modern hydrogen bomb. Every time that the weapon is invoked, its 'head' is never the same. It may be only obtained by Shiva.

Comparing these descriptions with the weapons of our ancient western gods it is immediately evident their total dissimilarity and difference. The thunderbolts owned by Zeus in Greece or the hammer of the Nordic god Thor are transformed into the *Vajra,* a metallic arrow able to free an incredible amount of energy or into the *Nagaastra,* a mortal metallic snake, at the moment of the impact. Jahve's pestilences in the Bible are transfigured, in the Indian sagas, in a metallic arrow able to cause the same effects of the modern bacteriological weapons, the *Asurastra*. Unlike any other known myth the Indian Astras seem to require a material counterpart, like a support to generate their destruction.

Usually, as it is possible to see from the chart above reported, the divine weapons are in fact invoked under the form of "arrows", often of metallic nature, but they may sometime to be thrown under any other form.

While we read again the effects of these divine weapons, the phrase pronounced by the author of *2.001 Odyssey in the space*, resound in our mind, "*every technically advanced technologies is indistinguishable from magic*".

Astra's nature

The divine weapons, also called *Divya Astra*, described and narrated in the two Indian epics of the *Mahabharata* and *Ramayana*, play a central and ubiquitous role inside these two texts as well as in the mythological and religious Hindu pantheon.

Anyway it is interesting to observe how in some cases the only way to get such weapons was to reach the 'Upper heaven', where the Deva lived, away from our planet. There was Arijuna's specific case who obtained his weapons also travelling on the Himalaya while, in the chapter 27 of the Ramayana[130], Rama received some divine weapons called *Upasamsamhaara Astras*, in all respect similar to the modern missiles.

Here there are two extract from the Ramayana in which Rama acquires the knowledge and the utilization of the Astras from some wise men, "...*Tadaka attacked with great impetuosity, but Rama defended himself and after a short fight hilled him. After that Rakshasi had exhaled the last breath, Visvamitra hugged Rama and taught him how to use some celestial weapons. Rama with these became much more strong against any enemy*"[131].

Much more interesting is the next step, "... *the three people decided to visit Agastya, one of the most powerful wise man that the history records. When they saw him, understood how he had been able to accomplish so many prodigious business. They admired him sitting in his yoga position called of the lotus, he was covered in ash, and he was shinning with an intense spiritual energy. After offering him the homage of duty, Rama wanted to spend time talking to him and to enjoy his company and to assimilate his deep spiritual consciousness. At the end of the interview, Agastya gave him several celestial weapons and started to teach him to their use*"[132].

It is during the battle that the powers of these weapons are expressed at their maximum extent. Suggestive is the story reported on the XXXV chapter of the Balakanda of the Ramayana where it is narrated how the wise man

---

[130] Valmiki, *Ramayana*, Bala Kanda, Chapter 27 & 28.

[131] Valmiki, *Ramayana*, Bala Kanda, Chapter 12, the forest.

[132] Valmiki, *Ramayana*, Aranya Kanda, Cap. 32, encounter with Agastya.

Viswamitra obtained the status of Brahmana, and Vasishtha's friendship, not after having had a conflict with him that saw the use of the most powerful Astra, "... *after a long time Shiva appeared. So Viswamitra asked him his graces. Some arcs and some arrow with their mysterious mantra. He obtained them and equipped with these divine weapons haughtily he attacked Vasishth's hermitage. At Viswamitra's arrival he threw the first weapons generating fires, the ascetics, the birds and the animals that were living in the forest started to flee in fear also if Vasishtha told them to stay calm. So Viswamitra threw once after the other his formidable weapons against Vasishtha. Vasishtha had with him the only things that usually a brahmana own, a bag, a stick, and a cup for the water. Viswamitra used the missiles revealed by lord Shiva. The soporific missile, one intoxicant, those that are producing unbearable heat, those that dry everything, the missile that disintegrates every things, those that breaks everything as a thunderbolt and one fatal as the death. Vasishtha was angry and destroyed all Viswamitra's weapons that were entering by the stick one after the other inside the bag. At this time Viswamitra used the most powerful weapon known: the Brahma-Astra. The Gods and the Demigods that were rushed to see that remained in suspense. Vasüshtha issued from every pore of his skin an unstoppable divine energy so that the present gods asked him to stop this flow. The wise man stopped himself after to have destroyed the last weapon of the king*"[133].

The report of this incredible battle is for sure conditioned of the mythographical transposition that the history have had over the millennia but, in its drama, he still let us to read the power of the weapons used, Astra whose description diversifies totally from to any other weapon till then observed and known in a myth. Anyway, the matter has been unluckily neglected as much by the researchers and academics. This at least till when a scientific study, also this one fail to be noted, has given the right dignity to the weapons of the Indian divinity clarifying a fundamental point on their nature.

On *Mahabharata's* VI book, the *Bhismaparavan*, we find described these extraordinary divine weapons with the following words "...*it had as rays some flaming missiles; with the wind that was made by its weapons, with the thunder made by the clang of his vehicle, with the flames that came out from the big weapons... Bhisma was for his enemies like to the fire of the end of a cosmic age. He fallen in the middle of row of chariots, he came out shortly after... he attacked impetuously the middle of the Panduide's army... with six very fast missiles with a tremendous clang, that were similar to the sun and shattered every opposing defense*"[134].

And much more "...*hearing the noise that his thrown weapon issued, similar to the thunder of the lightning, all the creatures crouched. Fourteen thousand Cedi, Kaci and Karusha all of them fighters with their carts, famous, noble families, ready to die, determined not to go back, everyone with their decorated banner with gold, attacking Bhisma they van-*

---

[133] **Traduction of Massimo Taddei, source www.duepassinelmistero.it**

[134] *Mahâbhârata, Bhismaparavan* (Book of Bhisma), chant 114.

ished in the battle as in the Death that was waiting for them with its mouth open, directed to the other world with all their carts, horses and elephants. And then we saw everywhere, o King, carts with broken axis and harness, with the crashed wheels by hundreds and thousands. About broken carts with all their armors, about warriors flew away from their carts, about broken arrows and armors about axes, scimitars, clubs, pieces of carts, quivers, swords, arms, heads with still off the earring, gloves for fingers protection, banners culled and broken arches, and all of these were scattered on the ground"[135].

During these battles the heroes may to achieve the victory almost exclusively using the Astra. It is also, inside these same stories that the divine weapons reach the acme of their descriptive developing bringing in the foreground many of the oldest India's practices and most of all practices and beliefs forgotten during the next millenniums.

Does not exist any other mythological corpus in the world where the concept of a divine weapon has been developed and has become as complex as in the Indian[136] epic. Therefore, without a correct understanding of the Astra, it is impossible to appreciate and fully understand the importance and the value of these epics as well as the traditions relate to them. Much more fundamental it is however a matter revealed on 2.000 by Jarrod Whitaker, a professor of Eastern Religions at the Wake Forest University, who brought to the public attention the concept of *tejas* or "fiery energy" without its understanding would not be possible to make a comprehensive and exhaustive examination of these weapons.

According to Whitaker inside the Indian mythology was incorporated a recurrent ensemble of principles regarding *Tejas's* nature, some concepts that directly and essentially tied to the nature and function of the divine weapons.

Before to penetrate inside this new horizon of research however, it is important to observe how the concept of 'war machines' was evolved and developed outside of the Indian 'epic' context and how had left only some small hint of its existence inside the technical and classical works of the Indian wisdom.

War machines

A text impossible to find, but anyway fascinating, that collects some of the weapons and some of the divine machines used in the past, was published on 1.979 by G.R. Josyer, the man who only six years before had published the first translation in English of the Vymanika Shastra. In his work titled *'Diamonds,*

---

[135] *Mahâbhârata, Bhismaparavan* (Book of Bhisma), chant 102.

[136] Jarrod L. Whitaker, *Divine Weapons and Tejas in the two Indian Epics*, IndoIranian Journal 43: 87–113, 2000. *Kluwer Academic Publishers. Printed in the Netherlands.*

Mechanism Weapons of War Yoga Sutras'[137] Josyer, as per its subtitle, gives a comprehensive overview of what will be the topics covered in the book, *'about the old Indian carts, the weapons and the realization of artificial diamonds'*.
Analyzing the book it is especially interesting to stop on the chapter devoted to the Yantra[138] where it is reported a long list with their technical characteristics and the materials used to build them.

IMMAGINE

Copertina del testo di G.R. Josyer *'Diamonds, Mechanisms Weapons of War Yoga Sutras'* compendio di quattro libri dettati dal Pandit Subbaraya Shastry (©Enrico Baccarini)

A close examination of the techniques used in antiquity is nothing short of staggering. *"The following carts are made with a metal called Veera. It is an alloy formed from the dissolution and the melting of three metals called Kshwinka, Arjunika and Kanta (magnetic), respectively in three, five and nine parts, all that is called Veeraloha namely metal Veera. When it is subjected to some Shastrics processes cannot be destroyed by the fire, by the air, by the water, by the electricity, by the cannon, by the gunpowder or similar. It will become very strong, light and of the color of the gold. This metal has been designed specifically for these carts"*.
Then follows a detailed list of the Yantra realized with the metal Veera. In the ancient Indian past these instruments were described even by technical and engineering texts, which provided minute technical details and the account of their magnificence.
It is interesting to notice how in several descriptions there is the name 'Ratha', the ancient terminology previous to the term vimana with which we used to identify the Gods's flying carts.
In the same way the list given by Josyer is certainly coming, in its original source, from the material dictated by Subbaraya Shastry and from the knowledge contained inside the *Yantra Sarvasva*, the treaty called *'Everything on the flying Carts'* of which the Shastry himself had been among the depositaries.
Where the book is listed, the mystic Bharadvaja is indicated as its compiler instead of Josyer, who appears as the modern curator. The same bibliographic information is indicated in the tables regarding the Vimanika Shastra.

---

[137] G. R. Josyer, *Diamonds, Mechanisms Weapons of War Yoga Sutras*, Josyer, 1979, International Academy of Sanskrit Research, 244 pagine.

[138] The word "Yantra' is derived from the Sanskrit *Yam*, literally *to control*, and has been widely used in ancient India to identify different realities. In this case the word Yantra is placed within the category that defines the 'machines' built in a remote past of this country for many different purposes.

In some cases Josyer has inserted in his translation some modern words to explain in a better way the meaning of what was described in the old text but also because sometime there were no words to express what was stated in Sanskrit.

Such 'poetic freedom' does not alter the original meaning present in the old language but perhaps, to us, enriches the contents and the meanings. We have made a literal translation of Josyer's text trying to avoid any alteration of what expressed in the original text.

Some descriptions may be somewhat 'suggestive' or picturesque, but we have to consider that they do not belong to 'our' western culture and even more they are coming from another Civilisation, according to a different evolution process.

They might have attained similar achievements to ours, but not using the same manners and the same extent. Here is a brief list of the 63 Yantra used in the old India, with some of their technical specifications and some of their peculiar characteristics.

Panchamukha Yantra - a cart with this name owns some doors to east, to southwest, to north and on the top. It weighs about 170 Ratals. It transports about thousand Ratals. With the aid of the electricity it can travel at five Kroshas per hour. It is used as a way of transport for men and for wars. Because of the carts is conducted by a spirit called Gaja it is called Gajaakarshanna Panchamukha Ratha.

Mrugaakasrshana Yantra - these carts are designed in the shape of animals as oxen, donkeys, horses, camels, elephants and so on.

Chaturmukka Ratha Yantra - this cart has some openings on four sides. It weighs 120 Ratals. It can be conducted with every type of oil (fuel, ndr) possibly what is coming from coconut shells, or through using the electricity. It travels at six Kroshas per hours. It is used to travel, during wars and to transport some objects.

Trimukha Ratha Yantra - this cart weighs 116 Ratals. It has doors, at the bottom, at the top and on one side. It can support a weight of 600 Ratals. It is conducted by the aid of an oil extract from the root knotted of the Simha-Krantha and from that one extracted on the stems of a type of grass. If these oils are not available we can use in their place the electricity. It is used for the same purposes of the above mentioned cart Cahkra-mukha-Ratha Yantra.

Dwimukha Yantra – it weighs 80 Ratals. It has some openings on east and west. It is governed by a wheel mounted with screws. It travels at 3 Kashas per hours. It can carry a weigh of 300 Ratals. It is used for the purposes referred to above.

Ekamukha Ratha Yantra – this cart has only one door. It weighs 48 Ratals. It has a weigh of 200 Ratals. It travels thanks to the aid of the oils extracted by the seeds of Kancha-Thoola and Sovlaalika or through the electricity. The speed is 1 Keosha per hours. It is used for the above mentioned purposes.

Simhaasya Ratha Yantra – this cart has the anterior side with an appearance of a lion. It owns two doors. It weighs 75 Ratals. It carries a weigh of 50 Ratals. It can travel both on the ground that on the air. It possesses the ability to expand and to contract. It is used for the purposes mentioned above.

Vyaaghraasya Ratha Yantra – it is modeled in the shape of a tiger, it owns wings, weighs 64 Ratals, carries 200 Ratals of weight. It travels in the air espanding his wings with the electric power but contracts its wings with the power of the steam. It is used for the purposes mentioned above.

Dolamukha Yantra – it is modeled in the shape of a litter, has two doors, weighs 50 Ratals. It travels to 3 Kroshas per hour. It is governed by the aid of the electricity and by an oil, namely from the Shilyusha extracted from the wine.

Kurmamukha Ratha Yantra – is modeled in the shape of a turtle. It owns two small doors, weighs 32 Ratals. It is used only to spy.

Ayah – Prasaarana Yantra – among those that are governed through the electricity this one particularly travels on the iron lines spread on the Ground. It can be constructed to contain from the 40 to 80 wheels. It seems something like a train, and weighs 4.000 Ratals. It can bring 25.000 Ratals. It travels at 3 Kroshas per hour with the power of the electricity. It is used to transport men and goods from a place to another.

Panchamukee Yantra – this cart has five sides. It weighs 115 Ratals, can carry 12.000 of them. It has another cart that allows to the five doors to open and to close by themselves. It is governed by the electricity. It travels at 4 Kroshas per hours. It is used for the purposes mentioned above.

Eka Chakra Yantra - it carries only a wheel. It is shaped as a trap. It weighs 105 Ratals. The movement is given by the wheels that are working as bellows. It travels at 3 Kroshas per hour.
Trimukhi Yantram – this cart has three sides. It contains three compartments that can be separated. It weighs 1.000 Ratals. It travels on the water. The three

compartments are arranged in such a way that the cart can travel with the second compartment if the first is damaged and if it is also the second, the third can safeguard the contents separating itself from the other if it becomes necessary. If the top is in a dangerous situation can rise in the sky and travel in the air. It is used for the purpose mentioned above.

Jrumbhala Yantra – this cart has an opening at the bottom. It is shaped as a closed umbrella. The coating is made by a thick and waterproof fabric produced by the juice of five trees or Pachavarga Kasheeree Vruksha. It weighs 42 Ratals. It carries 300 ratals. It can expand in the shape of a pavilion through a screw propeller housed inside. So it can contract in its original form through another screw. It looks like a flag. It is used for secret reconnaissance in order to spy. It can travel to 6 Kroshas per hour thanks to the electric power or with the aid of its wheels operating through its bellows.

Goodha Gamana Yantra – this cart can accommodate only three people. Its weighs is half of a Maund. It looks like an ordinary tower. It contains five keys. It can travel on the ground and in the air. Its motion is almost invisible. It can travel at 8 Kroshgas per hour with the power of a fuel oil called Sinjurika. It is used for secret travels.

Wyrajika Yantra – this cart is made by some glasses of abhraka or mica. It owns sixteen openings. It weighs 3 Ratals. It brings 5 Ratals. It appears as a shining light and so no one can understand that it is a cart. If someone has to go near it, the sparkling light produced by turning the internal key will kill him. It can travel both on water and on land. With the electric power of the sunlight can travel at 12 Kroshas per hour. It is used for travels, during war and for the dispatching money.

Indranee Yantra – this cart is constructed with paper, worked from grass that belongs to the Maunjavarga; the 3^a, 9^a, 11^a, 22^a, 30^a e 42^a type of grass known with the name of Pishangangamunja, PingalaMunja, Rajjumnunia and so on. This cart cannot be destroyed by fire or water. It is extremely light and strong. It can travel at 15 Kroshas per hour with the aid of wheels moved by the wind. It brings 100 Ratals.

Vishwaavasu Yantra – this cart owns two doors. It weighs 148 Ratals: it brings 3.000 Ratals. Thanks to the help of the steam can travel to 2 Kroshas and half per hour. It can go back and forth. It can be expanded or contracted. It contains seven keys. It is used for the purpose mentioned above.

Sourambhaka Yantra – this cart owns three floors. There are secret seats for 400 people in each of the three levels. The seats are not normally visible. The levels can only be perceived. It weighs 230 Ratals. It brings 36.000 Ratals. It can travel thanks to the aid of the electricity or of the steam, or with the aid of the spirits of the seventh type of wine. It can travel to 32 Kroshas per hour. It is useful to transport some men or some objects in war.

Sphotanee Yantram – this cart owns only one door, it weighs 50 Ratals. It brings 200 Ratals. It sails in the water, as a bubble water, some time it can rise over the water and sometime can dive under the water. It moves itself with the power of the steam or through the spirits of the Kanajala Kshaara. It travels at 4 Kroshas per hour. It is used for the marine espionage.

Kamatha Yantram – this cart is modeled as a turtle. It weighs 500 Ratals. It brings 8.000 Ratals. It owns two doors. It travels under the water. It is used for the purposes mentioned above.

Pervathee Yantram – this cart is modeled as a lotus. It owns four doors. It weighs 69 Ratals. It brings 800 Ratals. A pole is fixed in its central point and encloses some keys that let the cart to expand and to contract as the lotus does. Trough the steam or the electricity it can travel to 24 Kroshas per hours. It used to travel on remote islands.

Thaaraamuckha Yantram – this cart contains a face of seven keys that are shining as a stele. It has twelve doors, it weighs 2.000 Ratals. It brings 25.000 Ratals. If it is activated the first among the seven keys, a melodious music is accompanied with any type of musical instruments, will be heard by all the people that are in its inside: if it is activated the second one it will be possible to see a dramatic view or a scenario of action: if you active the third, a gentle stream of water will flow through the occupants, so that they can make use of the liquid as they wish: activating the fourth, some tables with flowers, perfumes, camphor, bananas etc. will be available for all the occupants so that they can worship the God: activating the fifth, some trays with excellent food will be offered to them and while they are eating, activating the sixth, the dishes will turn through some wires: activating the seventh, some beds will be available for everyone. If the keys will be repositioned as they were before, everything will vanish. It can travel with the aid of the steam or of the electricity at 4 Kroshas per hour. It is used for the purpose mentioned above.

Rohinee Yantram - this cart is modeled on the shape of a hollow bamboo and it owns the same color. It weighs 3.000 Ratals. It brings 50.000 Ratals. It contains 500 compartments where the explosive powder, the projectiles, the weapons etc. can be preserved. However through some external firebreaks nothing may be burned or damaged, as the fire will be suppressed by the nature of the metal with which is composed. With the help of the steam or of the electricity it may run at 6 Kroshas per hour. It is used especially during the wars.

Raakaasya Yantram – A glorious light as the light of the moon will come out from the cart, every three hours. This light will light up till a distance of 64 Kroshas within which everything that will be inside its beam of light will be clearly visible. It weighs 1 Ratal. Inside the cart there is a wheel that continuously runs on the right as the sun. It can travel on the land, on the water and on the air. It is useful to discover some objects in the distance. With the help of the spirits of the wine of the sixteenth type it can travel at 4 Kroshas per hour on the Hearth, 8 Kroshas over the water and 12 Kroshas in the air.

Chandramukha Yantram – this cart has its front as a lunar disk: it is dark in its central part and brightening all around. It weighs 400 Ratals. It owns 16 doors. It brings 16.000 Ratals. It owns five floors and sixty eight cylinders. These cylinders are very useful during the saturation of the five type of smoke, of the seven powers (electric or of energy, ndr) of the thirty two types of powders and of the forty eight types of gas. When they are inside these cylinders, cannot cause any damages. This cart can travel trough some paths dug inside the ground. It moves thanks to the help of the spirits of the wine of the 13° type. It reaches a speed of 16 Kroshas per hour. It is used in war.

Anthaschakra Ratham – this cart is modeled on the shape of a retort rod of a litter. This rod looks like the two poles angularly bent of a crusher and will be always folded on it. There are screwed 32 wheels. This cart has to be fixed on the ground. It is used to transport elephants, camels, horses, men, vehicles etc. or to bring them from distant places. All this is made through the screws inside it. The cart must be placed on the fifth circle of the war camp.

Panchanaala Yantram – this cart is assembled by combining five cylinders. In each of the five cylinders there are some distilling machines. These are used to produce not only some oils, energy, etc. but also rope, powders and so on. It weighs 230 Ratals. It travels at 3 Kroshas per hour with the aids of the 9° class of the wine.

Thanthree Yantram – the anterior part of this cart looks like a trap of cables. Inside the cart, in its center, there is a magnetic wheel. Behind it there are some faithful representations of lions, tigers and some others wild animals, all of them made of cables. On its front there is a magnifying glass of the 103° class. Acting on the keys, these lions, tigers and so on, made on iron can be made to roar or jump up to some individuals who approach them so that no one will approach. It weights 80 Ratals. It carries a thousand of Ratals.

Through the power of the energies of the 3th class of the wine it can travel at 4 Kroshas per hour. It is useful during a war.

Veginee Yantram – this cart is modeled on the shape of an umbrella. It can run very quickly turning the screws at the junction of the wheels. It can accommodate only three people. It can travel at 8 Kroshas per hour.

Shaktyudgama Yantram – it is a cart that spreads light in the sky. It owns five floors. It contains some large glass jars (containers) in each floor. On the first floor, the glass jars are full of tar mixed with coal. On the second, the glass jars will be full of sea foam, or foam, with the extract of the pond. The jars of the third floor will be filled with the five essences of Pranaksharas's oil. Five spheres along with the mercury are mounted in those of the fifth floor. The cables of these five jars are joined according to the Shastric principles. The jars of the first floor must be filled with the electricity and through this process the containers will be loaded on the other floors. Through this trend the light can be spread in the sky. This cart weights 32 Ratals. It is used to build airplanes.

Mandalaavartha Yantram – this cart is modeled on the shape of a whirligig. It owns six sides and sixty-four screws inside. It weight 68 Ratala. It carries 8.000 Ratals. It turns as a whirligig around armies and crowds of people. It can spin around three times, making a distance of two kroshas per hour and thanks to the help of the electric energy and of the eleventh class of the spirit of the wine. It is useful during war and people's mutinies.

Ghoshanee Yantram –it is modeled as a huge snake. It owns three coatings and twenty-four faces. It is filled with electricity. It owns 148 cylindrical rooms to store poisonous gasses. It can make a noise, acting through internal screws, just as thirty-two lightning. It releases poisonous gases while it is travelling. It is possible to hear the sound that had been made at a distance of 14 miles a quarter. The people that will be found in its vicinity will die because of the mortal effects of the deafening noise and the poisonous gases. Those that will find over eight kroshas from it, will faint. It weights 116 Rtals. It carries 6.000 Ratals. It can travel at 6 Kroshas per hour thanks to the electricity and the spirit of 13° type of the wine.

Ubhayamukha Yantram – this machine owns the same symmetry in both sides. It owns inside a fresh water stream above which arises another flow of tar. In the middle there are some oils belonging to seven varieties. It contains inside 71 keys. Working on these keys the poisonous gases, the powers or anything else that is dangerous for the life, will be wiped out in the range of twelve (about) miles around the cart and the atmosphere will be purified. It weights 48 Ratals. It carries 108 Ratals. It travels at 5 Kroshas per hour with the help of the electricity or the energy of the 27° class of wine. It is used to purify the atmosphere when and where you need.

Thridala Yantram – this cart is modeled on three leaves of Bilwa patra… and has two compartments. The first is squared, the second is triangular, and the third has a hexagonal shape. Every one of these compartments own two doors. Every compartment has some Peshanee Yantras. A Peshanee Yantras is a cart that Grinds grain, such as wheat, into powder. This cart is driven by electricity.

Thrikuta Yantram – this cart owns two towers, as the peaks of a mountain. Each of these towers measures one hundred (bahu) yards in height. Each tower owns 32 keys inside. There are some cylinders in each key. Upon the towers there are some flags and wheels. In front there are some equipment to measure the cold. This cart reports the weather, the wind, the sun, the light, the rain, the lightning, the falling of stars and some other future phenomena.

Thripeetha Yantam – this cart owns three bases. In the first there is a cart with three heads as an elephant, but it owns two trunks on each head. In the second there is an apparatus with three heads; each head owns two trunks of Vyali's animal. In the third there is an apparatus with three heads, each of them has the appearance of a rhinoceros with some fangs. The three bases can be mounted together and if it is necessary can be separated. The first of these Yantras can stop the flow of the water of a river, suck its water and modify the direction of its flow. The second can break into pieces the mountains and thereby to create a passage. The third is able to dig a hole into the ground, to suck the water from below and to throw the same through the fangs posed above its head. It weighs 6.000 Ratals. It carries 80 Ratals. It travels and works thanks to the steam, to the electricity and to the energies of the 23° class. This cart is used to build bridges and streets in the water and some galleries through the mountains and rocks.

Vishwamukha Yantram – this cart is very spacious. There are inside twelve cylinders containing magnifying glasses. These cylinders are very big and are mounted in a way that it is possible to orient in any direction, as it may be necessary. It weighs 1.800 Ratals. It brings 40.000 Ratals. In its inside there are two floors that can be combined or separated using some buttons. It travels at 12 Yojanas thanks to the help of the spirits of the 32 typology and thanks to the steam or to the electricity. The top can be separated and can hovering in the sky. By fixing to it the cylinders, it covers an area in the sky of 24 Yojana where the forests, countries, seas, towns etc… become clearly visible and it is possible to obtain images of the same. It is used in travels.

Ghantaakaara Yantram – this cart appears as seven Almirahs are fixed between them. Inside it there are several types of cables, the essence or Reavaka of the 16° type of magnet and also some others Dravakas are present. Inside, in each of these Almirahs, there are also two metal or white brass bells built so to produce a very disturbing sound. We can know with the waves produced all the news from the world. It is used in the collection of information and pictures.

Vishthrithhsya Yantram – the cart owns a mouth completely open. It weighs 76 Ratals. It carries 120 Ratals. In front of this cart there are five buttons that look like towers. In the first tower there is a pot of Chandra Kantha's stone. This stone vase, as soon as the moon rises, exudes water up to fill. The same water is used by the occupants of the cart to drink. The other towers attract the powers of the clouds, of the stars and so on. It travels at 3 Yojanas per hour thanks to the energies of the 14° class or to the electricity. It is used for travelling.

Kravyaada Yantram – this cart owns three sides. It weighs hundred of Ratals. It carries 10.000 Ratals. With the help of the steam can travel at 9 Yojanas per hour. It is used to travel and to transport goods.

Shankhamukha Yantram – it is a cart that contains a perforated device with five faces similar to a shell called Shankha Mukha Yantram. There are some keys to expand or to contract the cart when or where it is necessary. It weighs a thousand of Ratals. It is used to build wells, deep holes or to drill mines. It can dig 213 bahus or yards in an hour. It is used for the purposes contained in the description.

Gomukka Yantram – this cart is modeled as the face of a cow. It weighs 80 Ratals. It carries 700 Ratals. There is a continuous flow of running water inside its mouth. It travels at 2 Yojanas per hour with the help of the energies of the 20° class. It is used for provisioning the water.

Ambaraasya Yantram – for the viewers this cart looks like the sky. It weighs 180 Ratals. It carries 2.400 Ratals. It is used to transport elephants, camels, and so on. It travels at 3 Yojanas per hour using the electricity and the steam.

Sumukha Yantram – this cart shows the wonderful likeness of a crab. It weighs 118 Ratals. It carries 1.150 Ratals. It can travel through the use of the energies of the 14° class, through the steam or the electricity. It travels at 2 Yojanas per hour on the earth, 4 Yojanas in the air and 3 Yojanas on the water. It is used to travel and to transport goods from a place to another.

Thaaraamukha Yantram – the spheres, which are made with the metal product from the falling stars, are called Thaaraamanies. A cart that contains these spheres is called Thaaraamanies Yantram. In its inside there are three big cylindrical pillars. It is also possible to find inside this cart a smaller cart that contains some draavakas or acids, electricity, some crystals and so on. At the bottom of the three pillars above mentioned, there are some buttons. Using the first button we produce a bright light as a rainbow. Instead with the second button we will irradiate a shining light as the sun when is covered by the clouds. Moving the third it will be released a smoke similar to the dew. When this cart sails on the sea it is possible to take some images of all the carts or animals that are moving up or under the sea surface. It is used to discover some objects that are both above and below the surface of the sea.

Manigarbha Yantram – this cart owns a circular or round shape. Inside this cart there are some spheres called Souraka, Paavaka, and so on that attract the heat of the sun's rays. It weighs 64 Ratals. It carries 70.000 Ratals. It contains 12 faces to let the sun's rays to enter. It travels at 3 Kroshas per hour through the energies of the third class. It is used to travel and to capture the heat of the sun's rays.

Vahinee Yantram – this cart owns 16 buttons and 12 metallic cylinders. It is 32 feet high for 11 of circumference. There are under it 48 drilled equipments. There are 96 wheels that throw away the mud excavated, 22 keys that let to dig the stones and there are 12 devices to draw water. This cart can be securely anchored to the ground. The water so sucked flows like rivers. This cart can dig the ground till 82 thousand feet deep. It is used to dig the ground and to draw the water.

Chakranga Yantram – this cart is modeled on the shape of a trap. Throughout the structure there are some wheels with stones. Moving a wheel full of wind this one will go out, moving another one some water will be released. In this way there are some wheels which if activated will release fire, steam, poisonous gasses, dew, energy, colors and so on. Through the rotation of the wheels it travels at 2 Kroshas per hour. It is used in different ways.

Chaitraka Yantram – this cart is modeled on the shape of a scorpion. Inside there are 24 joints. There is a button at each joint. Each key is differently numbered and colored. Based on the key which will be activated will be produced music, conversations, melodious instruments, images, and many other wonders. Those who approach the cart will be captured (in the sense of "to be photographed", ndr) both in their appearance and in their mind. It is used in Bhedopaya or to conquer the enemies with deception.

Chanchupata Yantam – this cart has the shape of a bird with open mouth. It owns four wings. There are five buttons in each of these wings. From its open mouth some cables are connected to the ground. Till these cables will be connected to the ground this will acquire a peculiar power whereby the people, if standing in this area, will be numbed. Acting the keys that are connected to the wings the people that will be in the infected zone they will feel weak or the ground will crack, depending on the role that will be required to the keys.

Pingaaksha Yantram – this cart has the appearance of a litter. This cart has all over its body many green eyes. There is a button for all these eyes. The cart must to be fixed firmly on the top of a mountain. It is 60 feet high for 14 for the circumference. It can also be placed in a town when this is surrounded by the enemies. The keys in this cart are made and fixed through some cables that are extending below the surface of the earth for 24 miles all around the area. Inside the cart some buttons are arranged and numbered for all these external keys. Working on the first button we will operate on a particular key and the doors of the fort will close. Working on another the ditches will be filled with water. In this way, working on the other buttons, according to the default order it will be possible to create with each buttons some fantastic phenomena such as huge river of fire, flooding water, cyclones, etc. This cart is used to defend some towns or villages against stronger opponents.

Puruhootha Yantram – this cart owns the appearance of a Mrindanga, or a musical instrument. It is 25 feet high and measures just as many feet in circumference. Inside the cart there is another Shabda-spota Yantra. When the button is activated a terrible noise bursts equal to the simultaneous roar of 63 lions. It used for the nature of its job.

Ambareesga Yantram – this cart is modeled on the shape of an upside down terracotta pot. It is 46 feet in high and 23 in its circumference. It owns in each of the four sides some keys that are looking like the feet of a turtle. It travels on the water at 6 Kroshas per hour thanks to the help of the Chakra Bhastrika. It is used to search the things on the earth and under the surface of the sea bringing them to the light.

Bhadraashwa Yantram – its shape is modeled on that of a horse. It owns a tail of 38 feet in length. It weights 54 Ratals. It gallops as a horse with the help of the spirits of the 32° class. It has the speed of three horses. There are on the top some keys with three faces. It moves as a horse in a circular way, when it is put into operation through the keys. It covers a distance of 12 Yojanas per hour. There will be issued some bright sparks as he gallops, shining of light that will dissipate all the dew or the fog that is covering that zone and will clean all the atmosphere. It is used in the places and in the times of the dew where and when the same dew obstructs the view.

Virinchi Yantram – its shape is like a globe. Around it there are 32 cables of 80 feet in length and 40 of circumference, both in the front that on the back of the cart. There are three keys for these cables. Operating on the first key the cart is charged with powder and projectiles. It shoots operating on the third key. It destroys in pieces the mountains in a measure of 24 feet for each shot. It is used to build tunnels in the mountains and among the rocks.

Kuladhar Yantram – this cart is modeled on the figure of a crow. It owns three beaks as those of the crows. In its inside there are some electrical machinery. On its top there are some keys similar to some small boxes where there are some round buttons. When this cart is fixed on the rocks and starts to work, it extracts with the help of its beaks some slabs of stone of the desired size. It is used to cut the stones.

Balabhadra Yantram – this cart is shaped on a metal boiler tipped. It is 64 feet long and 16 feet large. On both sides there are 16 fixed plows of 16 feet for 4 of wide. Each plow owns two winds. At its beginning and its end there are some rotating screws. In its inside there is some electrical energy or a steam boiler. Upon these cart there are 24 buttons. Under these buttons there are some wheels. On a side there are 32 screws. The cart starts to plow the land when the buttons are pressed. It moves at 3 Yojanas per hour. It plows an area of 3 Yojana from 64 feet, per hour. The depth of the mud raised from the ground is of 3 feet. It is used to cultivate the land.

Shaalmali Yantram – this cart has a square shape and a white color such as that of Shireesha's acaria. On the top there are sixteen buttons, each designed for a specific purpose. Turning the first key, a pair of hands appears like the trunks of elephants that can support a weight of hundred Ratals. Acting the second button such weight will be placed wherever is needed. The other buttons are intended to lift weights from deep water, to place the pieces of stones, of wood or similar, or to arrange out of the water for bridges under construction. It can bring down weights from an altitude of 200 feet.

Pushpak Yantram – it has the shape of the rising moon while it is forming. It is equipped with many frames that are suspended to it. There are many frames on each side and 8 of them suspended in the middle. On the right side there are some carts similar to pigs, while on the left side there are some carts to saw. In the middle there are some wheels screwed suspended to the chains. There are also two wheels. This cart can be placed in a place where the wood will be cut and sawn. If the first buttons of the top wheel is activated, the above mentioned cart similar to pigs will descend one by one. Activating the second one, the cart will fall on the trunk of the tree and will shake it and will cut it with a tremendous noise producing a great amount of smoke and fire that spreading around for 16 miles, burns all the waste on the soil and cleans the air. By the action of the fire on the trees, the oils etc. will be extracted and stored in containers at the base of the same trees. The heat of the fumes produced by the fire make all the trees in the area soft as plane trees. The leaves fall. Operating on the third key some carts descend and wonder in the place exhaling tremendous breaths. Thanks to these strong winds the dust of such area will be swept away and the land will be deforested. In the same way, if we will operate the button on the left side, from their cribs the saws will drop down one by one. Turning the first screw of this wheel the saws can arrive ready on the side where the trees will be cut. Operating on the third screw, the saws will come back to their place and a pair of hands, as trunks of elephants, will descend. The pairs of hands will reap the pieces of wood that have been sawn. This cart weights 180 Ratals. It can move into the forest thanks to the power of the steam. This cart had to be fixed on the ground. It can saws 3.200 Ratals of wood per hour. It used to saw and to cut large quantities of wood.

Ashtadla Yantram - it is a machine modeled on the shape of a lotus containing 8 petals. Under each of these petals you will find some containers. In each of these containers you will find eight things that are smoke, light, water, steam, air, Rushakam, Vishasaram, Manjusham Katusaram that are described in the Meghotpati Prakaranam. In the center of the lotus there is a key, in which there are eight screws for the eight petals. By actuating each screw what that is connected to the petal will be brought to the top and will form a cloud. Pressing the central button of the steam it will be irradiated as the sunlight. As soon as the heat of these fumes will act on the clouds previously formed, it will start to rain. This machine is specialized in producing the rain.

Souryayana Yantram - this machine looks like a column of 116 feet in height and 58 in circumference. At the top there is a sieve containing some holes and it is made of glass of the 96th class called Somapa. From this sieve, contained in the pillar, there are twelve machines in order and above the sieve there is a glass cover of the 97th class called Somasya Darpana. Above this there is a wheel with a glass cover called Kumudinee containing some rays made of glass class 98th and called Chandrika Darpana. In the twelve points of the machine there are twelve screws above and twelve lower screws. Turning the screw the contents of the machine will rise in the proportions as for electricity, cold fluid, Shaitya Drava, Sudha Mushee, Soonruta, Pushkalee, Pranada, Dravinaamrutha, Sooraneee, Jambaalee, lulita, Vaachaklavee, Gacyoosha.

Through some cylindrical tubes that are attached to the wheels of the sieve these energies will flow touching the glass of the top cover. Turning the screw for electricity the wheel starts to turn 1192 rpm. So the energy of the rays of the moon called Someeya will be attracted to this wheel and fall through the sieve.

Thus this energy will fill the container below in the form of gas that must be kept tightly. Its use is that, when the limbs such as the head, hands or feet, of a person are cut will be resettled in the right place and the body will be kept in a special container. The body should be wrapped in a blanket of bark of a plant called Vaarshneeka Valkala.

When to that body will be injected with the aforementioned gas Somadrava for 5 Rajanikas, the body will be resurrected. This must be done within five minutes after the wound was caused. It is used to put in place the limbs immediately, or resurrect those killed in the manner described above.

The Tejas astras, and the energy weapons

A fundamental error seems to have always accompanied the Western interpretation of the ancient divine Indian weapons, something that has only recently been brought to light by some scholars as Jarrod L. Whitaker[139], Associate Professor of Eastern religions at the Wake Forest University in North Carolina.

Whitaker in his analysis focuses on a fundamental fact, a 'detail' that radically subverts the understanding of these instruments of offense, a factor that totally changes the attribution 'nuclear' allocated to the ancient Indian text descriptions found up to now.

Whitaker showed, in a search lasting several years, as the term in which they were always identified the Astras was the Sanskrit word *Tejas*. In its literal meaning of the term is "*bright, shining, radiant energy, fire*". The divine weapons must therefore be interpreted as something different than imagined until now, they should be read in a different light that connate deeply their characteristics by specifying them in a context that calls for an "energy" shining and destructive.

In its transliteration commonly accepted, both in the lexical etymology, the term *Tejas* assumes the meaning of 'energy' and the corresponding weapons become able not only to inflict harm but also to liberate, during their use, a force able to amplify the destructiveness and the power.

---

[139] Jarrod L. Whitaker, *Divine Weapons and Tejas in the two Indian Epics*, Indolranian Journal 43: 87–113, 2000. Kluwer Academic Publishers. Printed in the Netherlands.

This simple notation totally changes the concept that until now has been attributed to the Astras.

The term itself, in its cultural and historical evolution went to show, in the ancient Sanskrit language, the 'fire', one of the bench-mahabhutani, the five building blocks of the lowest aspect of the matter.

The divine weapons cannot be fully understood unless we analyse the concept of *Tejas* and its general implications.

Before this, however, we must keep in mind Whitekar's words about the present understanding about this subject, "... *Unfortunately, the scholars have never seriously studied the divine weapons in their details. When confronted with a weapon as the divine Agneya astra (weapon fire), the historians have generally explained its nature as an old musket or weapon gunpowder*"[140].

If, as we saw in Chapter III, the concept of 'firearm' was already present in the ancient India, the same certainly cannot be related with the descriptions of Astra present in the ancient sacred texts and correlate even to the more destructive power unleashed by the divine weapons that could cause an *ante litteram* rifle.

Generally the Tejas preserves its ancient connection with the fire and the heat but in its connection with an Astra the value and significance of this weapon change originating a connection with a destructive energy.

Paul Mangone states that the "...*tejas appears as a universal self-sustaining, independent energy from the gods and that obeys to its intrinsic laws to which the same god is facing*"[141].

There are no limits to the ways *tejas* can manifest their powers, they can be subtle and hidden as well as destructive and explosive, in any case they are tangible tools of death.

It is interesting to observe how in the few researches produced about the concept of *tejas* always rotate about a main pivot; this energy is multifaceted and changeable, it can take different forms while maintaining its original principle as shows us the *tejasvins,* a particular form of this same energy 'used' by the Rishis to threaten the 'planets' and the entire 'universe'.

As Whitaker always reminds "*the concept of divine weapon is fluid, they are energy weapons*" even if a more interesting element is provided by the ancient sacred

---

[140] Whitekar, op. cit., p. 88.

[141] Paolo Mangone, *The development of Tejas from the Vedas to Puranas*, in Proceedings of the VIIIth World Sanskrit Conference, Vienna, 1990, p. 138.

texts, "... *through the mantra a divine weapon is generally linked to a normal weapon (Sastra) - typically an arrow - which obviously makes it vastly more effective*"[142].

The union between the two realities creates a weapon of unusual power, in which the divine energy amplifies the power of an earthly weapon to turn it into an Astra.

Many weapons take their name from the presiding deity and manifest natural phenomena associated with the particular power that they reflect. The Agneya Astra, for example, or 'weapon of fire' belongs to the god Agni, god of fire, while the Astra Varuna, or weapon of the water, belongs the homonym divinity.

The transliteration of the effect on the weapon in relation to a corresponding deity seems to arise from the need to understand and classify these divine weapons within a religious and material context that will clarify and justify the objectives.

Often we assist, even at their anthropomorphism in sentient beings, or the Astra becomes a human or animal transfiguration of their destructive fury[143].

The *Tejas* is energy, a power that can be coupled and joint to a material weapon but also to be used in its purest form, the most destructive.

Whitaker's study identifies some fundamental points associated to the energy of the Tejas, some factors that are present in the Indian scriptures and associated with this form of energy. Let's see some of them.

The control of the divine arms
Any man or warrior, who acquires some divine weapons, has a corresponding quantity and quality of *Tejas* to control and to use during a battle.

The danger of divine weapons
Since the *Tejas* absorbs another *Tejas*, there is a danger intrinsic the divine weapons that, if not carefully controlled, can lead to catastrophic results.

Neutralization of divine weapons
When incited beyond measure, or made uncontrollable, the *Tejas* must be neutralized. This process can occur in two ways, a Tejas of higher power can destroys the lower or a dominant Tejas absorbs the energy of the secondary increasing its level.

The Puranic literature provides us with elements of great interest in understanding the genesis of this energy.

---

[142] Whitekar, op. cit., p. 90.

[143] Look at *Mahabharata*, 3.170.38-48 e 3.12.31.1.

Witnesses of this are the Vishnu Purana[144], the Markandeya Purana, the Matsya Purana, Padma Purana, the Bhagavata Purana or Harivamsa Purana.

We wanted to list the most important of them to show how this energy is widely debated within the oldest Indian sacred texts and how many, of these studies are profuse and deep.

The Vishnu Purana provides us, however, a mythological story of extreme interest. The creator of Astras utilizing the *Tejas* was Visvakarma the architect of the gods/Devas or one who had built the first Vimanas[145], the most important of which was the Pushpaka Vimana[146].

The text tells how the god, managed to win and harness the Sun in order to take it on the Earth after having also 'cut' part of its energy.

Visvakarma harnessed the sun with the Vaishnava Tejas, which fell violently on Earth and with such energy, the celestial architect built Shiva's trident[147].

In the Markandeya Purana is narrated how Visvakarma acquired the art of shaping the energy, he "... *began creating Astras for the other Gods and these beautiful objects had been made with the Tejas of the Sun for the pacification of the enemy*"[148].

The Padma Purana is even more explicit by stating "... *thanks to the friction of the foot of Sarva* (the ancient name of the god Shiva, ndr), *some sparks broken free from the Tejas and with these Vishwakarma built the Astra and the Celestial Chariots*"[149], the Vimana.

In the case of Narayana Astra, that was defined as the *ultimate weapon*, the Mahabharata is more explicit saying that the god Vishnu built this Astra by the

---

[144] Visnu Purana 3.2.8-12

[145] Adiparva, Ch. 66, vss. 26-31.

[146] Mahabharata, Vana Parva, Chapter161, Stanza 37.

[147] Visnu Purana, 3.2.8b-12.

[148] Markandeya Purana, 108.4b-5a.

[149] *Padma Purana*, 6.9.31b-33a.

tejas of the god Narayana[150], one of his manifestation, and "*his Tejas was so powerful to extinguish the Tejas of other Astra*"[151].

Whitaker, in his study[152], shows that the *Tejas* constitutes the constitutive and foundation principle of the divine weapons. Without this the *Astras* could not exist. Secondly, the researcher indicates the high destructive power that the *Tejas* takes place once in the form of weapon. This energy pervades all divine weapons becoming the founding principle of their power.

All mythical tales show how the weapons of the gods are a metonim of the *tejas*. Hits own use, and the power given off by it, are now also deeply connected to the amount of energy available. We will have a condition of *alpatejas* or limited energy, which can be transformed into *mahatejas* 'very tejas'. To understand this principle is interesting a legendary episode that tells how Vishnu had 'energized' the Vajra tejas of Indra, the weapon of lightning, to defeat the demon Vrtra[153].

The ancient texts are clear in asserting that many divine weapons possess inside them a power so destructive that they can kill thousands of people in an instant. The energy intrinsic in these *astras* is clearly described in an episode of the Mahabharata in which Arjuna acquired by some of their weapons and began to try them thoughtlessly to observe their effects.

The following passage contains the words of the sage Narada that address to the hero "*Oh Arjuna, Arjuna, you should not use the divya astras! They should never, in any way, be issued with a misuse ... within these Astras is a great demon, when these are protected according to the tradition they will bring great happiness but if they will not they will bring destruction all over the world worlds. Oh Pandava do not do that anymore!*"[154].

The destroyer demon (*Sumahan doash*) released by the Astras is so descturctive to devastate the 'worlds'. What are these negative effects (*niradhisthana*) mentioned in the ancient texts? The following passage reveals the exact nature of their misuse, the incident calls into question always Arjuna when he received the *Pasupata Astra*, equivalent to the *Brahmastra*.

---

[150] Visnu/ Narayana plays a key role in the genesis of the Hindu's universe, within him is, in fact, that the creation is reabsorbed and it is by him that everything has once again origin. In the Narayana Upanishad he is decanted as the vital principle that permeates and preserves the whole creation.

[151] *Mahabharata*, 7.170.40

[152] Op. cit., p.94.

[153] *Mahabharata*, 5.10.12-38 e 5.16.16.

[154] *Mahabharata*, 3.172.18-21.

The god Shiva addressing directly to Arjuna "... *I give you the great Pusapata Astra, which is my favorite! O Pandava you are able to control it, to thought it and to recall it. Even the great Indra does not know him, or Yama, or the king of Yakas, even Varuna or Vayu. How can humans know it? However, O Partha, it must never be hurled mindless against a human because if it is launched against someone with an insufficient Tejas, will destroy the whole world*[155]*. There is no one in the three worlds ... that is invulnerable to it*"[156].

To summarize what has been said on the *Tejas* and on the *Deva Astras*, we see that there are some fundamental principles by which this energy can and should be used. First of all the divine weapon possesses as a beating heart a *Tejas*, by the very high potential energy that pervades the universe.

Let us have a parallelism, this energy seems to be similar to that one contained at the atomic level or quantum also about the effects that could produce if properly controlled and directed.

The *Tejas*, and consequentially the weapons which make use of it, seem to follow a precise hierarchy and, if not properly controlled and triggered, this force naturally 'expands' until its container and all the surroundings are destroyed.

If you were to confirm such situation, the ancient texts state that it will radiate an event of a cataclysmic scale as described for the *Narayana Astra* and the only option to defuse such destruction will be to remove any possible source of energy from which the *Tejas* will draw its potential strength that amplifies the power.

This last aspect, as was pointed out by Whitaker, is extremely interesting. The *Tejas* seems, in fact, to be able to tap into the surrounding energy to boost its power but once harnessed and channeled into a weapon that divine power is somehow harnessed and controlled until to produce the desired '*effect*'. This is the 'hierarchy of power' mentioned above.

The divine weapons do not appear to be the result of a simple and fantastic mythographic creation but they have their own rationality and a precise and clear internal logic even more validated by the concepts contained in the ancient Indian physical. The treaties on the physical sciences, also called *Bhoutika Shastras*, in which is also contained the *Tejas*, claim that a subtle energy permeates all life forms, from animated to those inanimate, tying 'Everithing' to a common property that the charges and gives them substance, features and power.

---

[155] *Mahabharata*, 3.163.49.

[156] *Mahabharata*, 3.41.13-16.

The *Tejas* is only one aspect and one manifestation of this pervading energy but is a force that in time extremely remote was harnessed and used to make weapons by the very destructive potential[157].

The Visvakarma, the Lanka and the Dwarka

We introducing now a digression from Tejas and divine weapons of the gods. As we could highlight the figure of Visvakarma plays a predominant role in the creation of the divine weapons and in the construction of the first Vimana. Of equal value was his role as architect and for that reason two cities are attributed to him which have been a recurring element in all our discussion, Lanka and Dwarka.

According to the Indian mythology *'Sone ki Lanka'*, or *the golden Lanka*, was the place where the demon Ravana lived during the Treta Yuga. As narrated in the *Ramayana*, it was also the place where the demon held prisoner Sita, Rama's wife, after abducting her.

There is an interesting legend[158] behind the construction of the city of Lanka. When Shiva married Parvati, the god asked Viswakarma to build a beautiful building to stay; the divine architect built a golden residence for the couple.

Shiva invited Ravana for the ceremony to perform the ritual 'Grihapravesh' and, after the sacred ceremony, Shiva asked Ravana if he wanted something in return as *Dakshina*[159]. Ravana, overwhelmed by the beauty and grandeur of the building, asked Shiva the same building! Shiva was forced to accept and Lanka became the palace of Ravana.

Lanka was not the only city built by Viswakarma, the architect of the gods. Also to his ingenious ability we have to bestow the building of Dwarka[160], the capital of Krishna. As told in the Mahabharata, Krishna would have lived in Dwarka transforming it in his *'Karma Bhoomi'*, in other words its operations center.

---

[157] V. Raghavan, M.A., Ph.D., *Yantras or Mechanical Contrivances in Ancient India*, The Indian Institute of Culture Basavangudi, Bangalore, Transaction No. 10, February 1952.

[158] Swami Parmeshwaranand, *Encyclopaedic Dictionary of Puranas*, Volume 1, Sarup & Sons, 2001, p. 1388.

[159] By the term *Dakshina* is intended a practice of spiritual discipline in which the subject has to become aware that 'we must first give in order to receive'.

[160] W. G. Archer, *The Loves of Krishna in Indian Painting and Poetry*, Echo Library, 2007, p. 44.

The city was also the scene of the battle between the god Salva and the same Krishna on board of their Vimana, a clash that saw the use of numerous Astras and which reached its epilogue with the victory of Krishna.

We cannot forget the remains found submerged in the recent years off the coast of the city. The archaeological evidence appears to support the myth and give veracity to the existence of an ancient city for long time considered the fantasy of ancient authors.

In the final analysis we are in front of something that goes beyond the very concept of the classical myth presenting a new reality charged not only of deep inner meanings but also of scientific concepts and techniques, which have been converted in a legendary vision later on.

Gods's wars

The main tread of our analysis was the constant presence, in the ancient Indian epic texts, of the Vimana and their Astras, some divine weapons able to destroy and annihilate the armies as well as "*to make whole countries uninhabitable*".

We analyzed the most ancient Indian writings in search of answers and found that almost all of them talk profusely and in detail of incredible weapons, some works in which the authors are using such specific details to leave no doubt about their naturalness.

Adding to all this the 'reverse engineering' made in the recent decades in many university laboratories, the question becomes even more complex and difficult to diminish the critics as they attempted to make several Western scholars.

The mere mention of all the steps that have fallen in our analysis would require so many volumes to fill an entire library. The Selection presented to the reader is representative of the above and constitutes a useful tool to study and to understand the depth and power of the *weapons of the gods*.

"*The valiant Aswatthaman, standing up, and resolute in his chariot, he touched the water and invoked the weapon Agneya to which cannot withstand even the gods.*

Pointing it towards all his visible and invisible enemies the son of the preceptor, the slayer of enemy heroes, inspired by the mantras he fired an explosive column that opened in all directions and caused a bright light, like smokeless fire which was succeeded by a dense shower of sparks that completely surrounded the army of Partha. The four cardinal points, a haul that gaze could not embrace, were covered with dark, due to the cloud of the explosion.

The soldiers and Rakshasas Pisachas narrowed to one another by throwing loud cries. A violent and bad wind began to blow, the Sun itself did not give more heat.

The crows cawed terrified, the clouds roared in the sky raining blood. The Birds and The animals of the Earth gave signs of intense suffering. The elements themselves appeared perturbed. The sun seemed to turn in the opposite direction, the universe, flayed by the heat, seemed to have a fever. The Elephants and other animals of the Earth, flayed by the energy of the weapon, they ran terrified, breathing heavily and looking protection against that terrible force ... even the water began to boil for the heat and the animals that lived in that element gave signs of intense suffering and seemed to burn. From the four cardinal points, from that heaven of hell, fell a shower of sparks acute and fairs that hit swirling with the power of Garuda. Hit and burned by those weapons of Aswatthaman, which were full of energy and had the violence of the thunder, the enemy warriors fell like trees felled by a raging fire. Big elephants burnt by the weapon fell to the ground all around, throwing terrible trumpeting strong as the roar of the clouds. Other big elephants, flayed by the fire, started to run around, throwing screams of terror, like in the middle of a forest on fire"[161].

Another representative step describes the effects of the weapon Narayana and the devastation and the panic created among the ranks of the army.

"Sanjaya said, 'when the weapon called Narayana was invoked, a violent winds began to blow, with gusts of rain, we heard bursts of thunder, although if the sky was cloudless. The earth shook and the sea swelled in agitation. The rivers began to flow in the opposite direction. The tops of the mountains, or Bharata, began to crack. Several animals began to flee on the left side of Pandhava. Darkness fell and the sun became dark. Various carnivorous creatures began to fall on the ground, with joy (to feed of the corpses). The gods, the Danavas and Gandharvas, or the monarch, were filled with fear. Watching the tremendous agitation (of nature), they began to ask each another, shouting, which could be the cause. Actually seeing (the effects) of the fair and terrible weapon invoked by the son of Drana, all kings, frightened, and felt a great pain.

After Drona, the preceptor, was killed during a battle by Dhrishtadyumna, as the powerful Asura, Vritra, the pitcher of thunder, the Kurus, or Dhananjaya, is dismayed and gave up all hope of victory. Eager to save themselves, all fled from the battlefield.

Some drove wildly the wagons without the driver Parashni, and stripped of banners, flags and umbrellas and their equipment dispersed.

Others, in panic and private of the reason, fled precipitately, beating kicking horses of their chariots.

Others, on carts with yokes and the wheels and axles broken, fled plagued by fear.

Others, who rode on horseback, were taken away, almost completely dismounted.

Others, torn from their seats and nailed to the neck of the elephants by the spears, were taken away from their animals.

Others were trampled to death by elephants covered by arrows.

Others, deprived of weapons and armor, fell to the ground from their vehicles or mounts.

Others were cut by the carts or crushed by the legs of the elephants and the horses.

Others shouted the names of their kings and their children, fled in terror without recognizing each another, deprived of all energy from anguish.

---

[161] *Mahabharata*, Drona Vadha Parva, Section CCI.

Others, after putting the commanders, sons, brothers and friends in the wagons, were seen taking off their armor and wash them in the water. After the killing of Drona, the Kuru army, reduced in these terrible conditions, rushed away"[162].

Also *"Mahadeva is taken from anger, neither gods nor Asuras, Gandhavas or Rakshasa can have peace, even if they hide in the deep of the ocean. In those days, Daksha had gathered the necessary elements to make a sacrifice: Mahadeva, in his anger, destroyed that sacrifice. In truth became very hard on that occasion.*

Throwing an arrow with his bow, he threw terrible roars. The celestial were full of anxiety and fear. In truth, when Mahadeva was enraged and the sacrifice (in his corporeal form) ran away, the gods were extremely frightened by the sound of Mahadeva's arc and the sound of his hands. The gods and the Asuras fell down and made an act of submission to Mahadeva. All the waters swelled in agitation and the earth shook.

The mountains split open and all the cardinal points and the Nagas marveled. The universe, enveloped by thick darkness, we could no longer see anything. The splendor of all the stars, including the Sun, was destroyed.

The Rishis, filled with fear, became restless and eager for their good and for that of all creatures, celebrated some rites.

Surya was the one who ate the main offer. Once again Mahadeva headed towards the gods a rain of hot and sharp arrows that looked like tongues of fire mixed with smoke, or clouds with lightning. Seeing that rain of arrows, all gods, bowing to Maheswara, assigned to Rudra, a major part of the sacrifice. Before the valiant Asuras, possessed, in the sky, three cities. Each of these cities was excellent and great. One was made of iron, another of silver and the third of gold.

The city of gold belonged to Kamalaksha, the city of silver to Tarakaksha and the third, made of iron, had Vidyunmalin as lord. With all his weapons, Maghavat (Indra) was unable to make any impression on the cities"[163].

This description, taken from the Mahabharata, seems less precise in the exposition of the effects produced by the weapons but focuses mainly on the religious-ritualistic aspects of the dispute to the origins of the struggle. It is interesting, the reference to '... *hot and sharp arrows, which look like tongues of fire mixed with smoke, or to clouds or to lightning*', a battle fought not by men but by two gods, the choleric Mahadeva and Shiva.

Emblematic is also the next battle that is described both in the Mahabharata and in the Rigveda. As apparently dissimilar, the scholars have highlighted their correlation with the name of the city around which the fighting took place. In the first case Hyranyapura, and in the second case Hariyupa, different spellings of the city known as the Harappa belonged to the homonymous civilization of Mohenjo Daro was part. The reference to Hymavat instead is a personification

---

[162] *Mahabharata*, Drona Vadha Parva, Section CCI.

[163] *Mahabharata*, Drona Vadha Parva, Section CCII.

of the Himalayan mountains, the location that puts the events, together with the reference to Harappa, in the actual Sindh in Pakistan.

"*It is said that the same prince Maheswara, of a supreme power, has been met before by Partha at the foot of the mountains of the Hymavat! Stimulated by the head of the celestial creatures, he killed with a single wagon thousand Danavas, who lived in Hiranyapura! That son of Kunti is now allied with Vasudeva, of great intelligence. I think now he is able to destroy the three worlds, including the gods themselves*"[164].

While the second song reads "*In aid of Abhyavartin Cayamana, Indra destroyed the seed of Varasikha at Hariyupiya, struck the edge of the Vrcivans, and those who were behind fled in terror. Three thousand (warriors) covered with armor, looking for common glory on the Yavyavati, or very searched Indra, the sons of Vrcivan fell under the arrow, as vessels that are exploding, and both of them went to met their destruction*"[165].

The reference is toward the weapons that kill thousands of people at the same time as '*exploding vessels*', and is quite representative and paradigmatic.

We have selected the following passage because in his verses are described the defensive and the offensive weapons of an incredible power: "*Drona, full of great anger, struck Yudhishthira with the Vayavya. The son of Pandu, however, neutralized such weapon with his weapon that was like that. Seeing his weapon neutralized, the son of Bharadwaja, full of great anger and eager to slay the son of Pandu, tossed to Yudhishthira many celestial weapons, like Varuna, the Yamya, the Agneya, the Tvashtra and Savitra.*

The powerfully armed Pandavas, however, fearlessly repelled all the weapons of the Born-from-a-jar, that were or were about to come directly against him. Then the Born-from-a-jar[166], *intending to fulfill his vow and wishing, even for the sake of his son, to kill the son of Dharma, invoked in existence, O Bharata, the weapons Aindra and Prajapatya.*

But the first of the race of Kuru, Yudhishthira, with the behavior of an elephant and of a lion, with the wide chest and with the red and large eyes, soaked with an energy scarcely lower, he called in existence the weapon Mahendra.

With this he neutralized Drona's weapon, full of anger and wishing to destroy Yudhishthira, he called into existence the weapon of Brahma. As we were enveloped by a thick darkness, we could not see what was happening. Although all the animals were, o monarch, full of a great terror. Seeing Brahma's weapon pointed against him, the son of Kunti, Yudhishthira, o king, he neutralized him with his weapon of Brahma"[167].

---

[164] *Mahabharata*, Abhimanyu Cadha Parva, Section LXXV.

[165] *RgVeda*, Libro 6, Hymn 27, verses 5 and 6.

[166] The term 'Born of a Jar' refers to the ways in which women gave birth at that time, crouched on their own feet and in which the newborn baby was placed in a special earthenware jar.

[167] Mahabharata, Ghatotkacha Vada Parva, Section CLVI.

In the Ramayana we can identify, among the dozens present, a song equally representative "*Then Bharata, Rama's younger brother, angry, delivered a terrible spear. Caught in the snare of destruction, three hundred thousand Gandharvas were killed in an instant, blown to pieces by that hero. The inhabitants of the celestial regions were unable to remember another conflict so terrible, during which, in the blink of an eye, it was such a high number of dead warriors. Having slain the Gandharvas, Bharata, the son of Kaikey entered within those two opulent and magnificent cities and there, Bharata established Taksha in Takshashila and Pushkala and Pushkalavata, in the land of the Gandharvas, in the enchanting region of Gandhara*"[168].

This passage speaks of the death of three hundred warriors in a single moment, a reference rather unusual when compared to our mythological traditions and the corresponding weapons used. It seems rather to attend a weapon of mass destruction launched against an invading army.

Another item of interest is the reference to '*the region of Gandhara*' historically existed between the northern Pakistan and the eastern Afghanistan.

This information situates the events of the Ramayana where S.B. Roy, Davenport and many other scholars have identified the mythical Lanka with the city of Mohenjo Daro. Another factor that seems to confirm what until now exposed!

Of a different nature are the pieces that we are going to read. These verses seem to refer unambiguously to a nuclear explosion occurred in antiquity but, as has been commonly presented as a unitary text by many authors, in fact we are in the presence of a collage of excerpts from several different chapters of the *Mahabharata*, parts that in many cases relate to different events from each other.

"*... a single projectile charged with all the power of the Universe.*

An incandescent column of smoke and flames as bright as a thousand suns rose in all its splendor ... an vertically explosion with clouds of billowing smoke...

... the cloud of smoke, rising after the first explosion, opened in circular waves, such as the opening of a giant umbrellas ...

... it was an unknown weapon, an iron thunderbolt, a gigantic messenger of death which reduced to ashes the entire race of the Vrishnis and the Andhaka. ...

The corpses were so burned as to be unrecognizable. The hair and the nails fell out. The pottery broke without apparent cause, and the birds turned white. After a few hours all foodstuffs were infected ...

... to escape from this fire the soldiers threw themselves in the streams to wash themselves and their equipment".

Also in the Mahabharata, in the eighth book called *Karna Parva*, we find an ambiguous reference[169] to a 'glory' of 10,000 Suns.

---

[168] Ramayana, Uttara Kanda, Chapter CI.

[169] Mahabharata, Libro 8, Karna Parva, Section 34.

In the version that can be found in most of the texts published nowadays 'suns' have been reduced to 1000, however, probably due to an error of transcription, moreover it is arbitrarily associated with the power of a nuclear device. Actually the song indicted refers to a deity named Nila Rohita whose name can be interpreted as "red and blue smoke" and as a result of his anger he gave off "*a flash of a strong energy, which burns with splendor ... and wrapped by the superabundant energy of the fire, blazed with splendor*".

On the other hand it is also true that in the Drona Parva, as well as in the chronicles of the war of the Mahabharata in general, the weapons described are often quite conventional, like bows and arrows. "*Keeping the bow at full capacity bending, almost circular, no longer able to distinguish the movements. Take the arrow, reciting mantras, throw it and get another one had become all one. From that deadly weapon flowed an uninterrupted current of deadly darts, hitting with an inhuman accuracy*"[170].

The dichotomy between human and divine weapons reiterates clearly as in the ancient Indian world existed, and it was very clear, this division.

The differences between these two categories are structured, in both cases, with a level of detail such detailed as to leave no words. As we have repeated many times in the case of the divine weapons, the technical-scientific subtleties present in their descriptions seem to reveal at their base a real and concrete component, a background that is not the result of a mythographic invention but seems to draw on knowledge scrupulously preserved over the millennia. The reverse engineering experiments made in some Indian universities on the data contained in some ancient texts have instead established the credibility and reliability of some of these 'mythical tales'.

The description of an '*thunderbolt of iron*', for example, can be found eight times in the sixteenth book of the Mahabharata, the Mausala Parva. The use by Samva of this weapon was so devastating that "*... all the individuals of the race of Vrishnis and Andhakas became ashes ... resembling to a gigantic messenger of death*".

Away from the legends of any other civilization on the planet, India seems to have passed and kept precise accounts of those events that involved in the remote past of the humanity. The Astras represent the objective expression of something unique, the concrete manifestation of a technology that differed profoundly from the notions held in the historical past from any other civilization on the planet but also by the modern technological concepts.

Its essence left indelible traces in an initiatory chain of sages and ascetics who preserved part of this knowledge away from the clamor of the civilization handing down from master to disciple in the eternal wisdom chain of the Shruti.

---

[170] Mahabharata, Drona Parva, paragraphs 113-118.

CHAPTER V - The Vimana literature

The Rig Veda

The Rig Veda is the oldest holy text of the human history.
Generally it is believed that it was put in writing to the second millennium BC, with a use of the Vedic dialect in the composition of the Holy Scriptures that lasted until about the fifth century BC when it began to emerge the classical Sanskrit.
Many scholars, both Indian and European, disagree to this dating believing that the Vedas are even older. There is an example the studies made from the Italian Michelguglielmo Torri which states *"The two strengths of this theory - that is, the origin of the autochthonous Aryans that is the Vedic civilization, nda - referring to the fact that, it being understand to the indication of the 1000 BC as the completion date of the composition of the hymns collected in the Rig Veda, is not at all certain what is the start date. This could be much older than 1500 BC and date back to 3000, 4000 or even 7500 BC. The first element in support of this is taken by the astrology, that is that inside of the Vedas there is a series of astronomical references which, when decoded, suggest that the composers of the Vedic hymns have been living under a sky characterized by pattern stars and by characteristics some solar parables characteristics periods much older of the 1500 BC"*[171].
The same studies have been made on epics like the Mahabharata and the Ramayana providing extremely similar dates

Sanskrit literature

In the ancient Sanskrit epics, and even more in the ancient Veda, the worthy predecessors of the *Vimana* are so-called *'flying chariots'* used by the various deities. Indra had his own Chariot of the Sun while the other moved in the sky by flying chariots with wheels pulled by animals, usually horses.
In the RigVeda the Vimana are never directly mentioned but, in some verses[172], their presence is clearly evident through the description of the 'mechanical birds' that are able to fly.
For the Hindus, the historical evidence of these flying chariots is evidenced not only by the faith reposed in the sacred texts handed down over the

---

[171] Michelguglielmo Torri, *Storia dell'India*, Bari, Laterza, 2000, pag. 32.

[172] RV 1.164.47-48

centuries, these transpositions that in all cases have preserved an exceptionality of descriptions and terms unique in their kind, but also the certainty that before our time other civilizations walked on our planet, some cultures have been destroyed by huge cataclysms that brought humanity to have to start almost from scratch.

Swami Dayananda Saraswati[173], the most important and respected Indian Masters of the nineteenth century, described a Vimana starting from the RigVeda with the following words "*...it jumps in the space quickly with an instrument that is using fire and water ... containing twelve stamghas (pillars), a wheel, three cars, 300 pins, and 60 instruments*"[174].

A further step of the Rig Veda[175], which has not yet shown in this field of study, states further the capabilities of these aircrafts, "*O Asvini ... You have done for the son of Tugra [Bhujyu], between the bursting waters, that ship with animated wings with which to fly, on which ... you brought him forward. And he flew away with an easy flight surviving to the mighty wave. Four ships, greeted with joy in the middle of the ocean, driven by the Asvin, saved the son of Tugra, the one who had been thrown headlong into the waters ...* ".

In the text of the epic Ramayana, the god Rama, back from Ayodhya, used a flying machine called *Pushpaka Vimana* described as a 'boat'.

Also in the Ramayana, the *Pushpaka Vimana*, previously belonged to the demon Ravana, is described as follows "*The wagon Pushpaka, which resembles the Sun and belonged to my brother, was taken by the powerful Ravana; this so excellent flying chariot flys anywhere you want ... this wagon looks like a bright cloud in the sky ... and the King [Rama] took it, so the wonderful chariot commanded by Raghira, rose in the higher atmosphere* "[176].

This description is the first reference to a concrete and unequivocal Vimana distinguishing and differentiating from previous descriptions that referred generally to flying chariots drawn by horses.

In the Jainism in India we can find an interesting transliteration of the word Vimana. The reference is to the *Vimana-Vasin,* or *Deva Vaimanik,* literally the

---

[173] Maharishi Dayanand Saraswati (12 February 1824 – 30 Ottobre 1883) was an important scholar and theologian, reformer and founder of the Arya Samaj, a Hindu reformist movement.

[174] Cited by Mukunda, H.S.; Deshpande, S.M., Nagendra, H.R., Prabhu, A. e Govindraju, S.P. (1974). "*A critical study of the work "Vyamanika Shastra*", Scientific Opinion: 5–12, p. 5.

[175] RV 1,182, 5-6

[176] Dutt, Manatha Nath (translator), *Ramayana,* Elysium Press, Calcutta, 1892 and New York, 1910.

inhabitants of the Vimana, a class of deities who served the Tirthankara Mahavira[177] the last of the 24 prophets, for this religion, have succeeded in historical cycles in order to reveal the Jainism to the humanity. This prophet lived between roughly among the 599 BC and the 527 BC.

These 'gods of the Vimana' lived in the *Urdhvaloka* or the upper world[178].

According to the *Kalpa Sutra* of Bhadra-Bahu, the Tirthankara Mahavira twenty fourth emerged out the great Vimana Puspa-Uttara[179] while the 22 Tirthankara Arista-nemi emerged from the great Vimana Aparijita[180].

For Digambara's traditions, the nineteenth Mallinatha was the son of the king and queen Kumbha Prajavati and descents on Earth thanks to the Vimana Aparijita as his place of birth was in the Asvini Nakshatra, near the stars Castor and Pollux in the constellation of Gemini[181].

The fourth and fifth Tirthankara, named Abhinandana and Sumati-natha, traveled into the space aboard of their Jayanta Vimana[182] also called Vimana Sarva-artha-siddhi, belonged to the gods Jayanta[183], while the fifteenth

---

[177] Johann Georg Buhler (ed. by James Burgess): *The Indian Sect of the Jainas*. London: Luzac, 1903. p. 69 e Hermann Georg Jacobi, Jaina Sutras: Parte I, Forgotten Books, 2008, p. 169.

[178] Udharva Loka consists of 12 Dev Lok, 9 and 05 Greveyak Anutar Viman, which are the kingdoms of Vaimaniks or astral gods who are not freed. Above the Anutar vimans, at the summit of the universe, is located SiddhaSila, the realms of the gods freed also known as the Siddha, the perfect all-knowing being, revered by Jains.

[179] Saryu Doshi (trad. di Thomas Dix), *Dharma Vihara, Ranakpur*. Axel Menges, 1995. p. 11a.

[180] Enciclopedia Mewar alla voce "*Ranakpur, founding of*".

[181] Umakant Premanand Shah, *Jaina-Rupa-Mandana*, Volume 1, Abhinav Publications, 1987, p.159.

[182] Johann Georg Buhler (ed. by James Burgess), *The Indian Sect of the Jainas*. London, Luzac, 1903. p. 67.

[183] Johann Georg Buhler (ed. by James Burgess), *The Indian Sect of the Jainas* , London, Luza 1903, p. 74. See also Thomas William Rhys Davids, William Stede, *Pali-English Dictionary*, to the voice 'Vimana', Asian Educational Services, 2004.

Tirthankara, named Dharma Natha, who went to heaven on board of the 'Vijaya-Vimana'[184].

From the list just presented cannot escape even a Buddhist text, previously mentioned, called *Vimana Vatthu*, which in Pali language literally means "*Vimana's Stories*". This is the sixth book of the Khuddaka Nikaya and describes the splendor of the heavenly mansions belonging to the Deva, the heavenly gods, through 83 stories, divine residences conquered by them as a reward for the meritorious and heroic acts performed in their previous life.

Their stories would be directly learned from the wise men Moggallana and Vangisa by the Deva during a stay in their worlds and subsequently reported to the Buddha[185].

In this case, the term Vimana takes a more metaphorical meaning always staying within that area that wants it related to the divine beings, the Devas precisely, and their heavenly 'homes'.

In the Mahabharata there is also the case of Asura Maya and his Vimana, which measured eleven cubits in circumference and had '*four powerful wheels*'.

Besides the use of the 'blazing missiles', the poem recalls the use of other deadly weapons such as the 'dart of Indra', the *Indravajra*, maneuvered through an instrument that the ancient texts define a 'circular reflector'.

This weapon, when switched on, it produced a 'shaft of light' which focused on any target, consuming it immediately with his strength.

In a passage of the saga, the hero Krishna is chasing his enemy Salva in the sky when the Vimana of the latter, called Saubha, become invisible to the eyes of the god disappearing.

Undeterred, Krishna used a special weapon "*... I shoot an arrow that kills through the research of the sound*".

The Mahabharata[186] also mentions the "omniscient Yavanas" defined *sarvajnaa yavanaa* and sometime identified in ancient times with the Greeks, as the creators of some Vimana[187].

---

[184] Johann Georg Buhler (ed. by James Burgess), *The Indian Sect of the Jainas*, London, Luzac, 1903. p. 69.

[185] A commentary on this text was written by Dhammapala and is a part of Paramatthadipani, sometimes defined Vimalatthavilasini. The histories of the Vimana Vattu were connected, by Mahinda, to Ceylon in his first sermon at Anula and to his five companions. Mhv.xiv.58.

[186] Mahabharata VIII.31.80

[187] Clive Hart, *The Prehistory of Flight*, Berkeley, 1985.

The Vedas and the other inhabited worlds

Studying the ancient India and its atavistic mysticism imposes, to us Westerners, a necessary humility that is the consciousness of dealing with something totally alien to our way of relating to the spirituality and to the daily life.

Even more fascinating is the understanding that this people reached, about the world, ourselves and of science, already several thousand years ago, dozens of centuries ahead of the most innovative hypotheses and discoveries of the contemporary research.

Contrary to the common opinion, the true Hinduism is neither polytheistic or monotheistic religion but is a *henotheist*. The different gods and the worshiped Avatar, must be considered as different forms of the One, the Supreme God, Brahman, events that are used by the gods to make themselves more accessible to the man.

This distinction allows us to understand that there are significant differences between gods themselves, although if they are the emanations of the One, and higher beings considered semi-divine for their skills and abilities.

There are therefore major gods identified in 33 gods, defined *trayastrimsas,* and the Deva, literally '*one who radiates light*', identifiable in a myriad of celestial beings in charge of the most varied tasks in the higher worlds and on the Earth.

These beings are, all of them, considered a divine plan but believed bodily in their physicality. Their residences are the *Loka,* literally superiors 'Worlds'[188], understood in the subdivision of the universe in many inhabited worlds outside of our. From this logical and religious presupposition, the Hinduism has always considered the Earth as one of the many inhabited worlds, one of the countless planets where life had developed.

For us Westerners, the concept of life on other planets seems to be born in the modern times as a result of the advancement of the science and of the technology of our species, in fact already from the ancient Greece with the school atomist Democritus and Leucippus, it began to talk about '*plurality of inhabited worlds*'[189].

Such formulations, however, saw a sharp setback because of the rigid conceptions of the Aristotelian thought. Over the centuries, many thinkers tried to get back carriers of this research finding, however, almost always, the strong hostility of the prevailing status quo.

---

[188] AA.VV., *Dizionario Della Sapienza Orientale Buddhismo Induismo Taoismo Zen*, Edizioni Mediterranee, 1991, p. 240.

[189] Annibale Fantoli, *Extraterrestri storia di un'idea dalla Grecia ad oggi*, Carocci, 2008.

Among the few who managed to establish itself we find some philosophers as Giordano Bruno[190] or Isaac Newton but with some results that led the first at the stake while the second to hide his ideas until after his death

Reading the Vedas, whose first 'written' draft is officially dated about 2.000 BC, we find many references that "*life exists on other astronomical bodies in the space*" while in the *Bhagavad-Gita* we learn that "*there are countless universes and countless planets within each universe, and each planet is inhabited by different varieties of populations*".

To understand how this concept is deeply ingrained in the Indian culture have also recovered a step of an ancient sacred text, the *Srimad-Bhagavatam*, which tells how the mystical Kardama Muni was usually "*fly through the different planets as the air that propagates without brake in all directions*"[191].

During these trips it is told that Kardama Muni was able to cross '*the three materials universes of our world*', the lower planetary system known with the name of Adholoka, the planetary system called median Madhyaloka and the superior universe, called Urdhavaloka.

Puzzling is the assertion in the Padma Purana, one of the most important texts that constitute the major eighteen Puranas. The text was put in writing between the eighth and eleventh centuries AD but was previously one of the most ancient oral traditions of India.

Inside the wise Pulastya, one of the Seven Sages of the first age or Manvantara, he speaks to Bhishma of the essence of the religion and its importance. At certain time of the treatment it is discussed the number and the different types of forms of life that exist in the Universe.

In the Padma Purana it is stated that in the cosmos there are 8.400.000 forms of life, 900.000 of which are aquatic, 2.000.000 consist of trees and plants; 1.1 million are small living beings like insects and reptiles; 1.000.000 volatile; 3.000.000 are animals and finally 400.000 are human[192] species so forms of intelligent and of sentient life!

The only thing we can do is to take note of these numbers.

Always the *Srimad-Bhagavatam* tells how Kardama Muni met during his travels some beings belonging to higher planets, invisible to the man, called

---

[190] Author of *De Infinito Universo et Mundi*.

[191] Third canto, paragraph 23, verso 41.

[192] See also the work entitled '*Hinduism and Science*' di T.D. Singh, Director of the Bhaktivedanta Institute and President of the Vedanta and Science Educational Research Foundation. Paper presented at the "*Science and Religion: Global Perspectives*", 4-8 June 2005, Philadelphia, PA, USA.

Vaikunthalokas literally *'the planet or world beings Vaikuntha'*[193], the home of Vishnu.

The sage Sri Swami Sivananda says[194] that in the Vaikuntha Loka the Muktas, the pure souls, move through the use of the Vimana.

Sri Sivananda continues by asserting that the Vaikuntha Loka is crowded by Vimana of Vaidurya (cat eyes), of emerald and gold color, that are visible only to those *'who have bowed to the feet of Vishnu'*.

Since the plans are immaterial is curious to see how these 'beings' must still move through some 'material' flying machines that, incidentally, are the Vimana!

In the Mahabharata, Book Sambhava Parva section 93, it clearly speaks of other worlds (*lokas*) and ships (*vimana*) able to travel in the Universe.

Ashtaka said: "*Those five ships* (Vimana, ed) *that are appearing in the sky, that are shining like gold, for those who are? Climbing aboard, men can reach the eternal regions of the sky?*".

Yaiati replied: "*These five ships, glorious and blazing like fire, will transport you in the heavenly places*".

Ashtaka said: "*O King, Hop on one of these, and retreat in the sky. We can wait, we will arrive later*".

Yaiati replied: "*We can go together. In truth, all of us have earned the sky, observ, the glorious celestial path is becoming visible*".

Vaishampayana continued: *So, those excellent monarchs, boarded on the ships, heading towards the sky, reached the places provided for them.*
So it was, that the illustrious king Yaiati, for his great works, saved his descendants, going up to the heaven. Leaving the Earth, he held the Three Planetary Systems (trilokas) *with the fame of his actions.*

Also in the Hindu cosmology we often refer to 14 Loka[195], seven belong to the highest worlds, the *vyahrtis*, and seven to the lows, the *Patalas*.

---

[193] Srimad Bhagavatam Canto 4 Chapter 12, Verso 36. Further descriptions of *Vaikunthalokas* are found in the *Brahma Samhita* and in the *Bhagavata Purana*. Another reference can be found in Srimad Bhagavatam at verses 1.3.28.

[194] Swami Sivananda, *What becomes of the soul after death?*, The Divine Life Society, 1979.

[195] In their enumeration are Satya-loka, Tapa-loka, Jana-loka, Mahar-loka, Svar-loka, Bhuvar-loka, Bhur-loka, Atala-loka, Vitala-loka, Sutala-loka, Talatala-loka, Mahatala-loka, Rasatala-loka, Patala-loka.

The concept of Loka, in the Vedas, indicates not only a material world but also has a more spiritual connotation that can co-exist regardless of its spatial presence and fulfilling a mystical immaterial meaning[196].

The Indian beliefs in the recurring cycles of life lead us then to the description that is made of multiple worlds in *Sampark*[197], mutual contact, between them.

In this respect it is interesting a publication released in 2002 in Hindi, and translated into English in 2007, by Penukonda Narasimha Rangan.

The author, an electronics engineer, published a book in two volumes[198] titled *Mahásamparka*[199], *The Great Contact*, in which the scholar[200], after carefully analyzing the Rigveda and the Mahabharata, asserted that 5.000 years ago, an extraterrestrial race would come to our planet landing in the town of Badarinath, Himalayas, becoming responsible for the dissemination of the culture of the Rishis from which would be born the Rigveda, and that would be the protagonists of some of the events described in the Mahabharata.

Returning to the ancient Indian scriptures it look like in their coding religious are hidden some knowledge highly evolved comparable to the modern ones.

As the Hindu[201] cosmology is geocentric, then as opposed to the actual arrangement of the planets in our solar system, their level of astronomical knowledge was a milestone unmatched for centuries.

According to this view, also, every planet in the universe has its own specific role and all are part of an upper level subject to the same universal laws.

The planets are considered to be living things in constant evolution and in which, each represents a stage in the evolutionary journey.

Starting from the worlds with no life it forms the evolutionary line of progresses in the development of the consciousness from the unconscious to

---

[196] Soiver, Deborah A., *The Myths of Narasimha and Vamana: Two Avatars in Cosmological Perspective* State University of New York Press, Nov 1991, p. 51.

[197] In Sanskrit it is indicated by the word *sampark* which means "contact" as in the case of *Mahasamparka* namely "the great contact".

[198] The first titled 'Tarka', The Book of Reason, the second 'Sadhya', The book of the Possible.

[199] P. N. Rangan, *Mahásamparka*, Prism Books, Bangalore, 2007. (ISBN978-8-172-86537-5/8 and 86537-2).

[200] In 2004 Rangan was the author of a work based on the first chapter of *Vaimanika Prakarana* of Bharadwaja, which is the *Vimanika Shastra*, in which he tried to explain scientifically the 32 secrets needed by the pilots to command a Vimana.

[201] See the *Surya-Siddhanta: A Text Book of Hindu Astronomy*, Kessinger Publishing Co, 2007, edition edited by Ebenezer Burgess.

the conscious traveling through different worlds and allowing to the man to acquire gradually some faculty ever higher until his trip will not be concluded.

According to Hinduism this planetary chain exists throughout the universe and obeys to some precise rules, so every life form in the universe is subject to the same rules that involve a path toward the perfection.

Beside the countless material universes, the Hindu cosmology also talks about a spiritual world without borders where the 'living entities', purified, live in the full knowledge of the life and of the ultimate reality.

From Vedic literature we learn of the existence of many races created by the supreme deity, the main are the *Gandharvas,* the *Devas* (demigods) and the *Asuras* (demons).

Srila Prabhupada explains the features of the Devas and the Asuras in the comment to the Srimad Bhagavatam (3:10:17), "... *the inhabitants of the higher planets are called Devas, or celestial beings, because they are all devotees of Lord Vishnu. All the devotees of the Lord, Vishnu, are Deva, demigods, while all others are Asura*".

The Devas, gods responsible for the material phenomena, are appointed to preside over the different sectors of the material universe and to ensure that they function properly.

Then we find eight governors[202] of the planetary systems, subordinate to Brahma, under whose supervision act 33 million of Deva, each of them is responsible for managing a particular department of the Universe.

Interestingly, the classification into 'areas' but even more their exorbitant number. Researching the etymology of the Sanskrit term Deva as we have already observed it means literally "the one" or "that which emanates light" which will indicate, in its mythological-religious transliteration, a celestial being[203].

In the most recent times to us among the leading sages who have spoken of other inhabited worlds there is Bhaktivedanta Srila Prabhupada, who in his "*Easy journey through other planets*"[204], he asserted as the Hinduism, from its beginnings, had considered the cosmos as a garden bursting of life.

---

[202] Indra, Agni, Yamaraja, Nirriti, Varuna, Vayu, Kuvera, Siva.

[203] This term refers to entities capable of performing enterprises higher than those of human beings from which derives their 'divinization'.

[204] His Divine Grace A.C. Bhaktivedanta Svami Prabhupada, *Viaggio facile attraverso altri pianeti*, The Bhaktivedanta Book Trust International, 2006.

Srila Prabhupada always says[205] that the Moon would be inhabited by highly elevated beings "*might live in an underground and sublunar world*" and in touch with the demigods.

Another famous Indian master was Paramahansa Yogananda, yogi who taught the Western world how to practice the Kriya Yoga and the awakening 'of the divine consciousness in the man'.

In the book *Autobiography of a Yogi* is possible to find an illuminating phrase, "*... there are many astral spheres filled with such beings. Their inhabitants use some astral vehicles or masses of light to travel from one planet to another faster than the electricity and the radioactive energies*"[206]. The interesting fact is that this text was written before 1947, the date on which officially began the UFO phenomenon, which is, before any modern contamination that might have influenced Yogananda.

During our research we came across in a[207] hymn of the Rig Veda that has left us astonished. The first line of the text speaks of *'generations of gods'* operating in the cosmos and on our planet. The second line instead, more interesting, says that we could *see* again the descendants of these primordial beings, when these specific hymns will be "recited again".

Besides the cryptic call, the interesting and curious data is that we talk about 'future generations of the gods', then the descendants of the ancient visitors who will return to Earth. We are facing a generational continuation of contacts with beings that sporadically occur on our planet to bring their knowledge and wisdom?

Further analysis of the Vedic cosmography show how 14 planetary systems are identified, and the Earth is positioned in a central region[208].

Srimad Bhagavatam says that the peculiarity of our planet lies in the fact of being a very special place where the material and spiritual evolution would be far more favorable than in other planets, especially regarding the resolution of karmic debts and then the extinction of negative legacies from the past lives.

The *Vaikuntha*, the spiritual sky, would be unattainable by men but is traversed by the Deva with their flying machines, the Vimana.

---

[205] Chant 4, chapter 22, text 54 of the *Bhagaat Purana*.

[206] Paramahansa Yogananda, Autobiografia di uno yogi, Edizioni Astrolabio, p. 378.

[207] RigVeda 10.72.1

[208] *Srimad Bhagavatam*, Chant 5, chapter 26, verse 8 - 36.

Another peculiarity seems to be reported in some ancient Indian traditions. In ancient times the people of Siddhaloka[209] would be able to travel in the cosmos without any kind of mechanical assistance; its movement would have been possible thanks to the 'sound vibration' supported by the *'concentration of the magnetic field of the mind'*.

This technology would be able to carry not only the material but also the mind of the man who had addressed this trip.

According to other tradition the great yogi Kardama Muni would have instead created a giant Vimana with which he traveled for years with his wife Devahuti, to visit the different planets in the universe.

It is also said that in previous eras of our planet the Devas[210] regularly visited the Earth with their Vimana but also the Rakhsasas, evil peoples headed by beings as Ravana and Salva.

The Vimana of Salva, described in the tenth canto of the *Bhagavatam*, used a complicated system refraction of the light to be invisible. Instead, the Cannibal God Tripura was traveling on our planet aboard a huge Vimana "like a city", which was destroyed by Shiva in person because it was causing death and destruction among the hermitages of the sages.

In the higher worlds are also the Pitris, literally "fathers", considered ancestral beings whose creation took place separately from the human race. Their task was to awaken in man the ancient vital qualities including those emotional, psychological and spiritual.

In the higher worlds are also the Pitris, literally "the fathers", considered ancestral beings whose creation took place separately from the human race[211]. Their task was to awaken in the man the ancient vital qualities including those emotional, psychological and spiritual.

The *Pitris*, or *Pitaras*, are considered the fathers of the mankind and were the last to arrive on Bhuloka, the Earth, preceded by the *Arupa*, of a high level so devoid of bodily form, which succeeded the *Deva* or elementals, also called *Rupa*.

---

[209] Douglas Renfrew Brooks, *Meditation revolution: a history and theology of the Siddha Yoga lineage*, Muktabodha Indological Research Institute, 2000, p. 196 e 211-212.

[210] The Devas, the celestial beings, were divided into two main categories, the "unborn", said Ajata, and those evolved by humanity, the Sadhyas, then brought in the celestial spheres.

[211] The term is also applied to human ancestors believer's usually up to the seventh generation.

In the text known as Mahanirvana Tantra[212] or *Tantra of the Great Liberation*, the god Shiva exposes to his wife Parvati the different techniques of meditation to be able to move over the influences of the nature and the degeneration of the current era, the Kali Yuga, and so be able to raise to a wider awareness of the Self, the fundamental condition to win the cycle of birth and death.

In his introduction, a particular recall has caught our attention, a section entitled "*The inhabitants of the Worlds*", here's the translation.

"*The worlds are inhabited by many types of beings, ranging from the highest Deva (of which there are many classes and degrees) to the most simple animal life. The scale of beings proceeds from bright manifestations of the Spirit to those in which (corporeality, ed) is so veiled that it seems almost to disappear in its coating material. There is only one light, one Spirit, whose manifestations are varied. A flame enclosed in a clear glass does not lose much of its luster. If we replace the glass paper, or some other substance still more opaque, the transparency of the light is darkened. A metal coating is so dense as to exclude the view of the light that still burns inside with an equal splendor. In fact, each of these forms that are veiling are the Maya*".

According to the most ancient Hindu wisdom different types of beings are living in the cosmos, some entities sometimes so high as to be imperceptible to our senses, and sometimes they put on a closer level to our so that they can be perceived as intangible shadows.

Then we find beings like us, sometimes different in their characteristics but still with human features.

In all cases, the fundamental concept that the *Tantra of the Great Liberation* teaches us is that in all these appearances "*There is only one light, one Spirit, whose manifestations are varied*".

The above is just a hint to the endless references in the Indian world of the other inhabited worlds.

To add some more would be redundant if not to further consolidate an idea already structured that for this ancient people we were not alone in the universe.

The Kumaras and the Narada Muni, the travelers of the Worlds

If the Vedic literature seems to have pervaded since its very deep soul by the knowledge of the life in the cosmos, it seems to be even more interesting to analyze the few cases where we come from the general to the particular.

---

[212] *Mahanirvana Tantra, Tantra of the Great Liberation*, translated by Arthur Avalon (Sir John Woodroffe), 1913.

In the pantheon of Hindu deities the best known deity for being able to travel in space is Narada or Narada Muni[213]. This is to be identified as a divine sage who plays a prominent role in many Puranic texts, especially in the Bhagavata Purana[214].

Narada Muni is portrayed as a Brahmin, a monk, with the power to travel among the planets and the distant worlds together with his *vina* (a stringed instrument) to awaken the love of Vishnu.

In the Srimad Bhagavatam we find a reference of great interest, a verse which speaks not only of how Narada Muni traveled in the space but that there were also some other pilgrims of the universe that flanked him "*From different parts of the universe came here great sages as Atri, Cyavana, Saradvan, Ariṣṭanemi, Bhrgu, Vasistha, Parasara, Visvamitra, Angira, Parasurama, Utathya, Indrapramada, Idhmavahu, Medhatithi, Devala, Arṣṭiṣeṇa, Bharadvaja, Gautama, Pippalada, Maitreya, Aurva, Kavaṣa, Kumbhayoni, Dvaipayana and a great personality as Narada*"[215].

It is interesting to notice as some of these figures are among the wisest and most revered Rishis of the ancient India, the fathers or the initiators of some millennial arts, but between all of them there is one in particular that draws our attention, *Bharadvaja,* who according to the tradition composed the treatise known with the name of Vymanika Shastra.

Is emblematic that a treaty of Astronautics was composed by a wise man clearly indicated to as a '*traveler of the universe*'!

In the Hindu religion Narada[216] is also called *Triloka sanchaari* because they are able to travel in the "three worlds" or the Swargaloka, the heavenly world, the Mrityuloka, the "world of death" that is the Earth and the Patalloka, the lower world.

Equally fascinating is the discovery of other figures that can move in the cosmos with a very precise message, which is to preach the essence of the Vedic scriptures on behalf of Vishnu. This same *essence* has the connotations not only religious but also technical and scientific, which led to the natural question whether in addition to a religious teaching these beings do not attend the worlds also to civilize their people.

---

[213] The Vedic texts celebrate Narada Muni as one of the twelve mahajanas or great authorities of eternal truth.

[214] The details of his life and his teachings are narrated in the Narada Purana, in Padma Purana and throughout the Srimad Bhagavatam.

[215] Srimad Bhagavatam, 1.19.9-10.

[216] Narada Muni also owns the title of Sruta-Dhara ie one who can remember everything and forever.

In the Indian mythology four figures are equally emblematic known as the *Kumara* or *Catursana*. The two names are equivalent, as antithetical, meaning the first *'the boys'* and the second *'the four old'*. Best known, for their looks, as the Kumaras in the Hindu tradition these deities are the four sons of Brahma, the primordial being, or Sanaka (the Old), Sanatana (the Lord), Sanandana (the Gaudente) and Sanat-Kumara (Always the boy).

The impasse that originates from their contrasting name is resolved by the fact that at the origins of the time they were born from the mind of Brahma himself but, in their immortality, preserved the appearance of children between 5 and 6 years. Always they have been considered great sages having chosen a life dedicated to the study and to the chastity, and preaching the essence of all Vedic scriptures from planet to planet[217], traveling in space.

It is curious to note how in the UFO literature have been many reported some cases in which the described beings are in all similar to the Kumaras or very small in stature, but extremely long lived and very wise[218].

Among the most illuminating passages about their cosmic wanderings of Srimad Bhagavatam we read, "*As soon as the great sages: Sanaka, Sanatana, Sanandana and Sanat-kumara reached the above mentioned Vaikuntha in the spiritual world through their efforts mystic yoga, sensed an immense happiness. They saw that the spiritual sky was lit from airplanes decorated in heaven, piloted by the best devotees of Vaikuntha ...*"[219].

The theologizing observable in these cases is probably the product of a culture that deified these beings.

As in the *cargo cults* the plane becomes the manifestation of the divine so you can think that beings from other worlds, and physical characteristics similar to our own, could have suffered the same fate.

A common thread seems to tie these beings that were traveling in the cosmos with two extremely important figures of ancient India, men who through their work demonstrate the importance of the knowledge that *travelers of the planets* had donated to the human race.

An interesting hint must be duly made to the language since the beginning of this land has been a key element in its history and its culture.

The *Devanagari*, literally the *'Writing of the City of the Gods'*, is a syllabic script used in many regions of India.

---

[217] Srimad Bhagavatam, 7.1.37.

[218] Among many cases we remember the case of Bucine (Arezzo) that saw involved Rosa Lotti Dainelli or case Friendship with Saju. In the latter case, a particular type of aliens had characteristics similar to children between 5 and 6 years.

[219] Srimad Bhāgavatam (3.15.26)

The alphabets of the Devanagari appears around the eighth century, as a further evolution of the Brahmi script, whose presence is attested in the region of the Indus Valley since the fifth century BC, probably an Indian adaptation of the Semitic scripts, penetrated into India through Mesopotamia.

The Samarangana Sutradhara

The Samarangana Sutradhara is a manuscript written in Sanskrit around 1100 AD by King Bhoja Paramara Dhar[220], divided into 83 chapters, for a total of 1144 pages.
The text is the most famous and important compendium of classical indian architecture, called Vastu Shastra, dealing also about city planning, house and temple architecture. It dictates rules for sculpting, the canons of painting as also the different positions to be taken with the hands or the body and the right pose to be taken with the legs.
From the time of Panini, the reformer of Sanskrit in the sixth century BC, until the period in which reigned king Bhoja, in India flourished many great universities such as those of Taxila, Valabhi, Dhar, Ujjain, Visala, etc.
The legacies of the ancient knowledge were passed down within the rigid pyramidal caste preserving during the time and constituting the best of Indian knowledge.
The historical records, however, can also provide us an explanation of the reasons that led to the destruction of a part of this knowledge.
From the second century A.D. it began an incessant looting by foreign people and tribes. Two centuries after the attacks began, in waves, by hordes of Arabs, Turks, Afghans, etc.
All universities and all the centers of wisdom known as the Temples, the Viharas[221] and the Bhandaras, containing books and other priceless treasures of the Indian heritage, had to support the fire and fury of the predators.

---

[220] According to Dr. Bhatia Prathipala Munja, grandson of Maha Rajadhiraja Kaviraja Sista Siromani Dharesvara, Sri Bhojadeva was first and foremost a man of great culture, a versatile scholar, a wiseman. He has been credited works in almost every branch of knowledge. According to Ajada, who wrote a commentary named Padakaprakasa about Sarasvati Kanthab-Harana, Bhoja wrote 84 works. Prabhacandra Suri Prabhava Carita refers to works of Bhoja in different branches of knowledge. Among the works attributed to him, there is a comment on the Patanjali Yoga Sutras, the Sarasvati Kanthabharana (grammar) and the Sarasvati Kanthabharana (Poetics), the Srngara Prakasa (Poetics and Dramaturgy), the Tattva Prakasa (a manual of asexual Saivas), the Brhadrajamartanda (Dharmasastra grantha), the Rajamrganka (medicine), and a work of astronomy dated to 1042 AD.

[221] Word for the Sanskrit language and Pali language the Buddhist monasteries where monks took refuge during the season of monsoon rains.

In this dark firmament of devastation and uncertainty, many traditions and knowledge were lost but others were preserved in secret through the legacy of oral tradition, the Sruti.

In parallel fits the effort of king Bhoja that, in the twelfth century, tried their conservation by a new written compilation.

Many discoveries attributed to Afghans, Turks, Arabs and many other people of the Middle East were actually inherited from the literary works of ancient India. We can find an example in what we call the "Arabic numerals", which in fact are "Indians", but that have come to us through the Islamic culture.

Among the most interesting works come down to our days of King Bhoja, we find the aforementioned *Samarangana Sutradhara*, a treatise of architecture in which a chapter was devoted to the machines, the Yantras, as well as the measures to be used in case of war.

It is precisely the chapter XXXI to have brought upon themselves a great interest among the researchers because it is in art and in the defensive tactic of war that were described the machines and the instruments of all futuristic and timeless for the period when the volume was written.

Entitled *Yantravidhanam, On the preparation of the mechanical countermeasures,* this chapter is surprising because it puts its focus on the construction and operation of certain types of aerial machines that use different methods of propulsion.

King Bhoja used the generic Sanskrit word yantra, instead of the more familiar Vimana, even if the same will be used at least once in the text, stating that he had drawn his knowledge from some manuscripts that were very ancient already in his time.

Reading the full text appears perfectly clear and obvious to the reader the presence of some aircraft, flying machines, which are described in detail. The following passages are limited to those aircraft capable of moving in the atmosphere but the Samarangana Sutradhara also describes flying machines capable of moving in the *Suryamandala*, the solar sphere, as well as others can go further in the *Nakṣatra mandala*, or the 'stellar sphere'.

Here are some excerpts and translations from the Sanskrit of the twelfth century.

"*The structure of the Vimana have to be built solid and durable, like a great flying bird of light material. Inside one must put a mercury engine with under it its iron heating apparatus. By means of the power latent in the mercury that sets in moving the turbine control, a man sitting inside may travels to a great distance in the sky. The movements of the Vimana are such that it can vertically to ascend, vertically to descend, and to move slanting forwards and backwards. With the help of the machines the human beings can fly in the air and the heavenly beings can descend on the Earth* "[222].

At this point the text becomes even more interesting and at the same time even harder to understand. It is first of all clear that the essential elements of what is described as the 'propulsion system' are deliberately vague or completely ousted.

The reason is explained in the next paragraph of the text which states that "*...the details on the construction of the Vimana are misunderstood by virtue of a necessary secrecy, and not for ignorance. These details of construction are not mentioned because you should know that if they were publicly such machines could also have used in a wrong way*"[223].

The fear that such technology could fall into the wrong hands is already widely evident in this paragraph. King Paramara Bhoja writes his text around 1100 AD but he himself says that the material to which he had drawn was already very old at his time.

Let's step back. Why an architectural treatise contains some references to these 'mythical' flying machines?

The obvious fact that emerges from what has already presented is that we are facing a historical *impasse*. When King Bhoja wrote his encyclopedic text these aircrafts no longer existed, but the knowledge at the base of their operation and their construction was encoded and transmitted by memory of the wise men, through the *Sruti*, up to reach a written form in the Samarangana Sutradhara.

As we have seen, the text is a treatise of architecture in which there is an entire chapter in which retraced the techniques of warfare and the consequential countermeasures and tactics you can adopt. In its subsequent evolution the term Vimana would assume the more prosaic meaning of 'sacred place', 'place of the gods', attributing and then identifying the object of our interest in the same divine and religious connotations.

The coexistence of these two sides, the architecture of the book and the purely military chapter, are justified by the eclecticism of its author and the vast erudition that has historically attributed to him. His works are still one of the greatest cultural and wisdom representations that the ancient India has given to the humanity.

---

[222] Translated from Sanskrit into English by Sudarshan Kumar Sharma, *Samarangana Sutradhara of Bhojadeva: An Ancient Treatise on Architecture* (With An Introduction, Sanskrit Text, Verse by Verse English Translation and Notes in Two Volumes), Parimal Publications, 2007, Isbn 8171103022, cap. XXXI, pp. 183-184.

[223] Translated from Sanskrit into English by Sudarshan Kumar Sharma, Sudarshan Kumar Sharma, *Samarangana Sutradhara of Bhojadeva: An Ancient Treatise on Architecture*, op. cit., cap. XXXI versi 95-100.

It is in this context that, by treating the art of war, the Samarangana Sutradhara becomes an encyclopedia of military art in which are treated all the possible techniques, tactics and equipment for use in war. From this point of view was probably the need to quote the ancient Vimana and the will to explicitly as possible roles and functions.

We could refer to a large number of books but those that we have selected at least provide us with a general vision of what these aircraft and what were their peculiarities and weapons.

"*At the critical moment the bullet fire must be issued, which will make the fact possible. These bullets in time expand themselves accompanied by a thunder, because of the medium that expands themselves. This resulting expansion acts like the power of an elephant in a circle without end*".

Later in the text we find a paragraph in which the wood is mentioned as a potential material for the construction of these surprising flying machine to which directly follows a description of the propulsion system that refers to a form of fuel similar to gasoline.

"*The realization of the machine is strongly desired ... using the light wood to build a great aerial car of a massive structure. In the middle there is the container of the liquid consumed by the engine that burns slowly during the complete combustion*".

Immediately after, there is a list of the possible movements and maneuvers feasible for the pilot. Some of these maneuvers may be considered quite similar to those carried out in the modern airplanes while others are similar to those of the hovercraft and of the most recent British fighter vertical takeoff Harrier.

"*They are famous the technical mastery in the following maneuvers: the vertical ascent, vertical descent, back and forth, the descent and the normal ascent, the inclination and the progress over long distances through adjustments of the running parties... the resulting sound in the air and the roar of the thunder may be similar to the trumpeting of a frightened elephant but can also be similar to a musical tones*".

It is interesting to note that in this step we try to give voice to explain, the sound produced by the Vimana comparing it to that of an elephant scared, therefore is extremely sharp and strong, is also to musical tones, so sweet and melodious.

This peculiarity differs from a multitude of myths in which the 'chariots of the gods' were silent or with the sounds heavenly.

There has passed the comparison but in this case seems to be in front of a turbine, or a propulsion system, that various its intensity depending on the state in which it is located, by the sweet musical tone to 'trumpeting of an elephant scared'. As if to remind us that an engine that attains power for its take-off!

Reading the Samarangana Sutradhara we find further evidence that what is described is actually something quite different from a simple poetic, allegorical and mythographic representation "... *their machines sparkling in every direction can travel in any direction that their imagination can reach. From their great heights they can see exciting dances, theatrical dramas, as well ritual ceremonies. Their yantra have become popular among the royal dynasties and in various nations. In this way the Elected Anime of someone are flying, while those of a lower level are walking. All these friends have succeeded in the things that they wished more to do through their machines, this shows that human beings can fly in the air and not only in the Earth, and that some celestial beings can descend among the mortals when they come to visit the Earth*".

Some of these aircraft are described with winged structures similar to those of the modern airplanes but is also specified that this type of 'machines' cannot go back or ascend or descend vertically.

The description provided is completely specular to that of a modern plane able to move only in a straight line and, of course, not to be able to move in the same direction of its engine.

If the steps mentioned are not enough, we find further specific conformation of these items in the next step. The term "*two wings*" clearly appears in the text in conjunction with the description of what would seem a propulsion able to generate a '*whirlwind*'.

"*Therefore it can be placed inside the mercury engine and can be mounted properly under it the heating apparatus of iron. So men can fix the two wings driving the turbine in motion with the hidden pilot, through the power of the engine to mercury, can travel to a great distance in the sky*".

This step is extremely interesting and reserve profound implications. A mistake commonly made by the detractors of this issue is that it has given the notion of 'motor mercury' to Vymanika Shastra when in fact it is precisely in the Samarangana Sutradhara that this concept is labile expressed in the paragraph just quoted.

Added to this is also a notion from the deep technical value, ie, the presence under this of a '*heating apparatus of iron*'.

The case in question specifies the presence of a technology, that is still completely unknown, but certainly formed within a scientific precise context.

We are not in front of a myth of a celestial chariot of which is given a brief description; instead we are confronted with a technical explanation, gaunt but explanatory, a system able to lift the aircraft.

The additional presence of two wings that allow to drive '*the whirlwind in motion*' in which a 'hidden pilot' can travel through great distances is extremely significant. Do not forget that we are in the presence of a text from 1100 AD.

We believe that the steps so far mentioned are a description of an apparatus for flying made by someone who do not directly observed it but sensed it by consulting ancient documents which described the characteristics and tried, according to his simple and limited knowledge, to propose it as close as possible.

King Bhoja was in 1100 A.D. and, as he himself says, was inspired by ancient documents for the realization of his treatise.

What follows is a description of a very complex and larger Vimana in which is described a propulsion system consisting of four motors run by mercury.

This paragraph contains the only direct reference to the term Vimana in Chapter XXXI of the Samarangana Sutradhara.

"An extremely fast and agile Vimana can be constructed, as big as the house of the God-in-motion. Four containers resistant to mercury have to be installed in the internal structure. When these were heated by a controlled fire from iron containers, the flying machine develops the power of a thunder using the mercury, becoming a yantra highly fascinating. Also, if this iron engine with properly joints, it is filled with fluid [mercury?, ndr]*, when ascending or descending on the Earth generates a power with the roar of a lion".*

Konark and the temples in imitation of the Vimana

Another notation, in the Samarangana Sutradhara, is fundamental to understand the Vimana and the impact that these aircraft had in the Indian culture.

In Chapter 49, verses 1-20, we unequivocally states that the shape of the Vimana was used and imitated by the ancient Indian architects to build the temples and palaces in their own image and likeness, which is modeled on its aircraft flying antiquity.

Completely unnoticed by the modern industry sector, this data allows us to understand and read under a quite different perspective the shape of the indians temples and that inexplicable attraction and charm that they have always had.

By comparing these temples with some of the sketches in the Vimanika Shastra of the Pandit Subbaraya Shastry, as we will observe, for example, the Sundara Vimana greatly resembles the typical form of Indian temples!

In addition, as previously remembered, in South India the apical part of the temples is just called 'Vimana'.

There are also far more unusual structures built following the instructions from the Samarangana Sutradhara.

The most emblematic case and appropriate to what we just presented, is the temple of Konark, in the region of Orissa, a temple in which there is a huge Vimana built in stone and pulled by seven horses and twelve wheels finely decorated.

 The Temple of the Sun, located in the district of Puri, was built in the thirteenth century by the king Narasimhadeva transposing on the Earth the Chariot, Ratha, through which the god of the Sun Surya moved in the sky.

During its construction, according to the legend[224], on top of the main temple was placed a giant magnet, one was located in the basement of the structure while a statue of the god of the Sun, made of iron, it is said to be suspended in the air by four large magnets.

Again according to the legend these magnets were so powerful that they caused serious damage to the commercial Portuguese ships that passed in the stretch of sea off, less than three kilometers from the Temple.

The compasses, attracted by the giant magnet, seems that were drawn toward the coast, bringing the boats to a certain disaster.

To save their expeditions, the Portuguese navigators removed the magnets from the temple which held a function of 'cornerstone' keeping in balance parts of it. Because of the imbalance originated, some walls lost their stability and finally fell down. As these traditions are deeply rooted in this territory from ancient times there is no mention of it in the historical documents.

A single element seems to corroborate this tradition, there are dozens of iron columns that still lie scattered across the area of the temple or still hold up part of its facilities.

It is a method of construction entirely unusual for that time, almost unique, but that shows how the knowledge preserved by Bhoja in the Samarangana Sutradhara came from a distant past in which the techniques of constructions pushed beyond all imagination.

As the modern criticism has focused only on Vymanika Shastra, trying to deny the reliability and the contents, the objective reality shows us that the sources that describe in detail the Vimana, their technologies and their pilots are much more suggestive.

We do not know what ancient texts drew King Bhoja but undoubtedly it came from different sources from Vymanika Shastra since the concepts contained in Samarangana Sutradhara differ almost completely from what is contained in the right dictated by Pandit Subbaraya Sastry.

The text of Bharadwaja also states that the pilots could not drive these aircrafts before they have learned 32 secrets to the base of their operation. These secrets, as we will see later, are some real guidelines on how to best use the Vimana and show us how it was not the 'myth' to drive the legend but some individuals trained to do their work.

The research of the sources, the verification of their contents, involved not only the ancient texts and their testimonies but tried to understand the feasibility of what once testified in these treaties. On the last decade we have been witnessed to a real literary and scientific countercheck and the results of which have stunned the worst detractor.

---

[224] Karuna Sagar Behera, *Konarak - The Heritage of Mankind*, Aryan Books International, 1996.

At the moment the Academy of Sanskrit Research[225] of Melkote, near Mandya, was commissioned by the Aeronautical Research Development Board in New Delhi, a study of one year on *'Non-conventional approach to Aeronautics'*, some aeronautical unconventional approaches, whose source was just the Vaymanika Shastra.

It is not yet time to analyze the incredible scientific wisdom contained in the Vimanika Shastra.

During the writing of this book, we came in contact with several Indian, Pakistani and international scholars that have tried to refute or confirm the incredible knowledge preserved in this text and the claims that we have released pose a new light on the ancient Vimana and the story itself of our civilization.

The Ramayana and the Vimana

It was attributed to the sage Valmiki, the final draft of the poem goes back to the I-II century AD: it is in fact earlier to the final draft of the Mahabharata, but it is believed that its original form can be traced back to the IV-III century BC (*Maurya* era), or even to the VI century BC.

The Ramayana, like the Homeric poems, can be considered a source and a compendium of all knowledge and cultural patterns of civilization reflected in its pages.

The ramaic epos has an educational role fully satisfying his paradigmatic task, since it is a repository of collective knowledge. Nevertheless this "hereditary sediment", transmitted by oral tradition, cannot be considered as a encompassing heritage, but rather as the result of a gradual layering and overlapping of historical, mythical, anecdotal and geographically material over the centuries that has been incorporated in an organic collection become a synthesis and a symbol of cultural, religious and philosophical of an entire civilization.

The poem presents the story of Rama, the seventh avatar of Vishnu, and his wife Sita. Rama is the crown prince of the kingdom of Koshala, a vast empire between India, Pakistan, Iran and Afghanistan, in the classical texts known for being the birthplace of the "*Seven Cities of the Rishi*" commonly attributed to the Indus Valley Civilization.

Rama is unjustly deprived of his kingdom and exiled from his capital, Ayodhya, and forced to live for 14 years in exile with his wife Sita and his brother Lakshmana, first near the hill of Citrakuta, where was the hermitage

---

[225] Academy of Sanskrit Research - http://www.sanskritacademy.org/

of Valmiki and many otherwise men, later in the forest Dandaka, populated by Rakshasa demons.

During their wanderings Rama, Sita and Lakshmana will come in the ashram of the mystic Bharadvaja located in the town of Chitrakoot, in the Madhya Pradesh. It will be however in Dandaka that Sita will be kidnapped by Ravana, the Rakshasa demon king, to lead her to Lanka.

Rama and Lakshmana will ally with Vanara, population of half men-half ape shaped people, among them stands the figure of the brave Hanuman. According to the poem and tradition Vanara build a bridge of stones and sand to connect the southern tip of India with Lanka; this figure, however, seems not to coincide with the geography described in the Ramayana that places most of the places object of the poem in the northern of India.

Valmiki continues his story[226] saying that the Rakshasa demons were accustomed to attack at night because that way their magic power was multiplied. The epic continues as referring to the infernal horde commanded by Ravana that killed and eaten thousands of Vanara by virtue of their property to become invisible and ubiquitous.

Rama and Hanuman were inside of this terrible battle but "*suddenly there came a great wind that shook the mountains, and we saw a flame of fire that floated in the air*". With these words it describes the arrival of Garuda[227], an anthropomorphic deity featured as an eagle whose Sanskrit name means "*one who brings a great weight*", thanks to his intervention the two heroes will have save their life.

The battles that Rama will dispute along with his army will take place both on the land and in the sky through the use of flying chariots, the Vimana, and the weapons by the incredible power, the Astras.

"*The Rakshasas used black Vimana that were also used against the armies to bomb them. He saw in the sky a mass of scarlet clouds: from there emerged blinking many blazing and roaring missiles; in particular from a giant Vimana fell many winged gold weapons and thousands of lightning, with very strong explosions, and many hundreds of flaming wheels. Makarna, to defend himself from the Rakshasas, launched a missile against the formation of aerial bombardment of Rakshasas, which was reduced to ashes and the great enemy ship rushed to the ground, but he managed to go on a part of the army of Karna, who remained killed.*

---

[226] Chapter VI of the Ramayana, known as, Yuddhakanda or *Book of the battle*, collects the most interesting descriptions about the use of flying chariots and weapons from the enormous destructive power.

[227] In the Vedas is the oldest reference to Garuda, with the name Shyena. In its description it is stated that he would have brought, the Amrita, on Earth from Heaven the water of eternal life, or, according to other traditions, the water of knowledge.

The Rakshasa forces had the shape of large mounds still in the sky, like a mass of clouds in the blue sky, surrounded by a rainbow. Some of these were of the Vimana with some decks and equipment according to the rules, and when they flew away it seemed a flock of birds. In particular, the golden Vimana of the Rakshasa, when landed, it looked like a mass of antimony (shiny metal, ndr) *with a beautiful form on the surface of the Earth. Not all the flying chariots had a circular shape that was the most efficient. The Vimana Elephant class, had furniture, windows, numerous apartments, and could carry other Vimana and had an enormous size* "[228].

This shocking report is a fragment of the war fought between Rama and the demon Ravana, a battle that seems to add up all the power in his description attributed to the technology of the Vimana and the destructiveness of the weapons contained inside them[229].

A story that, at least since the XVI century, has found a pictorial and exceptional transposition in South India, some frescoes discovered by the director Diego D'Innocenzo depicting an aerial battle fought between the Vimana.

As recalled by the guardian of the temple located near the largest center dedicated to Jagannath "... *this is a fresco of the Ramayana and is an integral part of the traditional art of Orissa. We realize this as frescoes on the walls of our temples and this kind of art started 500 years ago. Here we are seeing Rama and Laxman with some monkeys and many other soldiers. Behind this design there is the demon Ravana while is going to attack Rama*".

The Ramayana continues narrating the victory of the 'Apes-men' Vanara on the demons Rakshasa. Ravana was killed in a duel by Rama that could well return triumphantly to his capital Ayodhya with his wife Sita, and to be crowned king.

Sometimes pictures are worth more than a thousand words!

As a symbol of victory Vibhasana, the repudiated brother of Ravana, he gave to Rama the Pushpaka Vimana, a giant flying fortress belonged to the god Bramha and built by Viswakarma, the deity patron of the craftsmen, the engineers and the architects.

In the Ramayana the Pushpaka Vimana is described in very interesting terms, "... *this chariot was moving by itself, it was all shiny and painted; it had two floors and lots of windows, many rooms and many flags; while it was flying issued a melodious sound that sounded like a whisper,* "but also" *The carriage of flowers that resembles to the Sun and belongs to my brother, was brought by the powerful Ravan; this magnificent flying carriage travels wherever you want ... it looks like a bright cloud in the sky ... and the King himself*

---

[228] Romesh C. Dutt, *The Ramayana and Mahabharata*, Libro X, Yuddha, 1899.

[229] Uttar parsa matha, Puri, Orissa. North side next to the Jagannath temple.

(Rama) went inside, and the magnificent carriage, at the commanding of Raghira, got into the upper atmosphere[230]".

The Pushpaka was created by Maya for Kubera, the god of the health, but later it was stolen, together with the island city of Lanka, by the demon Ravana.

Looking in the Ramayana, we find an interesting notation on some internal elements constituting the Vimana, "... *the substance of their coaches is gold; their pillars are made of metal, and shine in the sky like lightning...*"[231] referring to the First book they refer to the shape of one of these objects, "*Come to us with your well built triangular machine and made by three columns and three wheels ...*"[232].

The aircraft is described in a singular way: each part is made of gold and is studded with precious stones such as sapphires. Inside there are some secret rooms "*bright as silver*" while on the top floor of the aerodynamic floor stands a white silky flag depicted with golden lotus flowers. Its floor is covered with beautiful crystals and on its inside there is a throne studded with sapphires while its speed is compared to that of the mind.

The Pushpaka Vimana could accommodate many people in it, it is said that many Vanara could find hospitality (Sloka 13) as well Vibhashana along with his ministers (Sloka 25)[233] and the whole army!

So "... *the chariot rose up in the sky, sailed on its way by air, while the monkeys, the bears and the surviving monsters were inside at ease*" and all the Vanaras, the Rukshyas, and the Rakshasa hosted inside could find a comfortable accommodation without giving discomfort each other (Sloka 27).

On command of Rama, the aircraft automatically lifted into the air and reached the great heights with a great sound (Sloka 1).

All these descriptions seem to be much more similar to a modern airplane than to a legendary aircraft fruit of the mind of the sage Valmiki.

In addition to the above, the description mentioned in the Ramayana Rama's journey seems to witness us the depth knowledge of the places that they were going to fly. On the way back to his capital the King recognizes the overflown territories listing them to his wife Sita.

---

[230] Dutt, Manatha Nath (translator),Ramayana, Elysium Press, Calcutta, 1892 and subsequently New York, 1910.

[231] *Ramayana*, Book V, 5.062.07.

[232] *Ramayana*, Book I, 1.118.02.

[233] Valmiki Ramayana published by Geeta Press, Gorakhpur, Uttar Pradesh, India.

Among the many, he mentions the Mount Kailash and its pyramidal shape, indicated as the place "*visited by the men of the sky*" and used as a reference point as a result of its form.

To date, curiously, the Mount Kailash by the Hindus is considered a holy mountain instead the name *Sita* is the river that would flow in *Shambhala*, the legendary underground kingdom dear to the Buddhist tradition.

Rama is to know a wide geographical area of two thousand miles but especially the description given in the epic text is that one of the places as they would really distinguishable from watching them from the top!

The question, as it may seem apparently secondary, is actually extremely complex. We may know very well a place at the ground level but when we are taken to a height of a few miles our landmarks will disappear completely and we would be facing to a new perspective and perception in which it would be extremely difficult to navigate and to find some points of reference.

The objection that could be made is that Valmiki has 'created' this report of Rama through a fantasy but if we compare the epic text with a real 'aerial' observation of the journey undertaken by Pushpaka Vimana we soon realized that they coincide perfectly.

It is a mere coincidence? We do not think so!

How could the author of the Ramayana to be so aware of these characteristics?

The only plausible explanation is that really 'someone' has traveled flying the route and this observation is then gone to integrate into the epic tale.

The same could be assumed for the description of the Pushpaka. The scholar David William Davenport is certain of that, "*the ancient authors have definitely seen... but due to the poverty of the language, or lack of terms needed, the image that they give is deficient from the technical point*" for that was implemented in the form of the epic story and embellished with some poetic descriptions.

In more recent times the same condition occurred with the Native Americans, when they entered in contact with the white man and our technology, describing 'thundering rods', 'long knives', and 'iron horses', in this case a transpositions of the rifles, the swords and the trains.

www.ingramcontent.com/pod-product-compliance
Lightning Source LLC
Chambersburg PA
CBHW062211220526
45471CB00009B/3162